HANDBOOK ON TOURISM AND CONSERVATION

RESEARCH HANDBOOKS IN TOURISM

Series Editor: Robin Nunkoo, *University of Mauritius*

This timely series brings together critical and thought-provoking contributions on key topics and issues in tourism and hospitality research from a range of management and social science perspectives. Comprising specially-commissioned chapters from leading authors, these comprehensive *Research Handbooks* feature cutting-edge contributions and are written with a global readership in mind. Equally useful as reference tools or high-level introductions to specific topics, issues, methods and debates, these *Research Handbooks* will be an essential resource for academic researchers, practitioners, undergraduate and postgraduate students.

For a full list of Edward Elgar published titles, including the titles in this series, visit our website at www.e-elgar.com.

Handbook on Tourism and Conservation

African Perspectives

Edited by

Joseph E. Mbaiwa

Professor of Tourism Studies, Okavango Research Institute, University of Botswana, Botswana

Oluwatoyin D. Kolawole

Professor of Rural Development, Okavango Research Institute, University of Botswana, Botswana

Wame L. Hambira

Associate Professor of Environmental Management, Botswana University of Agriculture and Natural Resources, Botswana

Emmanuel Mogende

Lecturer of Natural Resource Governance, Department of Environmental Science, University of Botswana, Botswana

RESEARCH HANDBOOKS IN TOURISM

Cheltenham, UK • Northampton, MA, USA

© Joseph E. Mbaiwa, Oluwatoyin D. Kolawole, Wame L. Hambira and Emmanuel Mogende 2023

All rights reserved. No part of this publication may be reproduced, stored in a retrieval system or transmitted in any form or by any means, electronic, mechanical or photocopying, recording, or otherwise without the prior permission of the publisher.

Published by
Edward Elgar Publishing Limited
The Lypiatts
15 Lansdown Road
Cheltenham
Glos GL50 2JA
UK

Edward Elgar Publishing, Inc.
William Pratt House
9 Dewey Court
Northampton
Massachusetts 01060
USA

A catalogue record for this book
is available from the British Library

Library of Congress Control Number: 2023945114

This book is available electronically in the **Elgar**online
Geography, Planning and Tourism subject collection
http://dx.doi.org/10.4337/9781839106071

Printed on elemental chlorine free (ECF)
recycled paper containing 30% Post-Consumer Waste

ISBN 978 1 83910 606 4 (cased)
ISBN 978 1 83910 607 1 (eBook)

Printed and bound in the USA

Contents

List of contributors		viii
1	Introduction to the *Handbook on Tourism and Conservation* *Oluwatoyin D. Kolawole, Joseph E. Mbaiwa, Wame L. Hambira and* *Emmanuel Mogende*	1

PART I TOURISM AND CONSERVATION IN PROTECTED AREAS

2	Tourism and conservation in protected areas: on collective efficacy *Jabulile Happyness Mzimela and Inocent Moyo*	10
3	The prospects of forest-based tourism for marginalised communities *Joyce Lepetu, Gofaone Rammotokara and Hesekia Garekae*	25
4	Human–elephant conflict: implications for rural livelihoods and wildlife conservation *Kenalekgosi Gontse, Joseph E. Mbaiwa and Olekae T. Thakadu*	37
5	Residents' perception of ecotourism development at Tachila Nature Reserve, North-East District, Botswana *Unabo Tafa and Joseph E. Mbaiwa*	54

PART II TOURISM AND CLIMATE CHANGE

6	Tourism and climate change adaptation in protected areas *Kaarina Tervo-Kankare*	73
7	Tourism and climate change vulnerabilities: a focus on African destinations *Kaitano Dube*	86
8	Tourism and climate change: consequences, adaptation and mitigation *Esraa A. El-Masry*	101
9	Evaluating climate change communication for sustainable environmental conservation in the tourism sector *Sharon Tshipa and Olekae T. Thakadu*	115
10	Perspectives on the effects of environmental change in northern Botswana and its implications for CBNRM *Maduo Mpolokang and Jeremy Perkins*	130

v

vi   *Handbook on tourism and conservation*

PART III   SUSTAINABLE AGRITOURISM

11   Potentials and challenges of sustainable agritourism in Fortín, Veracruz, Mexico   147
Karina Nicole Pérez-Olmos, Noé Aguilar-Rivera and Carlos Enrique Villanueva-González

12   'Negotiating with the juggernaut': on agritourism and the paradoxes of market-driven conservation   164
Mikael Andéhn and Patrick J. N. L'Espoir Decosta

13   Micro and small-scale culture-based tourism initiatives as a livelihood option for rural women in Kenya   176
Rita Wairimu Nthiga and Beatrice H. O. Ohutso Imbaya

14   Environmental impact of rural tourism   189
Gondo Reniko and Oluwatoyin D. Kolawole

15   Towards agritourism development in Zimbabwe: growth potential, benefits and challenges   204
Rudorwashe Baipai, Oliver Chikuta, Edson Gandiwa and Chiedza N. Mutanga

PART IV   DESTINATION COMMUNITIES AND NATURAL RESOURCES CONSERVATION

16   Commodification of nature and territorialization: conservation, local communities and Botswana's international cooperation   223
Kekgaoditse Suping

17   Community-based natural resources management and poverty reduction   237
Israel R. Blackie

18   Information communication technologies and community-based tourism organisations   249
Siamisang Sehuhula

19   Assessing the role of the central government and communities in alleviating poverty through ecotourism   264
Agnes Tshepo Nkone and Thekiso Molokwane

20   Sense of place and tourism in cultural landscapes   279
Joseph E. Mbaiwa and Gladys B. Siphambe

21   Co-management of world heritage sites for community benefit   300
Olekae T. Thakadu, Wame L. Hambira, Gaseitsiwe Smollie Masunga, Barbara N. Ngwenya, Abigail Lillian Engleton, Dandy Badimo and Ineelo Mosie

Contents vii

PART V   CONCLUSION

22   The interlinkage between tourism, environmental conservation, and
      natural resource management: a synthesis   329
      Oluwatoyin D. Kolawole, Joseph E. Mbaiwa, Wame L. Hambira and
      Emmanuel Mogende

Index   335

Contributors

Noé Aguilar-Rivera is a Research Professor at the Universidad Veracruzana Mexico in the Faculty of Biological and Agricultural Sciences in Córdoba Veracruz. His areas of research are environmental management in agribusiness, sustainable development, and circular economy. He has actively participated in teaching at the postgraduate level, thesis direction and developing and supervising several multidisciplinary projects on sustainability. He is currently a member of journal editorial committees such as *Sustainability Science*, *Discover Sustainability*, *Sugar Tech*, and *Frontiers in Sustainability*.

Mikael Andéhn is a Senior Lecturer at the School of Business and Management, Royal Holloway, University of London, UK. He has written on subjects such as tourism development, critical marketing, organization in online milieus, and the interplay between brands and places.

Dandy Badimo is a Research Scholar in Sustainable Tourism at the Okavango Research Institute, University of Botswana. His research interests are in community livelihood, impact assessment and evaluation, and tourism employment.

Rudorwashe Baipai is a Lecturer in the Tourism and Hospitality Department at Manicaland State University of Applied Sciences, Mutare, Zimbabwe. Her research interests are in sustainable tourism development with a particular focus on agritourism, sustainability, and community livelihoods in developing countries.

Israel R. Blackie is a Lecturer in the Department of Sociology at the University of Botswana. His research interests are in pro-poor development research such as intersection of state and local people and their institutions on common pool resource conservation, management, and rural livelihoods.

Oliver Chikuta is a Program Director at Modul School of Tourism and Hospitality Management at Nanjing Tech University Pujiang Institute, China. His research interests are in nature-based tourism, sustainable and heritage tourism, customer service excellence, and tourism marketing with a particular focus on universal accessibility in tourism.

Kaitano Dube is an Associate Professor of Tourism Geography at Vaal University of Technology in South Africa. He is a National Research Foundation of South Africa rated researcher who also works as an independent research contractor at the University of South Africa. He has written extensively on tourism, Covid-19, climate change, sustainability, and tourism aviation.

Esraa A. El-Masry is a Lecturer in Marine Geology at the Department of Oceanography, Faculty of Science, Alexandria University, Egypt. Her research interests include the advance of remote sensing and GIS and their applications in assessing the impacts of human development and climate change on coastal areas within the framework of coastal zone management.

Contributors ix

Abigail Lillian Engleton is currently Deputy Permanent Secretary (Tourism) in the Ministry of Environment and Tourism. She worked previously as a National Coordinator, UNDP Global Environment Facility Small Grants Programme, Gaborone, Botswana.

Edson Gandiwa is currently the Director Scientific Services at the Zimbabwe Parks and Wildlife Management Authority. His research interests are in community-based natural resource management with a particular focus on wildlife conservation and capacity building.

Hesekia Garekae is a Lecturer in the Department of Environmental Science at the University of Botswana. He has a PhD in Environmental Science, Rhodes University, South Africa. His research interests intersect forests, livelihoods, and conservation.

Kenalekgosi Gontse is a PhD student in the Department of Tourism and Hospitality, University of Botswana. His research interests are in human–wildlife interaction, tourism, and wildlife conservation.

Wame L. Hambira is an Associate Professor of Environmental Management at the University of Agriculture and Natural Resources in Botswana. Her research work is centred on global environment change and its socio-economic impacts, particularly on the tourism sector and tourism dependent communities. Her work also includes identifying potential response trajectories that inform environment and climate change policy in line with the national, regional, and global sustainability agenda.

Beatrice H. O. Ohutso Imbaya is a Senior Lecturer in the Tourism Management Department, Moi University, Kenya. She holds a PhD in Tourism Management from the same university and her research interests are in tourism entrepreneurship, community and rural tourism, and tourism education. Her teaching career spans over 30 years at tertiary and university levels. She is a member of Kenya Institute of Management, Ecotourism-Kenya, Association for Tourism and Leisure Education and Research, Africa, Tourism Professional Association, and Toastmasters International.

Oluwatoyin D. Kolawole is a Professor of Rural Development at the Okavango Research Institute (ORI), University of Botswana. His research interests are in social change/diffusion studies, natural resources management, rural development and the intersection of local and Western knowledge/innovations in sustainable, low external input (LEI) agriculture.

Joyce Lepetu is an Associate Professor in the Department of Range and Forest Resources at the Botswana University of Agriculture and Natural Resources. She has a PhD in Forest Science, University of Florida, USA. Her research interests are in forest conservation and livelihoods, Protected Area management, agroforestry, and climate change adaptation.

Patrick J. N. L'Espoir Decosta is an Associate Professor at the Research School of Management, Australian National University. His research interests are in tourism management and marketing with an increasing focus on Indigenous tourism and sustainability.

Gaseitsiwe Smollie Masunga is a Research Scholar in Landscape Ecology at the Okavango Research Institute, University of Botswana. His research interests revolve around wildlife conservation, community-based natural resource management, ecological disturbance factors, and landscape processes.

x *Handbook on tourism and conservation*

Joseph E. Mbaiwa is a Professor of Tourism Studies and Director of the Okavango Research Institute (ORI), University of Botswana. He is widely published in the areas of tourism development, rural livelihoods and biodiversity conservation in the Okavango Delta. He contributed to the preparation for the listing of the Okavango Delta as the 1000th UNESCO World Heritage site.

Emmanuel Mogende is a Lecturer in Natural Resource Governance at the Department of Environmental Science, University of Botswana. His research interests are in the political economy and political ecology of natural resources with a particular focus on wildlife conservation in Southern Africa.

Thekiso Molokwane is a Senior Lecturer of Public Administration in the Department of Political and Administrative Studies, University of Botswana. He has worked for the Government of Botswana, Lecturer at Limkokwing University of Creative Technology, and Postdoctoral Fellow in Public Administration – North West University, Vaal Campus RSA. His research interests include public policy, new public management, and public sector reforms with focus on public–private partnerships.

Ineelo Mosie is a Senior Technician in the Monitoring Unit at Okavango Research Institute, University of Botswana. His research interest is in fisheries management with particular focus on sustainable fisheries management and conservation in the Okavango Delta.

Inocent Moyo is an Associate Professor in the Department of Geography and Environmental Studies and Acting Deputy Dean of Research, Innovation and Internationalisation in the Faculty of Science, Agriculture and Engineering at the University of Zululand, South Africa. He researches borders, migration, political ecology, and the political economy of the informal economy in the Southern African region.

Maduo Mpolokang is a trainee Environmental Assessment Practitioner under Aqualogic Pty (Ltd), Botswana. He holds an MSc in Environmental Science from the University of Botswana. His research interests include tourism and community development, climate change, and environmental sustainability management.

Chiedza N. Mutanga is a Senior Research Fellow in Sustainable Tourism at the Okavango Research Institute, University of Botswana. Her research interests are in sustainable tourism development, with a particular focus on nature-based tourism, as well as Protected Area tourism and community livelihoods.

Jabulile Happyness Mzimela is a Lecturer and registered PhD student at the University of Zululand (South Africa) in the Department of Geography and Environmental Studies. Her PhD research focuses on the psychological factors that influence adaptation behaviour in the context of climate change and variability. Her research interests include human geography and environmental management.

Barbara N. Ngwenya is retired Professor of Applied Anthropology, Okavango Research Institute, University of Botswana. Her research interests and publications focus on human health in wetland and savannah ecosystem services, in particular access to and control over provisioning (water, food, and medicine) and cultural recreation (spiritual and tourism) including ways in which these affect and are affected by social and institutional relations. Professor

Contributors xi

Ngwenya is well grounded in client-centred participatory research methods including culturally sensitive documentation of indigenous knowledge systems and traditional health and healing practices in southern Africa.

Agnes Tshepo Nkone is a Social Studies teacher and a member of Botswana Social Studies Association. She is a regional Chairperson of the Social Studies Cluster in Kweneng District, Botswana. She holds a Bachelor of Education Secondary (Social Studies) degree and a Master of Public Administration degree specializing in Natural Resources Management.

Rita Wairimu Nthiga is a Senior Lecturer, Department of Hotel and Hospitality Management, Moi University, Kenya. She holds a Master's degree in Hospitality and Tourism Management from Kenyatta University, Kenya and a PhD in Tourism, Conservation and Development from Wageningen University in the Netherlands. Her research interests are in sustainable tourism and hospitality, conservation and development, hospitality operations management and contemporary issues in tourism and hospitality, as well as tourism policy planning and policy analysis. She is currently the Chair of the Department of Hotel and Hospitality Management, Moi University and a member of the Tourism Professional Association (TPA).

Karina Nicole Pérez-Olmos has a PhD in Agricultural Sciences from the Universidad Veracruzana in Mexico, where she developed the research thesis on 'Agritourism in the multifunctionality of the landscape'. Topics related to sustainable development are part of her research interests such as rural tourism, geographic information systems, food security, and biodiversity conservation.

Jeremy Perkins is an Associate Professor in Range Ecology at the Department of Environmental Science at the University of Botswana. He undertook his PhD research on Kalahari cattle posts in 1988 and returned to Botswana in 1992. Research areas include the coexistence of wildlife and cattle, Community Based Natural Resource Management (CBNRM), climate change and sand rivers.

Gofaone Rammotokara is a Technician in the Department of Range and Forest Resources at the Botswana University of Agriculture and Natural Resources. She has a Bachelor's degree in Range Sciences, Botswana University of Agriculture and Natural Resources. Her research interests include range ecology, ecological restoration and sustainability, economics of environmental issues, as well as the application of geographical information systems and remote sensing in forest ecology and management.

Gondo Reniko is a Senior Research Fellow at the Okavango Research Institute (ORI), University of Botswana in southern Africa. His research interests are in the interface between Western and Indigenous knowledge systems in the management of natural resources.

Siamisang Sehuhula is a PhD student in Natural Resources Management with the Okavango Research Institute, University of Botswana. His research interests are in environmental communication with a particular focus on public participation in environmental decision-making in Botswana.

Gladys B. Siphambe holds a PhD in Tourism Management from the University of Botswana. She is currently a Heritage Manager and a Projects Coordinator at the Department of National Museum and Monuments in Gaborone. She has been working with communities since 1997,

xii *Handbook on tourism and conservation*

assisting them to develop heritage tourism projects in their localities. Her main research interests are in cultural heritage tourism development, community mobilization, community perceptions towards tourism development, livelihoods sustainability, and biodiversity conservation.

Kekgaoditse Suping is a Lecturer of International Relations and Politics in the Department of Political and Administrative Studies at the University of Botswana. His research interests are diplomacy, international environmental cooperation, and natural resource politics.

Unabo Tafa is a researcher in tourism science, rural livelihoods, and natural resources management. In 2016, Mr Tafa produced an original research thesis titled 'Tourism development, rural livelihoods and land use conflict resolution at Tachila nature reserve'. The area of his research interest covers concepts of tourism development in Botswana, travel, hospitality, sustainable travel, and customer behaviour in tourism and travel. Mr Tafa's research output continues to contribute immensely to tourism development in the North East District of Botswana. He is currently undertaking a PhD and his research addresses politics in tourism development and the creation of peace parks.

Kaarina Tervo-Kankare works currently as an Environmental Expert, and was formerly a Lecturer at the University of Oulu (Finland) in the Geography Research Unit. Her main research interests cover sustainability, global environmental change, and tourism.

Olekae T. Thakadu is an Associate Professor (Environmental Communication) at the University of Botswana, Okavango Research Institute. His research focus is on environmental communication and education, risk communication, community-based natural resources management, knowledge translation and extension, and natural resources management.

Sharon Tshipa is a Research Scholar at the Leaders of Africa Institute, and a Development Research Fellow at Dataville Research LLC. Her research interests are in development research, development communication and science communication, with a proclivity for climate change.

Carlos Enrique Villanueva-González is a PhD student at the Faculty of Tropical AgriSciences of the Czech University of Life Sciences in Prague, Czech Republic. He has worked in research to determine the contribution that different local production systems provide to the conservation of biodiversity and the strengthening of livelihoods at the rural level in Guatemala.

1. Introduction to the *Handbook on Tourism and Conservation*

Oluwatoyin D. Kolawole, Joseph E. Mbaiwa, Wame L. Hambira and Emmanuel Mogende

INTRODUCTION

Generally, tourism sector operations are largely contingent upon both the natural and built environments. In this book, however, much attention is focused on nature-based tourism, which encompasses many variants of tourism including rural tourism, ecotourism and agri-tourism. While nature-based tourism has many economic benefits and is acknowledged as the fastest growing form of tourism (Kuenzi & McNeely, 2008), activities associated with it exert pressure on the natural environment upon which people's wellbeing and the success of the industry depend. The main goal of the book is, therefore, to underscore the mutually inclusive relationship between tourism and environmental conservation within the context of the United Nations' sustainability agenda. If carefully managed, active participation in tourism activities by all stakeholders including grassroots communities might promote a judicious and effective utilisation of resources; enhance environmental protection and justice; and help to achieve economic gains. Although not as simplistic as it seems, a concerted effort by all stakeholders in advancing effective governance of resources through relevant institutions might go a long way in achieving economic progress and environmental conservation.

The social and ecological systems upon which tourism depends cannot exist in isolation. Hence, tourism and conservation are inseparable and have been the subject of research for decades. As new environmental-related and governance issues emerge, there is a need to unearth pertinent and context-specific information to consider tourism planning and development in a scenario of climate variability and change. Indeed, many stakeholders such as tourism companies, communities, conservation organisations and national governments face a lot of challenges in their efforts to contribute positively to environmental conservation. Given that tourism utilises land resources among other competing demands, many managerial and institutional challenges associated with certain trade-offs naturally arise. Land, upon which tourism activities take place, is a finite resource and must be fairly accessed and utilised by all competing stakeholders if only to enhance social equity and the adequate ordering of ecological services. Therefore there is a need to develop some pragmatic sustainable development strategies that enhance good governance and management of natural resources wherever tourism activities exist.

While this volume assembles various perspectives and contributions from researchers working in tourism and conservation around the globe, it specifically addresses underlying issues of competing demand for land use, global environment change, social and environmental justice, and governance in the context of tourism and its roles in environmental conservation. This introductory chapter presents a general overview on the interactions between tourism and environmental conservation – the main theme on which the book hinges.

2 *Handbook on tourism and conservation*

THEORETICAL FRAMEWORK

The book's theoretical foundation derives from Ostrom's (2009) social-ecological systems (SES) framework, which underscores the embeddedness of natural resources within a complex and layered system comprising multiple subsystems at multiple levels. The multiple subsystems bear the semblance of a living organism whose constituent parts comprise cells, organelles, tissues and organs. The framework, therefore, analyses subsystems of a complex SES comprising a resource system, resource units, users and governance systems meant to produce outcomes at the SES level, including individual subsystems and their components as well as other larger or smaller SES (Ostrom, 2009). The need to develop a common framework to organise findings derived from different isolated, scientific disciplines in ecological and social sciences, which in the past had failed to 'cumulate' in finding a common ground on how to address complex environmental governance issues informed the development of the SES (Ostrom, 2009). The framework, which has had 'a long history of empirical research on the commons, institutions, and collective action', was later conceived as 'a general tool to diagnose the sustainability of social-ecological systems' (Partelow, 2018, p. 1; Ostrom, 2009).

Ostrom developed a framework of analysis on how the interactions between four first-level subsystems (including the resource system, resource units, governance system and users) affect each other to produce an outcome within social, economic and political settings and other related ecosystems. In this context, while the resource system connotes the totality of a geographical space comprising terrestrial and aquatic biodiversity upon which nature-based tourism relies, the resource unit describes individual species or kinds of flora and fauna found within the resource system. While on the other hand, the governance system comprises the organs of government, organisations and rules enforced to manage the resource system, the users are those who appropriate the resource system for their livelihoods and recreation. It is noteworthy that the use of resource system conservation (e.g., a wildlife management or controlled hunting area) is contingent upon the knowledge that people have about the system, its size, the mobility of a resource unit and the structure and level of governance (see Ostrom, 2009). In the context of our analysis, these variables are crucial when dealing with nature-based tourism and environmental conservation issues. For instance, the mobility of a resource unit (e.g., elephants' movement) could be influenced by climate change, and other related ecological and geographical factors. Understanding these dynamics could help in the implementation of sustainable tourism activities. From McGinnis and Ostrom's (2014) standpoint, people have the ability to make conscious choices either individually or collectively to affect the outcomes of any SES. Evidence abounds on how traditional societies employ diverse rules to manage resource systems on a sustainable basis in different sectors and regions (Acheson et al., 1998). People with common interests can work together and self-organise in developing institutional frameworks for enhancing the governance of natural resources within their locality (Partelow, 2018). To reiterate the power of the commons, institutions (both traditional and mainstream) and collective action (Partelow, 2018; Agrawal, 2001; Wollenberg et al., 2007) in natural resources management is a mere platitude. This book therefore seeks to draw lessons on sustainable tourism and conservation amidst emerging global environmental-related and governance concerns.

STRUCTURE OF THE BOOK

The *Handbook on Tourism and Conservation* is compartmentalised into four thematic parts. The four themes are Tourism and Conservation in Protected Areas; Tourism and Climate Change; Sustainable Agritourism; and Destination Communities and Natural Resources Conservation. The four themes comprising 20 chapters point in one direction – the intersection of tourism, environment and sustainable natural resources use. The concluding chapter provides a synthesis of the analyses and debates throughout the book.

Part I: Tourism and Conservation in Protected Areas

Part I analyses the role of protected areas (PAs) in natural resources conservation. It highlights issues surrounding PAs underscored by the International Union for Conservation of Nature (IUCN) as *Protected areas with sustainable use of natural resources*. Increasingly, partnerships between conservation agencies and the tourism sector are contributing to changes in attitudes around the issues of biodiversity conservation and environmentally responsible business practices (Bushell & Bricker, 2016). The use of concessions for tourism business as a conservation model in PAs is commonplace. Against the promise of conservation to communities in terms of environmental and infrastructural gains, more and longer contracts have been granted to more powerful concessionaires (Dinica, 2017). The questions of how communities can be empowered to enhance natural resource conservation through tourism in PAs and how tourism activities can further the advancement of conservation objectives are addressed in this section.

This part comprises four chapters. Chapter 2, entitled 'Tourism and conservation in protected areas: on collective efficacy', by Jabulile Happyness Mzimela and Inocent Moyo, employs a collective efficacy (CE) analytical framework to underscore the intersection between tourism and conservation in PAs. The authors' standpoint is that CE is crucial for individual-level cognitive mechanisms through which communities around and within PAs may influence their actions to enhance conservation of resources in the area. The next chapter entitled 'The prospects of forest-based tourism for marginalised communities' by Joyce Lepetu, Gorata Rammotokara and Hesekia Garekae assesses the potential of forest-based tourism in contributing towards sustainable livelihoods of resource-poor communities. The chapter draws its data from the representatives of the Village Development Committee (VDC), Conservation Trust, youth and women's groups. Based on its findings, the chapter concludes that forest-based tourism has potential in contributing to the success of any pro-poor tourism initiatives in the marginalised communities that are adjacent to PAs. Chapter 4, entitled 'Human–elephant conflict: implications for rural livelihoods and wildlife conservation', by Kenalekgosi Gontse, Joseph Mbaiwa and Olekae Thakadu is based on a study which investigated 119 active arable-land farmers and six key informants to determine the extent to which elephant crop damage negatively impacted on people's livelihoods and wildlife conservation at Khumaga in the Boteti sub-district of Botswana. Situated within the context of Social Exchange Theory, Joseph Mbaiwa and Unabo Tafa's chapter entitled 'Residents' perception of ecotourism development at Tachila Nature Reserve, North-East District, Botswana' employs both primary and secondary data sources to analyse rural people's disposition towards a development initiative known as Tachila Nature Reserve (TNR) designed as a tourist destination area for ecotourism purposes. Findings show that local communities in the district did not have any positive perceptions about the initiative and its conservation agenda because of the perceived deprivation

4   *Handbook on tourism and conservation*

meted out to them in relation to the seizure of their agricultural lands, which otherwise could have served as a means of livelihood for them.

Part II: Tourism and Climate Change

Part II focuses on the tourism–climate change nexus. Compared to the Global North, relatively little empirical research has been conducted in relation to climate change impacts on nature-based tourism in the Global South (Hambira, 2017; Hoogendoorn & Fitchett, 2019). Clearly, different forms of tourism are threatened in various ways by climate-related conditions. For example, while nature-based tourism and parks are affected by storm damage caused by tropical cyclones (Southon & Van Der Merwe, 2018), drought and aridity (Preston-Whyte & Watson, 2005), adventure tourism is at risk of extreme weather events such as floods (Giddy et al., 2017). Beach tourism on the other hand is vulnerable to sea level rise and its impact (Hoogendoorn et al., 2016). Under increased temperatures, changes in rainfall and humidity as well as wind speed and direction can have a severe effect on established cultural artefacts and heritage sites including newly discovered heritage such as rock art (Duval & Smith, 2012). As tourism is largely nature-based, especially in Africa, it is imperative to protect any tourism base against the recurrent phenomenon of global climatic change.

Part II comprises five chapters which address the effect of climate change on tourism. Kaarina Tervo-Kankare's chapter on 'Tourism and climate change adaptation in protected areas' reviews relevant literatures to unearth pertinent issues on adaptation in tourism in PAs and outlines the challenges related to the development of adaptation strategies and methods. Chapter 7 on 'Tourism and climate change vulnerabilities: a focus on African destinations' by Kaitano Dube uses primary, archival and secondary data to investigate the vulnerabilities of selected tourism destinations across Africa, with an emphasis on the vulnerabilities caused by global warming. Unarguably, the chapter's findings show that global warming poses a threat to several African tourist destinations. Esraa El-Masry's chapter entitled 'Tourism and climate change: consequences, adaptation and mitigation' employs the Tourism Climate Index (TCI) to assess the impact of temperature and precipitation increases on tourism activities in different tourist destinations in Africa. The chapter also presents a comprehensive and integrated approach for mitigating and adapting to climate change scenarios in relation to achieving sustainable tourism on the African continent. Sharon Tshipa and Olekae Thakadu's chapter on 'Evaluating climate change communication for sustainable environmental conservation in the tourism sector' employs the framing theory and SES framework to evaluate climate change communication for sustainable environmental conservation in Botswana's tourism sector. The chapter indicates that the coverage of climate change and tourism issues is trailing the general debates on climate change in the media. Chapter 10 by Maduo Mpolokang and Jeremy Perkins entitled 'Perspectives on the effects of environmental change in northern Botswana and its implications for CBNRM' employs semi-structured interviews to obtain information from wildlife tour guides working with safari lodge operators. The chapter analyses the widespread awareness of environmental variability and the threat it poses to sustainable tourism and effective wildlife management.

Introduction 5

Part III: Sustainable Agritourism

The third part draws readers' attention to the agricultural sector and how it could enhance nature-based tourism through agritourism. Agritourism is a major component of the modern patterns of rural development. It is an avenue that farmers use to augment proceeds derived from farm produce through Alternative Agrifood Networks (Ammirato & Felicetti, 2013). The convergence of the exploitation of agricultural resources in addition to the rediscovery of historical and natural heritage offers the opportunity to achieve sustainable development (Ammirato & Felicetti, 2013). Over the years, the agritourism subsector has received approbation from some scholars for its application of more environmentally friendly agricultural methods (Mastronardi et al., 2015). While agritourism could enhance environmental sustainability, it may also have a negative impact on the environment if it is not carefully managed and its activities are overtaken by the mainstream tourism activities. Nonetheless, some scholars have pointed out that there is little evidence on the environmental impacts of visitors, and which seemingly are left unconsidered by farmers, planners and tourism professionals (Kline et al., 2007). The question of the eco-friendliness of agritourism, therefore, arises.

There are five interrelated chapters in this part. Karina Nicole Pérez-Olmos, Noé Aguilar-Rivera and Carlos Enrique Villanueva-González's chapter entitled 'Potentials and challenges of sustainable agritourism in Fortín, Veracruz, Mexico' is an exploratory study that uses structured surveys and focus group discussions to unearth various potentials and challenges of sustainable agritourism development in Fortín. Among others, the identified potentials for agritourism development include agricultural resources and tourist services, while challenges associated with the subsector are those bordering on product development and marketing strategies, technical assistance, education and training, management and entrepreneurship, public policies, and safety and security issues. Mikael Andéhn and Patrick L'Espoir Decosta's '"Negotiating with the juggernaut": on agritourism and the paradoxes of market-driven conservation' offers an exposition on marketisation and the potential of agritourism in achieving sustainable environmental conservation while bearing in mind the risk of overexploitation and the subversion inherent in mainstream tourism. Rita Nthiga and Beatrice Imbaya's chapter on 'Micro and small-scale culture-based tourism initiatives as a livelihood option for rural women in Kenya' uses a qualitative approach and the community capitals framework (CCF) to examine the contribution of tourism-related initiatives at a village in western Kenya where some women's groups engaged in pottery production to enhance community livelihoods and how the initiative contributed both positively and negatively to environmental conservation. The chapter on 'Environmental impact of rural tourism' by Gondo Reniko and Oluwatoyin Kolawole used remote sensing and geographical information systems (GIS) to obtain data to assess land use and land cover change (LULCC) in Botswana. The authors' findings show that there was a significant shift in infrastructure development between 1990 and 2020 as witnessed in significant increments in built-up areas and a decline in forest cover, which was attributable mostly to increased tourism activities in the village. Rudorwashe Baipai, Oliver Chikuta, Edson Gandiwa and Chiedza N. Mutanga's chapter entitled 'Towards agritourism development in Zimbabwe: growth potential, benefits and challenges' employs a multi-case study and multi-stakeholder, qualitative approach to assess the growth potential of agritourism development on Zimbabwean farms. It also analyses the impacts of agritourism on people's socio-economic and environmental wellbeing, as well as the challenges associated with the subsector's development. Among others, the chapter outlines income generation,

6 *Handbook on tourism and conservation*

broadening of the tourism product base, promotion of environmental conservation, provision of affordable tourism options for local entrepreneurs, and preservation of local heritage and culture as some of the benefits conferred by agritourism.

Part IV: Destination Communities and Natural Resources Conservation

The last part of the book explores pro-poor approaches, which enhance all-inclusive community empowerment in ecotourism and natural resources conservation. The extent to which destination communities benefit from tourism has always been a concern for stakeholders (see Mbaiwa, 2017). Tourism requires significant investments, which have always been out of reach for communities hosting these tourism businesses. Over the years, joint venture partnerships (JVPs) in ecotourism businesses have been encouraged through community-based natural resources management (CBNRM) but the desired results have not been fully realised as a result of unfavourable policies that tend to disadvantage local communities when juxtaposed with their business partners. Consequently, the affected local communities continue to decry their marginalisation or alienation by tourism investors. Ultimately, community stewardship of natural capital is brazenly compromised. This raises environmental and social justice questions in relation to the impact of tourism development on local communities. Political will is, therefore, critical in ensuring that the economic benefits derived from the tourism industry are accessed by the affected communities (Hambira, 2019).

This part comprises six chapters addressing conservation and community-based issues. Kekgaoditse Suping's chapter on 'Commodification of nature and territorialization: conservation, local communities and Botswana's international cooperation' employs secondary information sources and interpretative analysis to argue that international cooperation facilitated by wildlife resources in Botswana has been fraught with certain challenges, which include increased territorialisation for wildlife conservation that has earned Botswana some international recognition at the expense of local community. It also argues that increased institutionalised commodification of wildlife resources has attracted exogenous tourism enterprises that dominate and control wildlife conservation and tourism policies, among others. Israel Blackie's chapter entitled 'Community-based natural resources management and poverty reduction' makes an assessment on how local communities in Botswana supposedly deemed to be CBNRM beneficiaries have been alienated from accessing and utilising natural resources available in their local area. The chapter thus suggests that the government deliberately and systematically dispossessed subordinate social groups within the state in its bid to modernize. Siamisang Sehuhula's chapter on 'Information communication technologies and community-based tourism organisations' is premised on an SES framework to analyse the ICT structures, processes and outcomes in community-based tourism organisations (CBTOs). The chapter employs literature review and thematic analysis to determine the limitations imposed on ICT adoption by legislative, institutional and technological factors. In Chapter 19, entitled 'Assessing the role of the central government and communities in alleviating poverty through eco-tourism' by Thekiso Molokwane and Agnes Nkone uses a case study and qualitative approach to unravel the mixed feelings that arose from government inadequacies in alleviating poverty through ecotourism. Joseph Mbaiwa and Gladys Siphambe's chapter 'Sense of place and tourism in cultural landscapes' employs a qualitative approach and the concept of sense of place to analyse the effects of globalisation and commodification on tourism activities in the study area. While emphasising that culture provides a sense of place and identity for the people, the chapter concludes that the commodification and globalisation of cultural landscapes do not always impact negatively on cultural heritage but instead could promote the preservation and sustainability of a people's cultural leaning provided such local traditions are respected by the outsider-visitor. Finally, in Chapter 21, entitled 'Co-management of world heritage sites for community benefit', O. T. Thakadu, Wame L. Hambira, Gaseitsiwe Smollie Masunga, Barbara N. Ngwenya, Abigail Lillian Engleton, Dandy Badimo and Ineelo Mosie investigate the need for active multi-stakeholder involvement and participation in the conservation of biodiversity resources and protected areas management. They investigate a number of initiatives by international organizations which seek to ensure that local communities are empowered to co-manage environmental resources and derive benefits from the management and conservation of natural resources.

CONCLUSION

This *Handbook* is primarily put together to enhance holistic policy development in tourism and environmental conservation. As earlier noted, the book is generally premised on the interdependency existing between tourism and biodiversity conservation. It primarily underscores how nature-based tourism is contingent upon ecological factors and the intricacies inherent in their interactions. Thus, most of the chapters in the book emphasize the need to devise innovative frameworks to achieve this objective. All things considered, the ensuing thematic issues laid out in the compilation may, to a large extent, assist policymakers and institutions to implement tourism policies that place pertinent issues in tourism and environmental conservation on the same pedestal.

REFERENCES

Acheson, J. M., Wilson, J. A., & Steneck, R. S. (1998). Managing chaotic fisheries. In F. Berkes & C. Folke (eds), *Linking Social and Ecological Systems*. Cambridge: Cambridge University Press, pp. 390–413.

Agrawal, A. (2001). Common property institutions and sustainable governance of resources. *World Development*, *29*(10), 1649–1672.

Ammirato, S., & Felicetti, A. M. (2013). The potential of agritourism in revitalising rural communities: Some empirical results. In L. M. Camarinha-Matos & R. J. Scherer (eds), *Collaborative Systems for Reindustrialisation: PRO-VE 2013. IFIP Advances in Information and Communication Technology*, vol. 408. Berlin: Springer, pp. 489–497.

Bushell, R., & Bricker, K. (2016). Tourism in protected areas: Developing meaningful standards. *Tourism and Hospitality Research*, *17*(1), 106–120. https://doi.org/10.1177/1467358416636173.

Dinica, V. (2017). Tourism concessions in national parks: Neo-liberal governance experiments for a conservation economy in New Zealand. *Journal of Sustainable Tourism*, *25*(12), 1811–1829.

Duval, M., & Smith, B. (2012). Rock art tourism in the Ukhahlamba/Drakensberg World Heritage Site: Obstacles to the development of sustainable tourism. *Journal of Sustainable Tourism*, *21*(1), 134–153.

Giddy, J. K., Fitchett, J. M., & Hoogendoorn, G. (2017). A case study on the preparedness of white-water tourism to severe climatic events in Southern Africa. *Tourism Review International*, *21*(2), 213–220.

Hambira, W. L. (2017). Botswana tourism operators and policy makers' perceptions and responses to the tourism–climate change nexus: Vulnerabilities and adaptations to climate change in Maun and Tshabong area. *Nordia Geographical Publications*, *46*(2), 1–74.

Hambira, W. L. (2019). A review of community social upliftment practices by tourism multinational companies in Botswana. In M. T. Stone, M. Lenao & N. Moswete (eds), *Natural Resources, Tourism*

8 *Handbook on tourism and conservation*

and Community Livelihoods in Southern Africa: Challenges of Sustainable Development. London: Routledge, pp. 52–63.

Hoogendoorn, G., & Fitchett, J. M. (2019). Fourteen years of tourism and climate change research in southern Africa: Lessons on sustainability under conditions of global change. In M. T. Stone, M. Lenao & N. Moswete (eds), *Natural Resources, Tourism and Community Livelihoods in Southern Africa: Challenges of Sustainable Development.* London: Routledge, pp. 50–60.

Hoogendoorn, G., Grant, B., & Fitchett, J. M. (2016). Disjunct perceptions? Climate change threats in two low-lying South African coastal towns. *Bulletin of Geography: Socio-Economic Series, 31,* 59–71.

Kline, C., Cardenas, D., & Leung, S. (2007). Sustainable farm tourism: Understanding and managing environmental impacts of visitor activities. *Journal of Extension, 45*(2), 2RIB2, 1–5.

Kuenzi, C., & McNeely, J. (2008). Nature-based tourism. In O. Renn & K. D. Walker (eds), *Global Risk Governance.* Dordrecht: Springer, pp. 155–178.

Mastronardi, L., Giaccio, V., Giannelli A., & Scardera A. (2015). Is agritourism eco-friendly? A comparison between agritourisms and other farms in Italy using farm accountancy data network dataset. *Springerplus, 4*(1), 1–12. https://springerplus.springeropen.com/articles/10.1186/s40064-015-1353-4.

Mbaiwa, J. E. (2017). Poverty or riches: Who benefits from the booming tourism industry in Botswana? *Journal of Contemporary African Studies, 35*(1), 93–112.

McGinnis, M. D., & Ostrom, E. (2014). Social-ecological system framework: Initial changes and continuing challenges. *Ecology and Society, 19*(2), 30. http://dx.doi.org/10.5751/ES-06387-190230.

Ostrom, E. (2009). A general framework for analyzing sustainability of social-ecological systems. *Science, 325*(5939), 419–422.

Partelow, S. (2018). A review of the social-ecological systems framework: Applications, methods, modifications, and challenges. *Ecology and Society, 23*(4), 36. https://doi.org/10.5751/ES-10594-230436.

Preston-Whyte, R. A., & Watson, H. K. (2005). Nature tourism and climatic change in southern Africa. In C. M. Hall & J. Higham (eds), *Tourism, Recreation and Climate Change.* Clevedon: Channel View Publications, pp. 130–142.

Southon, M. P., & Van Der Merwe, C. M. (2018). Flooded with risks or opportunities: Exploring flooding impacts on tourist accommodation. *African Journal of Hospitality, Tourism and Leisure, 7*(1), 1–27.

Wollenberg, E., Merino, L., Agrawal, A., & Ostrom. E. (2007). Fourteen years of monitoring community-managed forests: Learning from IFRI's experience. *International Forestry Review, 9*(2), 670–684.

PART I

TOURISM AND CONSERVATION IN PROTECTED AREAS

2. Tourism and conservation in protected areas: on collective efficacy

Jabulile Happyness Mzimela and Inocent Moyo

INTRODUCTION

Protected Areas (PAs) were first established in the nineteenth century in the United States for the sole purpose of long-term biodiversity conservation and protection against degradation (Jouzi et al., 2020; Moyo and Cele, 2021). During the colonial conquest, exclusionary Western environmental management epistemologies were extended into Africa resulting in the establishment of PAs (Moyo, 2023; Musavengane and Kloppers, 2020; Nsikwini and Bob, 2019). In South Africa, colonists and, later, the apartheid government enacted several Acts (the Natives Trust and Land Act of 1936; the Group Areas Act of 1950; and the Separate Amenities Act of 1953) to ensure that Black Africans were dispossessed of their native land (Musavengane and Kloppers, 2020), thus promoting the human–nature dichotomy (Moyo and Cele, 2021). In particular, the Natal Colonial Government took the first steps toward establishing a PA in the Drakensberg in 1903 by issuing a government notice (Ezemvelo KZN Wildlife, 2019). Between 1903 and 1989 the PA system was expanded and consolidated (Ezemvelo KZN Wildlife, 2019). Resource-dependent indigenous communities were resented (Carruthers, 2013) and conceptualized as a conservation problem (Nsikwini and Bob, 2019). Following that, their livelihoods were criminalized (Nsikwini and Bob, 2019) even though humans and nature are closely intertwined (Jouzi et al., 2020), and the natural environment has endured because of indigenous knowledge (IK) (Moyo, 2023).

Initially, PA establishment and management fell under the public authority's mandate, but in the 1980s, private actors were introduced under the neoliberalization policies, resulting in the commodification of nature for tourism (Gumede and Nzama, 2021; Hora et al., 2018). Subsequently, colonial game reserves were transformed into national parks and tourist destinations (Carruthers, 2013). PAs now theoretically serve a dual mandate of conservation and livelihood improvement through the provision of natural resources, ecosystem services and tourism activities (Stone et al., 2022). In reality, however, a win–win situation is not always achievable (Pan et al., 2022).

Increasingly, PAs are tourism destinations of choice and account for a significant proportion of the global tourism industry (Şakar et al., 2022). There are myriad benefits of PA-based tourism (Allendorf, 2022; Chang and Watanabe, 2021) delivered locally and globally (Chang and Watanabe, 2021). These benefits include biodiversity conservation; sustained ecosystem services; natural resource extraction; cultural, recreational, and aesthetic benefits; employment and/or income benefits from tourism activities (Allendorf, 2022; Stone et al., 2022); contribution to Gross Domestic Product (Baum and Ndiuini, 2020); and climate change mitigation (Chang and Watanabe, 2021) amongst other benefits. Notwithstanding, there are various pathways and mechanisms through which PA-based tourism exerts pressure on communities (Albrecht et al., 2022), particularly in the African context. For example, land dispossession and restricted access to natural resources adversely affect livelihoods (Allendorf, 2022; Moyo, 2023; Nepal et al., 2022).

Adaptive governance is required to accomplish the dual mandate. Accordingly, in the 1980s Black Africans began to engage in conservation through Community-Based Tourism (CBT) to fulfil the mandates (Musavengane and Kloppers, 2020). Literature regarding integrating tourism into the indigenous community nexus is constantly evolving (Stone et al., 2022). It is important to note that empowerment has become ingrained in the lexicon of tourism scholars concerned with how tourism might "work for development" (Scheyvens and van der Watt, 2021). In Africa, extant studies indicate that tourism development in PAs may empower one group while disempowering another. For example, Stone (2015) found that in Botswana's Chobe Enclave, CBT performance and environmental conservation are limited because community members experience empowerment and disempowerment. Similarly, in a recent study conducted in Manicaland Province (Zimbabwe), Gohori and van der Merwe (2022) discovered that community members perceived tourism impacts differently as they simultaneously empower and disempower the community.

The studies cited above provides insights into tourism development and community empowerment or disempowerment. Adaptive governance requires concerted effort and multiple actors, and this entails individuals coordinating with others and banding together resources for the group goal attainments (Thaker, 2012) – in this case conservation through tourism development. However, the effectiveness of any collective depends on 'collective efficacy' (CE), a psychological construct defined as "a group's shared belief in its conjoint capabilities to organize and execute the courses of action required to produce given levels of attainment" (Bandura, 1997, p. 477).

Against this backdrop, this chapter discusses the intersection between tourism and conservation in PAs by responding to the overarching research question: How can communities be empowered to enhance the conservation of natural resources through tourism in PAs? To respond to this question, the chapter is structured as follows: this introduction is followed by a theoretical framework that analyses the notion of CE and its utility. There is a discussion of the context or background of this chapter in terms of a survey of literature relating to community empowerment which is then followed by a discussion of the findings. In the conclusion, it is reiterated that CE is crucial for individual-level cognitive mechanisms through which communities surrounding and within PAs may influence their actions to enhance the conservation of resources through PA-based tourism.

THEORETICAL CONSIDERATIONS: ON COLLECTIVE EFFICACY

According to Bandura's Socio-Cognitive Theory (SCT), self and collective efficacy beliefs influence individuals' and collectives' thoughts, feelings, and actions in diverse contexts, including those relating to the natural environment (Bandura, 1997; Bonniface and Henley, 2008; Thaker and Floyd, 2021). Humans do not live in isolation; by nature they are social beings who perceive the significance of working in groups to accomplish common goals and find solutions to challenges that affect their quality of life (Bandura, 1997; Bandura, 1998; Mackay et al., 2021). Self-efficacy (SE) alone is not enough to explain group performance; CE, a group-level attribute, best explains agency in collective efforts (Bandura, 1997; Esnard

12 *Handbook on tourism and conservation*

and Roques, 2014; Filho, 2018). Hence, this chapter is squarely focused on CE, as the power of communities depends on CE perceptions (Bandura, 1997).

Bandura (1997, p. 477) defined CE as "a group's shared belief in its conjoint capabilities to organize and execute the courses of action required to produce given levels of attainment". The variance of defining CE based on the study domain is acknowledged in the literature (see e.g., Meares, 2002, p. 1604; Sampson et al., 1997, p. 919). However, Bandura first coined the phrase in 1986, and his conceptualization remains widely accepted among applied psychology scholars (Filho, 2018). In this chapter, CE is framed as an individual's belief in their community's ability (which has diverse backgrounds) to successfully undertake conservation actions in PA-based tourism and advance conservation tourism initiatives. CE may predict pro-environmental behaviour in networked communities (Cuadrado et al., 2022). That is, the more individuals and groups believe in the capacity of the community to conserve natural resources, the more they will engage in conservation behaviours not only individually but also in group terms.

As summarized succinctly by Bandura (1997, p. 478), "people's beliefs in their collective efficacy influence the type of future they seek to achieve, how they manage their resources, the plans and strategies they construct, how much effort they put into their group endeavour, their staying power when collective efforts fail to produce quick results or encounter forcible opposition, and their vulnerability to discouragement" (Bandura, 1997, p. 478). Research evidence demonstrates that CE perceptions are positively related to individual and group performance levels (Leonard and Leviston, 2017; Salinger et al., 2020; Velasquez and LaRose, 2015). This denotes that positive CE perceptions in a community correspond to a higher motivational investment of group members to band resources at their command to achieve the group's goal (Thaker et al., 2016, 2019). Several studies (e.g., Jugert et al., 2016; Thaker et al., 2016, 2019) demonstrate that individuals with high levels of perceived CE positively influence support for and/or engagement in environmental policies or behaviours.

In the case of South Africa, legislation imposes policies on communities in and around PAs to enhance natural resource conservation and tourism development. The public and private actors who exercise authority and control do not necessarily bestow power, resources, and privileges in acts of beneficence (Bandura, 1997). Communities must negotiate a share of benefits and control through collective effort and often through prolonged struggle (Bandura, 1997). Thus increasingly, there is a need for solutions and interventions "that call for a commitment of collective effort rather than litanies about powerlessness that instill in people beliefs of inefficacy to influence conditions that shape their lives" (Bandura, 1986, p. 453). Addressing beliefs that disempower individuals/collectives to act is a significant and pressing research area where the goal is to support natural resource conservation through PA-based tourism.

But what is the relevance of CE in conservation and PA-based tourism? Collective effort – whether bottom-up by nature or bottom-up and top-down collaboration(s) – is required for conservation and tourism development. Individual effort is marginal, and this necessarily requires or invokes CE to influence and encourage group participation (Velasquez and LaRose, 2015; Zheng et al., 2020). Researchers contend that CE is derived from the self-efficacy of group members (Bandura, 1997) and perceptions of social cohesion (Zaccaro et al., 1995). Self-efficacy and, therefore, CE antecedents include mastery experiences and past performance; vicarious experiences of observing others' performances; social persuasion; and affective states and the emotional tone of the group (Bandura 1997; Hoogsteen, 2020;

Pomfret, 2019). Accordingly, we expect an individual's CE perceptions to relate to their individual efficacy beliefs and group cohesion. The chapter focuses on empowering communities through PA-based tourism to enhance natural resource conservation and explores tourism activities further to advance conservation objectives in the South African context. To accomplish this, we highlight (see section after the methodology part below) that empowerment initiatives should be in the direction that will develop/strengthen CE for conservation and tourism development.

MATERIALS AND METHODS

This chapter is based on insights from three indigenous communities, namely AmaNgwane, AmaSwazi, and AmaZizi who live on the outskirts of uKhahlamba-Drakensburg Park (UDP), a PA in KwaZulu-Natal, South Africa (Figure 2.1). In total, 42 community members and three community leaders were purposively selected between 2018 and 2019. They were asked questions about community involvement in PA around conservation and socio-economic development. Data was analysed following the thematic approach as recommended by Creswell (2014). This entailed reading through and organizing the data for analysis, leading to the identification of describing, explaining, and connecting themes in the light of the overarching question that this chapter addresses.

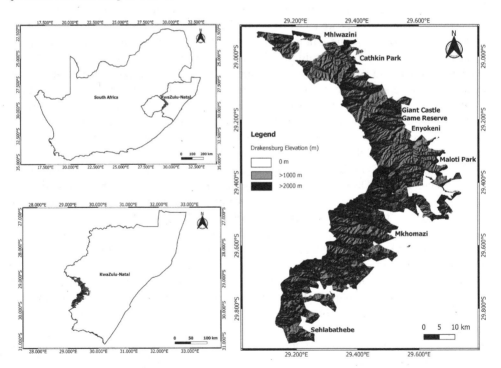

Figure 2.1   *Map of uKhahlamba Drakensberg*

PROTECTED AREAS, COMMUNITY EMPOWERMENT, AND TOURISM

The empowerment and participation of host communities in tourism destinations are critical for democratic policymaking (Jeong et al., 2018). Tourism development and sustainability cannot be assured without community empowerment (Joo et al., 2020). Hence, community empowerment has "become a mantra within the sustainable tourism literature" (Boley et al., 2017, p. 113). In this chapter, empowerment is defined "as the activation of the confidence and capabilities of previously disadvantaged or disenfranchised individuals or groups so that they can exert greater control over their lives, challenge unequal power relations, mobilize resources to meet their needs, and work to achieve social justice" (Scheyvens, 2020, p. 115).

Empowerment through tourism development enables communities to exercise control, and articulate their visions and actions (Gohori and van der Merwe, 2022). PA-based tourism is highly contested (Jeong et al., 2018). Empowerment, in this case, minimizes conflict between host communities and tourism development while also promoting the conservation of endowed tourism resources (Jeong et al., 2018). Community empowerment through tourism is based on endowed resources belonging to the community that is conserved while preserving sociopolitical culture (Jeong et al., 2021) and promoting linkages to local economies while minimizing tourism leakage (Stone, 2015).

In the context of sustainable tourism development, a new empowerment and sustainable development framework has been proposed by Scheyvens and van der Watt (2021). Six dimensions of empowerment are included in this framework – psychological, social, political, economic, environmental, and cultural empowerment – and these determine the impact of tourism activities on local communities (Scheyvens and van der Watt, 2021). Several community empowerment strategies for sustainable tourism have been developed and operationalized (by tourism planners, researchers, and practitioners) as community participation, capacitation, livelihood diversification, ownership, partnerships, CBT, and community sovereignty (Stone, 2015). In this chapter, we discuss mainly CBT for disempowered communities. The CBT concept was conceptualized to mitigate threats in PAs while also improving the local community's livelihoods (Gohori and van der Merwe, 2022). In addition to empowerment, CBT considers several issues including sustainability, social justice, and self-reliance (Giampiccoli and Glassom, 2021). Through CBT, communities are involved in tourism activities and businesses and can therefore exercise control, engage in development and management, and receive a share of tourism benefits (Giampiccoli and Glassom, 2021; Gohori and van der Merwe, 2022). CBT is meant to change social relations of power and capacity by distributing rights and benefits between indigenous communities and the governance structures (Allendorf, 2022) and by promoting constructive and collaborative ways of working (Allendorf, 2022). The relationship between empowerment, CE, and conservation is a critical study gap in the nascent empowerment literature. Thus, "empowerment is a product, process, and outcome. It is a product of a process of collective action where people are put in a position where they can develop social capital and politically exercise power" (Gohori and van der Merwe, 2022, p. 83). Therefore, CE is required as it determines the effectiveness of any collective action. Neglecting the role of CE in empowerment for conservation may have a detrimental effect on conservation.

Community Empowerment to Enhance Natural Resources Conservation through PA-based Tourism

PA issues are attributed to conflicting interests between the community and PA management. The fortress conservation system interfered with the integral relationship between people and their land and instigated extensive tensions between the community and PA management (Gumede and Nzama, 2021). Furthermore, the system has proven obsolete and ineffective because conservation efforts require the agency and support of the community to be effective (Moyo, 2023; Stone et al., 2022). Hence, PAs have not necessarily gained acceptance because of their very foundation of marginalization and exclusion of indigenous communities (Nsikwini and Bob, 2019). Matteucci et al. (2022, p. 180) agree that "the transformative potential of tourism for communities has not yet been achieved because the ontological and ethical assumptions on which tourism development is based have remained unquestioned". That is, the colonial patterns of land ownership and governance systems persist (Moyo, 2023), indicating disempowerment.

Reconceptualizing PA-based tourism by addressing the costs of tourism on local communities and improving societal benefits can be regarded as political empowerment. Co-management is one way to redress and resolve contestations due to fortress conservation (Stone, 2015). Effective, meaningful, and productive engagement is the hallmark of a meaningful co-management approach to the conservation of biodiversity (Moyo, 2023). In South Africa, the co-management approach was introduced in the post-apartheid period to reconcile land claims and biodiversity conservation (Moyo, 2023; Ezemvelo KZN Wildlife, 2019). The main characteristics of co-management are pluralism, communication and negotiation, transactive decision-making, social learning, and shared action/commitment (Plummer and Fitzgibbon, 2004, cited in Moyo, 2023). On paper, co-management appears to benefit communities, but in practice, communities continue to be disadvantaged and excluded from PA management, preventing benefits from being realized (Moyo, 2023), and in the process disempowering communities. Affording communities the opportunity to manage PA-based tourism enterprises devolves power and empowers the community politically, fostering a sense of ownership while providing a diverse set of societal benefits (e.g., economic and infrastructural benefits (Gumede and Nzama, 2019). These benefits fall under social and economic empowerment for enhancing natural resource management. Globally, benefit-sharing arrangements have been adopted, demonstrating linkages between conservation, tourism, and livelihoods (Nepal et al., 2022).

In some instances, the government and private entities benefit while the local communities are marginalized (Gumede and Nzama, 2019) and therefore experience significant levels of disadvantage across a range of ecological, cultural, social, economic, and health indicators (Gumede and Nzama, 2020; Nepal et al., 2022). This supports the widely held view that PA benefits are shared at the national and global levels while the costs are borne by local communities (Allendorf, 2022). Therefore, benefit-sharing is one way for communities to experience the positive impact of these establishments and be persuaded to support the conservation of natural resources (Bhammar et al., 2021).

Prior to the fortress system, societies were able to conserve biodiversity through IK (Moyo, 2023). IK incorporates lived experiences as it is deeply embedded in a particular environment and considers interrelationships (Gumede and Nzama, 2019; Moyo, 2023). This is the context within which indigenous and scientific knowledge forms should be integrated for partici tory tourism development (Gumede and Nzama, 2019; Moyo, 2023). This would indicate psychological empowerment since it would demonstrate that communities have always been self-reliant and would represent a commitment to preserving and respecting indigenous knowledge systems (IKS). In light of this review, the study documents the three communities' disempowerment and empowerment opportunities to enhance natural resources conservation through PA-based tourism.

Tourism Activities Further the Advancement of Conservation Objectives

The relationship between conservation, tourism, and livelihoods is complex and inconclusive (Stone et al., 2022). Accordingly, there is considerable disagreement over the overall impact of tourism activities on conservation. Wolf et al. (2019) argue that a symbiotic relationship between PA-based tourism and conservation may be possible. For example, tourists can enjoy the natural environment while PA managers generate profit, which can be reinvested in environmental conservation (Wolf et al., 2019). Endowed resources are often located in impoverished developing countries. Therefore, supporting conservation activities through tourism in these countries can fulfil development goals while also making conservation efforts efficacious (Bhammar et al., 2021). However, Matteucci et al. (2022) caution that tourism growth in economic terms does not always translate to enhanced conservation; in fact, environmental impacts may be detrimental.

To advance conservation efforts through tourism, it is necessary to resolve conflicts between resource-dependent communities and PA/tourism development managers (He et al., 2020). Conflict may arise as a result of several issues including limited natural resource extraction within the PA; competition between indigenous groups and wildlife; unequal distribution of tourism benefits; and damaging tourism activities (Allendorf, 2022). Conflict and trade-offs between conservation and tourism activities and livelihoods must be unpacked first and then resolved (via policies, regulations, and administrative orders) to garner local communities' support for conservation initiatives and tourism enterprises.

Tourism development must be intrinsically linked to conservation in such a way that they cannot be 'divorced' (Stone et al., 2022). A disconnect between theory and practice has been noted in that stakeholders (government, environmental protection agencies, and communities) are well-versed in conservation and engagement of indigenous communities as per the co-management approach, but execution is poor. Inequitable distribution of power and resources continues unabated in PA-based tourism because PA management benefits from the system, and is therefore reluctant to change the paradigm. For example, a study reveals that communities were never consulted during the establishment of PAs in Tanzania. Communities were given little opportunity to contribute because they were only informed about the PA once it was designed and established (Kegamba et al., 2022). In some developing countries, including South Africa, the participation of locals in nature-based tourism is often limited (Gumede and Nzama, 2019). Dejected individuals characterize some communities and this impedes effective collective action. Other community members have a disincentive for engagement because historically, they have not benefited or witnessed unequal benefit-sharing freeloaders benefiting. Understanding how tourism connects conservation and community development is critical for the sustainability of tourism and enhancing conservation efforts.

THE POTENTIAL OF TOURISM DEVELOPMENT IN UKHAHLAMBA DRAKENSBERG

Scheyvens and van der Watt's (2021) sustainable development and empowerment framework is used to explore conditions that may facilitate community empowerment with respect to PA-based tourism in UDP among the communities of AmaNgwane, AmaSwazi, and AmaZizi. It is important to note that empowerment dimensions may be interrelated, mutually reinforcing or eroding (Scheyvens and van der Watt, 2021).

Participation and Empowerment for PA Management

The paradigmatic shift to CBT is commendable, and beneficial outcomes have been noted in some areas. For example, Gohori and van der Merwe (2022) discovered that in Zimbabwe's Manicaland Province, through CBT a committee was elected to represent the community's interest in decision-making. On the contrary, the UDP communities claim that they were excluded when co-management plans were devised, and restrictions on access to and consumption of natural resources were imposed[1] (also see Moyo, 2023). Furthermore, the community members, including their traditional leaders, indicated that they are not involved in the management of the PA.[2] Yet CBT was conceptualized to devolve power from PA management to the local community (Nsikwini and Bob, 2019). This exemplifies the failure to reconceptualize colonial/apartheid governance systems in the PA that promoted the absence of Africans in the first place. That is, the community's power has been relinquished to PA management that advances their mandate, which does not necessarily address the community's needs. This finding supports Bandura's (1997, p. 477) assertion that "those who exercise authority and control do not go around voluntarily granting to others power over resources and entitlements in acts of beneficence". The findings demonstrate psychological, cultural, and political disempowerment and indicate non-compliance to sections 39(3) and 41(2)(e) of the National Environmental Management: Protected Areas Act 57 of 2003 where stakeholder involvement and support are deemed critical for effective PA governance (Ezemvelo KZN Wildlife, 2019).

It is important to note that participation typologies vary, and that participation does not necessarily equate to decision-making. For instance, the management structure of Kgalagadi Transfrontier Park supports the inclusion of Mier and Khomani San community views and opinions in the management of the PA and benefit-sharing arrangements (Lekgau and Tichaawa, 2021). However, the communities lack autonomy to make decisions and are instead consulted, which is problematic and limits their control (Lekgau and Tichaawa, 2021). Non-participation in decision-making is not exclusively ascribed to marginalization by PA management. The not-so-common viewpoint is that Black South Africans 'detest' anything to do with conservation because of historical land dispossession and exclusionary conservation ideologies implemented during apartheid (Musavengane and Kloppers, 2020). Furthermore, Kegamba et al. (2022) assert that participation may be hindered by the lack of knowledge and awareness whereby the community and PA management are unaware of the community's rights.

UDP community members further shared that they wanted to be involved in the governance and management of the PA believing that this would reduce conflict between the people and the PA, thereby fostering collaboration among stakeholders.[3] Literature reveals that the absence of community involvement in PA management jeopardizes the PA's sustainability

(Giampiccoli and Glassom, 2021) which may fuel people–park contestations. Hence, community participation in PA management is essential to enhance shared societal benefits and counter the negative impacts of tourism (Giampiccoli and Glassom, 2021). This necessitates political empowerment – the devolution of power to the community (Scheyvens, 1999).

PA managers need to create an environment conducive to collaboration and participation and responsive to community needs, allowing all concerned members to communicate their goals, plans, and progress. By doing so community members may shift their concerns about themselves to collective concerns. In this sense, "collaboration is not an assurance that tourism would be sustainable, however, absence of collaboration is tantamount to misunderstanding among stakeholders, crisis, and failure of sustainable tourism development at the tourism destination" (Eyisi et al., 2018, p. 38). Moyo and Cele (2021) suggested that communities must be trained to understand that they are partners and not subjects in PA management. This responsibility lies with PA management and academics.

Cuadrado et al. (2022) assert that CE beliefs for ecological behaviours are strengthened through participatory and collaborative activities, and consequently pro-environmental actions are demonstrated by individuals and society. When the situation at UDP is seen through the prism of CE, it becomes evident that the three communities need to mobilize to claim their rights as per the CBT approach, which includes the right to participate in decision-making and governance. The study reveals that interventions to develop/strengthen CE beliefs are critical to dissolve conflicts and promote tourism development and conservation. This suggests the need for the development of CE via the establishment of group associations to represent the community in decision-making/governance systems. In addition, CE may be developed by the community and in conjunction with PA management to mitigate contestations. The community's identification with the proposed association will encourage people to develop a sense of CE, which will motivate them to take collective action when appropriate (Politi et al., 2022). Additionally, community participation in PA management is likely to fuel CE and increase motivation for the dual mandate. Therefore, shifting the focus to perceived CE may be necessary for effective and adaptive PA governance.

Degree of Community Power

It appears that the UDP communities have no control over the distribution of benefits. Yet benefit-sharing is critical for the success of PAs, tourism businesses, and conservation (Bhammar et al., 2021). Overall, there is economic and psychological empowerment for the few that get jobs but disempowerment for the rest. The UDP communities expressed dissatisfaction with employment opportunities that were either seasonal or offered to people who came from other parts of the country.[4] The lack of employment opportunities means continued direct reliance on natural resources for livelihoods. This study is consistent with current literature, which states that locals barely benefit from PA-based tourism benefits but bear the costs of PA coexistence (Stone et al., 2022). The employment of migrants indicates tourism leakages and economic and social disempowerment. This concurs with Bhammar et al.'s (2021) study which reported that over 75 per cent of Bwindi Impenetrable National Park (Uganda) revenue is leaked. As evidenced by the literature, societal benefits motivate conservation behaviour, so these benefits must outweigh conservation costs. This is an egoistic perspective, but it works (Cuadrado et al., 2022). A synergistic relationship between conservation and societal benefits adds to beliefs of mastery. When individuals perceive their community as having the capacity to behave pro-environmentally, the more conservation initiatives they will partake in, their self-efficacy rises, and they act individually in collective terms (Cuadrado et al., 2022).

The UDP community members further shared that they want to own tourism business ventures or activities within the PA. They thought that if they were empowered through funding, they could initiate successful tourism ventures that they would own and control and not the current case where they are outsiders in what they considered a resource that must benefit the community in the first instance.[5] It is evident that we need to forge a way to achieve equity and legitimacy concerning the ownership of tourism enterprises in PA. As it is, only 5 per cent of businesses in the iSimangaliso Wetland Park are Black owned; the Black community is primarily involved in informal tourism (Giampiccoli and Glassom, 2021). In the context of UDP, this points to the need to develop employment opportunities for locals as this would empower them psychologically while enabling them to meet their economic demands and aspirations. Community members will support PA-based tourism if it assists with their own development (Scheyvens, 1999). The issue is that tourism positions are seasonal, menial, and limited such that the whole community cannot be offered jobs. Hence, we suggest a concessions policy to enable community ownership of tourism enterprises. The community could generate revenue through concessions and reinvest it in their enterprises. Gumede and Nzama (2019) contend that if locals have a sense of ownership of tourism enterprises, they will build self-reliance and be more inclined to commit themselves to such enterprises for extended periods.

PA management is in a position to assist community members in developing innovative tourism opportunities (e.g., via training, knowledge dissemination) and capitalizing on them while conserving natural resources. This would address the current low community participation in tourism enterprises and strengthen the community's authority and control. Community empowerment is not entirely the PA management responsibility; empowerment can be bottom-up. Hence we support the call for "the transformation of the tourism sector" (Giampiccoli and Glassom, 2021). The Makuleke community serves as an excellent example to emulate. This community (located in South Africa's Western Kruger National Area) founded the Makuleke Community Property Association in 1996, which operates the park in partnership with SANParks and Wilderness Safaris (Musavengane and Kloppers, 2020). This community association negotiated for 10 per cent of all park revenues and gives community members preference for employment opportunities (Musavengane and Kloppers, 2020). Furthermore, private actors are permitted to build and operate lodges on the condition that once the concession lease expires, the ownership is transferred to the community (Musavengane and Kloppers, 2020). This example clearly demonstrates that it is possible to prioritize locals when there are employment/income opportunities and that quality governance is possible when all stakeholders collaborate (Musavengane and Kloppers, 2020). If UDP communities are exposed to the Makuleke community's accomplishments, they will be inspired to believe that they, too, can improve via vicarious learning. Another study by Mabibibi et al. (2021) reveals that PA management indicated that local community members fill about 90 per cent of the Kruger National Park's general employee positions. These positions are permanent and therefore considered sustainable. Additionally, the Park plays a role in the establishment of community projects and ventures (including the Nkambeni Safari Camp and Mdluli Safari Lodge). In some of these projects, the community ownership is 50 per cent. In the preceding discussion, we recommended the development/strengthening of CE via the development of associations, to action group goal attainments. It is therefore suggested that communities

20 *Handbook on tourism and conservation*

leverage their CE beliefs to persuade PA management to give priority to locals when there are employment opportunities and to fund their tourism ventures.

Empowerment through Cultural Tourism Resources

The UDP community members opined that their participation in cultural tourism is limited.[6] Yet, they have cultural resources that they can develop and showcase for tourism purposes. This was particularly the case with the community of AmaZizi. In other words, the natural tourism resources in the form of the beautiful Drakensberg could be enhanced or augmented by the cultural resources of the people. The development of cultural tourism resources would mean the diversification of tourism enterprises and expansion of the cultural tourism industry as the environment will not be the only content of tourists' consumption of tourism products. Diversification can attract more tourists, increase revenue generation and reinvestment into the conservation of resources, and positively impact the community's economic fabric. Additionally, diversifying tourism resources to include cultural resources may reduce environmental impacts from mounting pressure on the environment (Bhammar et al., 2021).

As we expected, the community members shared the view that PA management was not supportive in this regard.[7] Cultural tourism development is an opportunity for collaboration as partners and not community members as subjects. This way the community would positively contribute to the conservation of the environment as they would view the success of the UDP as an endowed tourism resources enterprise as a foundation for their cultural tourism activities. The belief would be that, if the PA was not conserved and taken care of, it would be degraded and they would all lose out. If community members are not involved, they are likely to believe they have nothing to lose even if the PA is not conserved, whereas if they are involved, they are more likely to see the value of conserving the PA as its survival and preservation are the foundation of their activities and thus success and community development. The failure of PA management to support cultural tourism resource development can be classified as psychological and cultural disempowerment. The lack of support stems from the lack of participation in PA management. PA management cannot support these offerings if they are unaware of them, which we presume is the case because of the non-participation of communities in PA governance.

In arguing for cultural tourism resource development, we refer to the Makuleke community (located in South Africa's Kruger National Park). Through the Makuleke tourism initiative, this community has incorporated cultural tourism resources as one of their offerings (Mabibibi et al., 2021). This involves collaboration and connecting communities, government, and private entities in supporting conservation and tourism development (Mabibibi et al., 2021). In UDP efforts should be geared towards cultural tourism development. Therefore, in addition to engaging communities in decision-making, PA management should assist cultural resource tourism development by offering business skills to increase capacity and small grants to develop cultural tourism resources, a strategy suggested by Bhammar et al. (2021) to increase the community's involvement in and benefit from PAs.

By recognizing and developing cultural tourism resources and incorporating IK, the communities will be empowered psychologically via raised cultural awareness, self-esteem, and pride (Gohori and van der Merwe, 2022). Consistent with the literature on IK in tourism destinations, this study stresses the importance of incorporating IK in tourism. IK incorporates lived experiences as it is deeply embedded in a particular environment and considers interrelationships (Gumede and Nzama, 2019; Moyo, 2023). This is the context within which indigenous and scientific knowledge forms should be integrated for participatory tourism development (Gumede and Nzama, 2019; Moyo, 2023). This would be a form of psychological, cultural, and social empowerment. It would demonstrate that communities have always been self-reliant and represent a commitment to preserving and respecting IKS. Again, we propose associations to draw on CE to persuade PA management to invest in cultural tourism resources development both in terms of skills and financial resources.

CONCLUSION

PAs were exclusively established for conservation; nevertheless, they have evolved into vehicles that provide ecosystem services and aid in poverty reduction through tourism. The contribution of this study is premised on the recommendation to build/strengthen CE to yield new dispositions for advancing the multiple goals of tourism (conservation, tourism development, and community development). CE is one of the tenets of collective action intentions. Theoretically, CBT is a noble ideal; however, colonial/apartheid structures, ideologies, and inequities persist in UDP indicating poor uptake of CBT and community disempowerment. As a result, low community participation persists, and so do inequitable benefit-sharing and destructive tourism activities. Yet participation is legitimized by the policy frameworks in South Africa and by CBT. The lack of community involvement in PAs increases the costs borne by the community. The inclusion of the host communities in PA management and processes is the empowerment that is proposed in this chapter. Empowerment coalesces into conservation action; collective action necessitates CE. Empowerment initiatives should be in the direction that will develop/strengthen collective efficacy.

PA management can empower the communities through employment opportunities and investing in community tourism ventures. Tourism enterprises should also be diversified to include cultural tourism resources. Cultural tourism resources have not been fully integrated into the tourist experience in UDP. That is, tourism resources are not applied optimally because the community's capacity has been restrained. Community empowerment should be both bottom-up and top-down. All stakeholders should reconceptualize themselves and their relationships to one another and the natural environment in non-hierarchical ways to propel PA-based tourism forward. The salience of this chapter lies in recognizing that CE is a necessary construct that is crucial for individual-level cognitive mechanisms through which communities surrounding and within PAs may influence their actions to enhance the conservation of resources through PA-based tourism. CE may help researchers uncover how the community functions and how they can be assisted to act in unison for their mutual benefit.

NOTES

1. Interviews with community members in 2018 and 2019.
2. Interviews with community members in 2018 and 2019.
3. Interviews with community members in 2018 and 2019.
4. Interviews with community members in 2018 and 2019.
5. Interviews with community members in 2018 and 2019.
6. Interviews with community members in 2018 and 2019.

22 *Handbook on tourism and conservation*

7. Interviews with community members in 2018 and 2019.

REFERENCES

Albrecht, J. N., Haid, M., Finkler, W., and Heimerl, P. 2022. What's in a name? The meaning of sustainability to destination managers. *Journal of Sustainable Tourism*, 30(1), 32–51.

Allendorf, T. D. 2022. A global summary of local residents' perceptions of benefits and problems of protected areas. *Biodiversity and Conservation*, 31(2), 379–396.

Bandura, A. 1986. *Social Foundations of Thought and Action: A Social Cognitive Theory*. Englewood Cliffs, NJ: Prentice Hall.

Bandura, A. 1997. *Self-Efficacy: The Exercise of Control*. New York: W. H. Freeman.

Bandura, A. 1998. Personal and collective efficacy in human adaptation and change. *Advances in Psychological Science*, 1(1), 51–71.

Baum, T. and Ndiuini, A. 2020. *Sustainable Human Resource Management in Tourism: African Perspectives*. Dordrecht: Springer International Publishing.

Bhammar, H., Li, W., Molina, C. M. M., Hickey, V., Pendry, J., and Narain, U. 2021. Framework for sustainable recovery of tourism in protected areas. *Sustainability*, 13(5), 1–10.

Boley, B. B., Ayscue, E., Maruyama, N., and Woosnam, K. M. 2017. Gender and empowerment: Assessing discrepancies using the resident empowerment through tourism scale. *Journal of Sustainable Tourism*, 25(1), 113–129.

Bonniface, L. and Henley, N. 2008. A drop in the bucket: Collective efficacy perceptions and environmental behaviour. *Australian Journal of Social Issues*, 43(3), 345–358.

Carruthers, J. 2013. The Royal Natal National Park, Kwazulu-Natal: Mountaineering, tourism and nature conservation in South Africa's first national park c. 1896 to c. 1947. *Environment and History*, 19(4), 459–485.

Chang, L. and Watanabe, T. 2021. Dilemma faced by management staff in China's protected areas. *Land*, 10(12), 1–18.

Creswell, J. W. 2014. *Research Design: Qualitative, Quantitative, and Mixed Methods Approaches*. London: Sage Publications.

Cuadrado, E., Macias-Zambrano, L. H., Carpio, A. J., and Tabernero, C. 2022. The moderating effect of collective efficacy on the relationship between environmental values and ecological behaviors. *Environment, Development and Sustainability*, 24(3), 4175–4202.

Ezemvelo KZN Wildlife. 2019. *uKhahlamba Drakensberg Park: Integrated Management Plan*. Version 2, South Africa Ezemvelo KZN Wildlife, Pietermaritzburg.

Esnard, C. and Roques, M. 2014. Collective efficacy: A resource in stressful occupational contexts. *European Review of Applied Psychology*, 64(4), 203–211.

Eyisi, A. P., Lee, D., and Trees, K. 2018. Collaboration as a potential strategy for addressing socio-cultural impacts of tourism development: Insights from Nigeria. 7th Biennial International Tourism Studies Association Conference (ITSA) 2018, Tshwane, South Africa.

Filho, E. 2018. Team dynamics theory: Nomological network among cohesion, team mental models, coordination, and collective efficacy. *Sport Sciences for Health*, 15(1), 1–20.

Giampiccoli, A. and Glassom, D. 2021. Community-based tourism in protected areas: Elaborating a model from a South African perspective. *Advances in Hospitality and Tourism Research (AHTR)*, 9(1), 106–131.

Gohori, O. and van der Merwe, P. 2022. Tourism and community empowerment: The perspectives of local people in Manicaland province, Zimbabwe. *Tourism Planning and Development*, 19(2), 81–99.

Gumede, T. K. and Nzama, A. T. 2019. Comprehensive participatory approach as a mechanism for community participation in ecotourism. *African Journal of Hospitality, Tourism and Leisure*, 8(4), 1–11.

Gumede, T. K. and Nzama, A. T. 2020. Enhancing community participation in ecotourism through a local community participation improvement model. *African Journal of Hospitality, Tourism and Leisure*, 9(5), 1252–1272.

Gumede, T. K. and Nzama, A. T. 2021. Approaches toward community participation enhancement in ecotourism. In M. N. Suratman (ed.), *Protected Area Management: Recent Advances*. London: IntechOpen, pp. 201–217.

He, S., Yang, L., and Min, Q. 2020. Community participation in nature conservation: The Chinese experience and its implication to national park management. *Sustainability*, 12(11), 1–17.

Hoogsteen, T. J. 2020. Collective efficacy: Toward a new narrative of its development and role in achievement. *Palgrave Communications*, 6(1), 1–7.

Hora, B., Marchant, C., and Borsdorf, A. 2018. Private protected areas in Latin America: Between conservation, sustainability goals and economic interests. A review. *Journal on Protected Mountain Areas Research and Management*, 10(1), 87–94.

Jeong, E., Ryu, I., and Brown, A. 2018. Moderating effect of sense of community on the relationship between psychological empowerment and tourism policy participation of local residents. *Global Business & Finance Review*, 23(1), 36–46.

Jeong, E., Shim, C., Brown, A. D., and Lee, S. 2021. Development of a scale to measure intrapersonal psychological empowerment to participate in local tourism development: Applying the sociopolitical control scale construct to tourism (SPCS-T). *Sustainability*, 13(7), 1–16.

Joo, D., Woosnam, K. M., Strzelecka, M., and Boley, B. B. 2020. Knowledge, empowerment, and action: Testing the empowerment theory in a tourism context. *Journal of Sustainable Tourism*, 28(1), 69–85.

Jouzi, Z., Leung, Y. F., and Nelson, S. 2020. Terrestrial protected areas and food security: A systematic review of research approaches. *Environments*, 7(10), 1–15.

Jugert, P., Greenaway, K. H., Barth, M., Büchner, R., Eisentraut, S., and Fritsche, I. 2016. Collective efficacy increases pro-environmental intentions through increasing self-efficacy. *Journal of Environmental Psychology*, 48, 12–23.

Kegamba, J. J., Sangha, K. K., Wurm, P., and Garnett, S. T. 2022. A review of conservation-related benefit-sharing mechanisms in Tanzania. *Global Ecology and Conservation*, 33, 1–16.

Lekgau, R. J. and Tichaawa, T. M. 2021. Community participation in wildlife tourism in the Kgalagadi Transfrontier Park. *Tourism Review International*, 25(2–3), 139–155.

Leonard, R. and Leviston, Z. 2017. Applying a model of collective efficacy for understanding consumer and civic pro-environmental actions. *Socijalna ekologija: časopis za ekološku misao i sociologijska istraživanja okoline*, 26(3), 105–123.

Mabibibi, M. A., Dube, K., and Thwala, K. 2021. Successes and challenges in sustainable development goals localisation for host communities around Kruger National Park. *Sustainability*, 13(10), 1–16.

Mackay, C. M., Schmitt, M. T., Lutz, A. E., and Mendel, J. 2021. Recent developments in the social identity approach to the psychology of climate change. *Current Opinion in Psychology*, 42, 95–101.

Matteucci, X., Nawijn, J., and von Zumbusch, J. 2022. A new materialist governance paradigm for tourism destinations. *Journal of Sustainable Tourism*, 30(1), 169–184.

Meares, T. 2002. Praying for community policing. *California Law Review*, 90(5), 1593–1605.

Moyo, I. 2023. Beyond a tokenistic inclusion of indigenous knowledge systems in protected area governance and management in uKhahlamba-Drakensberg. *African Geographical Review*, 42(2), 141–156.

Moyo, I. and Cele, H. M. S. 2021. Protected areas and environmental conservation in KwaZulu-Natal, South Africa: On HEIs, livelihoods and sustainable development. *International Journal of Sustainability in Higher Education*, 22(7), 1536–1551.

Musavengane, R. and Kloppers, R. 2020. Social capital: An investment towards community resilience in the collaborative natural resources management of community-based tourism schemes. *Tourism Management Perspectives*, 34, 100654.

Nepal, S. K., Lai, P. H., and Nepal, R. 2022. Do local communities perceive linkages between livelihood improvement, sustainable tourism, and conservation in the Annapurna Conservation Area in Nepal? *Journal of Sustainable Tourism*, 30(1), 279–298.

Nsikwini, S. and Bob, U. 2019. Protected areas, community costs and benefits: A comparative study of selected conservation case studies from Northern Kwazulu-Natal, South Africa. *GeoJournal of Tourism and Geosites*, 27(4), 1377–1391.

Pan, X., Yang, Z., and Han, F. 2022. Exploring the historical evolution of tourism-environment interaction in protected area: A case study of Mt. Bogda. *Journal of Geographical Sciences*, 32(1), 177–193.

Politi, E., Piccitto, G., Cini, L., Béal, A., and Staerklé, C. 2022. Mobilizing precarious workers in Italy: Two pathways of collective action intentions. *Social Movement Studies*, 21(5), 608–624.

Pomfret, G. 2019. Conceptualising family adventure tourist motives, experiences and benefits. *Journal of Outdoor Recreation and Tourism*, 28, 1–25.

Şakar, D., Aydin, A., and Akay, A.E. 2022. Essential issues related to construction phases of road networks in protected areas: A review. *Croatian Journal of Forest Engineering: Journal for Theory and Application of Forestry Engineering*, 43(1), 219–237.

Salinger, A. P., Sclar, G. D., Dumpert, J., Bun, D., Clasen, T., and Delea, M. G. 2020. Sanitation and collective efficacy in rural Cambodia: The value added of qualitative formative work for the contextualization of measurement tools. *International Journal of Environmental Research and Public Health*, 17(1), 1–18.

Sampson, R. J., Raudenbush, S., and Earls, F. 1997. Neighborhoods and violent crime: A multilevel study of collective efficacy. *Science*, 277(5328), 918–924.

Scheyvens, R. 1999. Ecotourism and the empowerment of local communities. *Tourism Management*, 20(2), 245–249.

Scheyvens, R. 2020. Empowerment. In A. Kobayashi (ed.), *International Encyclopedia of Human Geography*, 2nd edn. Oxford: Elsevier, pp. 115–122.

Scheyvens, R. and van der Watt, H. 2021. Tourism, empowerment and sustainable development: A new framework for analysis. *Sustainability*, 13(22), 1–19.

Stone, M. T. 2015. Community empowerment through community-based tourism: The case of Chobe Enclave Conservation Trust in Botswana. In R. van der Duim, M. Lamers, and J. van Wijk (eds), *Institutional Arrangements for Conservation, Development and Tourism in Eastern and Southern Africa*. Dordrecht: Springer, pp. 81–100.

Stone, M. T., Stone, L. S., and Nyaupane, G. P. 2022. Theorizing and contextualizing protected areas, tourism and community livelihoods linkages. *Journal of Sustainable Tourism*, 30(11), 2495–2509.

Thaker, J. 2012. Climate change in the Indian mind: Role of collective efficacy in climate change adaptation. PhD dissertation, George Mason University, Fairfax, VA, USA.

Thaker, J. and Floyd, B. 2021. Co-benefits associated with public support for climate-friendly COVID-19 recovery policies and political activism. *Journal of Science Communication*, 20(5), 1–24.

Thaker, J., Howe, P., Leiserowitz, A., and Maibach, E. 2019. Perceived collective efficacy and trust in government influence public engagement with climate change-related water conservation policies. *Environmental Communication*, 13(5), 681–699.

Thaker, J., Maibach, E., Leiserowitz, A., Zhao, X., and Howe, P. 2016. The role of collective efficacy in climate change adaptation in India. *Weather, Climate, and Society*, 8(1), 21–34.

Velasquez, A. and LaRose, R. 2015. Youth collective activism through social media: The role of collective efficacy. *New Media & Society*, 17(6), 899–918.

Wolf, I. D., Croft, D. B., and Green, R. J. 2019. Nature conservation and nature-based tourism: A paradox? *Environments*, 6(9), 1–22.

Zaccaro, S. J., Blair, V., Peterson, C., and Zazanis, M. 1995. Collective efficacy. In J. E. Maddux (ed.), *Self-Efficacy, Adaptation, and Adjustment: Theory, Research, and Application*. New York: Plenum, pp. 305–328.

Zheng, D., Liang, Z., and Ritchie, B. W. 2020. Residents' social dilemma in sustainable heritage tourism: The role of social emotion, efficacy beliefs and temporal concerns. *Journal of Sustainable Tourism*, 28(11), 1782–1804.

3. The prospects of forest-based tourism for marginalised communities

Joyce Lepetu, Gofaone Rammotokara and Hesekia Garekae

INTRODUCTION

Globally, tourism is poised as a viable strategy for economic growth, more especially in developing regions which are still battling with poor socio-economic conditions (Chen et al., 2018; Mugizi et al., 2018; Safa et al., 2021). In developing regions, tourism is an important source of foreign exchange and employment opportunities (Manzoor et al., 2019). Before the advent of the Covid-19 pandemic, tourism was a significant contributor to the global economy. For instance, in 2019, the share of the travel and tourism industry to the global gross domestic product (GDP) was 10.4 per cent (WTTC, 2021). Still in the same year, about 334 million people were employed in the travel and tourism industry. However, the contribution of tourism to the global economy decreased by half in 2020. The sector contributed 5.5 per cent to the global economy and retained 272 million jobs (WTTC, 2021). This staggering decline in the significance of tourism to the world's economy came as a result of travel disruptions and associated restrictions put in place to counter the spread of Covid-19.

Notwithstanding the disruptions to travel and tourism, the sector provides economic, social, and cultural benefits to the host destinations and the nation at large. In economic terms, tourism contributes to foreign exchange and creates employment opportunities (Manzoor et al., 2019; Safa et al., 2021). Tourism can also stimulate local economic development at the host destinations (Mugizi et al., 2018), e.g., starting business enterprises to produce products for the tourism market. This further creates employment opportunities across the value chain. Therefore, tourism is pivotal to economic diversification. Against this backdrop, tourism is essential for sustaining rural livelihoods, more especially for marginalised communities who are excluded from the mainstream economy. The tourist expenditures at the host destinations form part of the proceeds to the local economy (Chen et al., 2018; Mugizi et al., 2018). Since tourism is labour intensive, it presents various employment opportunities across the different stages of the value chain. Notably, tourism could be a significant source of employment for vulnerable segments of the society who may otherwise find it hard to secure a job (Manwa & Manwa, 2014). Moreover, tourism is also an important sector for combating poverty in communities living in and alongside protected areas (Manwa & Manwa, 2014; Manzoor et al., 2019). Besides, tourism instils a sense of pride among the host communities as they interact with tourists to educate them about their culture and lifestyle (Mogomotsi et al., 2018). The tourists also appreciate the host destinations' culture. Similarly, tourism could be pivotal in rejuvenating traditional and cultural practices which might be on the verge of disappearance. This is an integral component of cultural preservation.

In Botswana, tourism remains a significant contributor to the national economy. Tourism is the second largest contributor to Botswana's economy after diamonds (Lenao, 2015). In 2019, tourism accounted for 9.6 per cent of the country's GDP (WTTC, 2021). Similar to the global figures, the contribution of tourism declined to 5.3 per cent in 2020, owning to the widespread Covid-19 pandemic. Approximately 10 per cent of the employed population is from the travel and tourism sector (WTTC, 2015). The recorded number of international day visitors has been rapidly growing over time, from 619,582 in 1994 to 2,531,979 in 2010 (Department of Tourism, 2012). Despite the current shocks to travel and tourism, the sector is still poised as a viable strategy in diversifying Botswana's economy away from the mining sector, owing to mining's high volatility and dependence on non-renewable resources (Manwa & Manwa, 2014; Mogomotsi et al., 2018). Moreover, tourism is an integral component of local economies in rural areas, especially for communities contiguous with protected areas who suffer limited livelihood options due to competing land-use interests. In these areas, tourism outlets provide various employment opportunities such as in accommodation, transportation, travel and tours, and as wildlife guides. The sector also provides a market for local resources, being a source of additional income to many households. Tourism has sustained many households in the Okavango Delta, a region still subject to abject poverty (Mbaiwa, 2017). However, empirical evidence has demonstrated that the benefits accrued from tourism are heterogeneous, and often vary within and across communities in the same locality (Mogomotsi et al., 2018).

Despite the significance of tourism to rural livelihoods in Botswana, the country's tourism offering is still largely anchored on wildlife-based resources (Stone et al., 2017). This undiversified approach potentially limits maximising benefits from tourism. The lack of diversification means that tourism activities are concentrated in certain areas of the country which happen to be rich in wildlife resources such as the Chobe region. Consequently, this causes congestions which might be detrimental to the resource base (Mogende & Moswete, 2018). Furthermore, the undiversified state of Botswana's tourism could lead to other equally important resources being obscured in the sector offering, for example forest resources. Forests have the potential to diversify the country's tourism activities away from wildlife. Notably, forest-based tourism could be an answer to the long-standing call for active community participation in the tourism industry. The diverse forest resources, particularly in non-protected areas, could be explored in order to assist in decongesting tourism activities which have been heavily concentrated within parks and game reserves. Against this backdrop, this study assessed the potential of forest-based tourism in contributing to sustainable livelihoods of the marginalised communities in the Okavango Delta. This study portrays forest dependency as a viable intervention to promote forest-based interventions such as tourism activities that can benefit the community.

LITERATURE REVIEW

Human well-being has always been the result of benefits that the ecosystem provides (Power, 2010). It is well known that forests abound with diverse ecosystem goods and services which are crucial to livelihood and environmental sustainability. A purely economic distinction sees the forest ecosystem classified into three types: 'timber based' (timber), 'no timber based' (non-timber) and 'intangible' (intangible asset) (Rizio & Gios, 2014). The forest ecosystem provides key ecosystem goods and services, which are classified into four broad typologies: provisioning (food, water, fuel wood, timber, fibre), regulating (carbon sequestration, flood control, heat amelioration), cultural (recreation, education, aesthetic inspiration, tourism, spiritual values) and supporting services (hydrological functions, primary productivity) (Rizio & Gios, 2014). Central to this study are cultural services, particularly tourism and aesthetic inspiration. Among the aspects of the forest landscape, there is also considerable potential about the service rendered to relaxation and the need for silence that reflects an intrinsic relationship between nature, culture and local communities.

The natural surfaces that make up a landscape are subject to various activities that can damage or vice versa contribute to the conservation of what they represent. The relationship between common natural resources and sustainable tourism represents an essential combination for the development of forest tourism. In fact, the forest landscape can perform various functions in the field of tourism and the natural area to which it is a background and there is a need to manage the resources sustainably. In order to be adequate, forest management must guarantee the flow of economic and ecosystemic utility for the local community through the production of goods and services that are economic in return. Such management must not only represent expenses inevitably incurred for the conservation and production of goods and services provided by natural areas, but must be able to provide additional income to the local community. Multifunctional forest management considers multiple potential products and their sources of income. Indeed, one of the most important functions of adequate forest management in community forests, in addition to obtaining raw materials, is the conservation of the landscape and natural resources as a background for recreational activities and establishment of local social networks.

As noted by Sgroi (2020), in the specific case of forest landscapes, the link between local populations and the forest involves a process of continuous transformation. This process is the result not only of physical interventions, but also of the socio-cultural context of the local populations, which, through their interpretation and appropriation of the traditions and customs handed down from generation to generation, have gradually shaped the forest landscape. Conservation and management of the forest landscape in a given community is often the expression of the link between that community and its surroundings. Forest management must be multifunctional, it must take the state of the forest as a public good in consideration of the fact that it must be able to provide local populations with sufficient products and services and so that they perceive the intact natural surface as a source of income. Indeed, Shackleton et al. (2007) indicated that indigenous forests and savannas with plantation forests provide valuable benefits to rural communities and the broader society. Their study appraised the role of conservation and forestry in sustainable livelihoods and poverty alleviation strategies and plans. The latter relates well to the proposed tourism enterprises of Shaikarawe community studied in this research, which is home to various significant indigenous trees found in Shaikarawe community forest. The stated problem is that despite the abundance of indigenous species and resources in the study area, they have not reached their full potential utilisation. Tourism could help increase their marketability while synergistically improving local communities, as they also possess vital tourist attraction values in the study area. The indigenous plants and the tourism activities have potential value for influencing the preservation of traditional and local knowledge of significant biodiversity. Research has suggested that indigenous communities are not only affected by tourism but that they can offer returns through entrepreneurial initiatives (Sharpley, 2000; Porter et al., 2018; Ramaano, 2021). In order for the management of the forest area to be economically sustainable in the long term, it must provide sufficient income for the local population without affecting the interests of future generations (Chifon, 2010). Effective forest management requires the production of the goods and services that form utility flows for an ecosystem and its local population to be guaranteed (Sgroi, 2020).

METHODS

Study Area

This study was conducted in the Ngamiland District in north-western Botswana (Figure 3.1). Although the district is characterised by semi-arid to arid area conditions, it contains the world's largest inland delta, the Okavango Delta (NWDC, 2003). The Okavango Delta is rich in flora and fauna. The predominant ethnic groups in this district are Batawana, Bayei, Baherero and Bambukushu, but others include Bakgalagadi, Basubiya, Baxhereku and Basarwa (the San) (NWDC, 2003). Ngamiland District is home to 193,725 inhabitants, with an annual growth rate of 2.8 per cent, which is higher than the national average of 2.3 per cent (Statistics Botswana, 2022). The district is mainly rural, and most people are settled along the water routes and sources, which they depend on for drinking, livestock, veld (wild) product harvesting, fishing and as a means of transport (NWDC, 2003). The major economic activities in this district are tourism, livestock rearing, handicrafts, small to medium scale industries and arable agriculture (NWDC, 2003). Despite the low fertility of the soil, juxtaposed with erratic rainfalls and drought conditions in the area, a large proportion of the population depends on the land for their income or at least for subsistence because employment opportunities are very limited (NWDC, 2003).

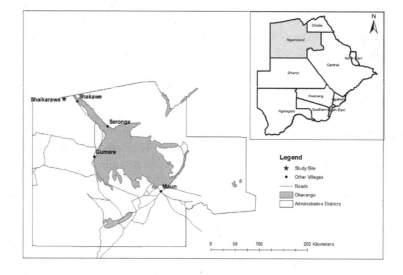

Figure 3.1   Location of the study area, Shaikarawe

Shaikarawe Community

This study focuses specifically on one site within the Ngamiland District, which is Shaikarawe settlement. Shaikarawe is a San settlement located along the Okavango Panhandle (Figure 3.1). The San people in this settlement are identified as Bugakhwe and have traditionally

The prospects of forest-based tourism for marginalised communities   29

relied on hunting and the collection of wild plant foods (VanderPost, 2003). Historically, the Bugakhwe have been savannah foragers (Bock, 2009). The population of Shaikarawe remains unknown but estimations approximate it around 300–400 people and many of them are living in extreme poverty, which is exacerbated by marginalisation and exclusion from the mainstream economy. The community also has a *kgotla*, which is the traditional place of justice, as well as organised community leadership, which includes their *kgosi*, Village Development Committee (VDC), and Shaikarawe Community Trust. The Botswana government and various NGOs have worked with this community on development issues and various development projects have been implemented. Some projects mentioned by the community included community gardening, beekeeping, and livestock.

Even though the settlement is not yet legally gazetted, the participants listed a few developments that were being done in the village, namely a garden, main *kgotla*, VDC offices, mobile stop clinic, primary school, cemetery and the proposed Community Forest Reserve.

The community has a proposed Shaikarawe Forest Conservation Area (SFCA) which was formally known as 'Shaikarawe Forest Reserve' which is located in Controlled Hunting Area (CHA) NG/1, about 12 km west of Mohembo and the Okavango River flood plain, and adjacent to the Namibian boarder. The SFCA is located on a piece of land measuring 7,666 hectares housing the Bugakhwe ethnic San group. This community initially conceptualised a 'Forest Reserve' in the year 2000 as a means of protecting the wild plant species, which they depended on, from unseasonal fires and over-harvesting by non-resident communities. This situation had led to forest degradation and destruction due to human contact activities from neighbouring villages. The construction of the Mohembo Bridge across the Okavango River has also brought negative environmental consequences for the existing forest reserve as the current contractors continue to mine road construction material from the conservation area. This situation has led to the reduction in the condition of the pristine nature of the forest reserve and hence the need to come up with mitigation strategies.

The community's original traditional territorial area measuring 7,666 hectares has been subjected to degradation over the years due to the slash and burn agricultural practices of adjacent populations residing east of Shaikarawe village. This means that since the SFCA is one of the last relatively intact forest reserves in proximity to the Shakawe-Mohembo West town planning area, its urgent protection is called for.

Previous mitigation attempts to assist communities towards conservation and management of forest reserve by other stakeholders such as 'Letloa Trust' or Trust for Okavango Cultural & Development Initiatives (TOCADI) have supported community initiatives since 2005 that have not borne any fruit.

Challenges facing the San community
Since the San in Botswana are no longer able to live as they did in the past, many have resorted to other types of livelihoods. Some are now engaging in subsistence farming and also enrolled in government projects aimed at stimulating food production. Despite the difficulties faced during this livelihood transition, there were noted successes for those who have been ploughing either seasonally and/or all year round (Le Roux, 1999). However, unfavourable climatic conditions in the area compromise agriculture and exacerbate its risk. Nonetheless, the San have been encouraged to build their capital and assets such as through rearing livestock in order to enhance their resilience to livelihood shocks (Le Roux, 1999). Livestock has remained an important asset for most rural households in southern Africa (Otte

30   *Handbook on tourism and conservation*

& Knips, 2005). Other San community have diversified their livelihoods to include additional contemporary strategies such as participation in tourism activities and craft production. For example, Shaikarawe community used to engage in annual cultural tourism activities with their Namibian neighbouring community before the Covid-19 pandemic. Despite the potential of tourism in uplifting rural livelihoods, the San community is beset with inadequate management and marketing skills necessary for implementing economically viable projects (Le Roux, 1999). On the contrary, the tourism activities often led to exploitation or misrepresentation of the San lifestyle and culture (Lebotse, 2009). Moreover, some San community members have experienced research fatigue. They opined that several researches has been conducted about them but the researchers rarely come back to disseminate their findings. Lack of benefits from the research was also cause for concern. The aforementioned concerns were also raised by the Shaikarawe community, hence this calls for active engagement of the participants from the outset and throughout the stages of the research project.

Data Collection and Analysis

The study draws from primary data which was elicited through focus group discussions (FGDs). The discussants were drawn from various groups and committees representing the Shaikarawe community. This includes Village Development Committee (VDC), Conservation Trust, youth, and women's groups. These groups were purposively sampled based on their position, as they are considered representative of the entire community. Four distinct FGDs were conducted. The composition of the FGDs ranged between 10 to 15 members, who were drawn from the aforementioned groups and committees. This sample size is within the recommended ranges for FGDs (Nyumba et al., 2018). The discussions were guided on the following issues: forest resource use, community empowerment, forest-based tourism opportunities, and challenges and threats to the forest. The discussions were facilitated by five people, one being the moderator who was responsible for asking questions and directing the proceedings, two people performed note-taking and their scripts were compared at the end of the discussions for quality assurance, while the remaining two were translators who were responsible for translating services when the need arose. Approximately, the discussions took one hour. The focus group data were first reviewed, and the responses then categorised into similar topics which were used to generate broader themes which guided the analysis.

RESULTS

Contribution of Forest Resources to Rural Livelihoods

Most of the participants expressed that the local community is heavily dependent on forest resources for their subsistence. Livelihoods and culture were the most frequently mentioned reasons for depending on forest resources. The participants indicated that forest resources contribute substantially to their livelihoods and well-being. Some of the participants maintained that forest resources remain the only source of subsistence, given the limited livelihood strategies they could pursue in their area. The adjacent forested area to the community is abundant with a diversity of plant species and provides food, medicine, raw materials, and aesthetic inspiration, among others. A variety of multi-use plant species were mentioned by

The prospects of forest-based tourism for marginalised communities 31

Table 3.1   The most frequently collected plant species (species sorted by frequency of mention)

Botanical name	Local name
Garcinia livingstonei	Motsaudi
Terminalia sericea	Mogonono
Baikiaea plurijuga	Mukusi
Dialium engleranum	Mohamana
Strychnos cocculoides	Mogorogorwana
Strychnos pungens	Mogwagwa
Eragrostis pallens	Motsikiri
Burkea africana	Mosheshe
Amaranthus thunbergii	Thepe
Schinziophyton rautenii	Mongongo
Diospyros chamaethamnus	Mokokosi
Ochna pulchra	Monyelenyele
Pterocarpus angolensis	Mukwa
Grewia flavescens	Motsotsojane
Drimia sanguinea	Sekaname
Euclea divinorum	'Motlhakola'

the participants during species listing (Table 3.1). *Schinziophyton rautenii* (Mongongo) and *Ochna pulchra* (Monyelenyele) were the most common species mentioned across all focus groups. These are valuable plant species used for producing oil used for cooking by most of the households in the area. In regard to culture, the participants opined that subsistence on forest resources is embedded in their cultural practice. Traditionally, the Basarwa tribe are renowned hunters and gatherers who have survived on wild resources since time immemorial. Moreover, the participants were of the view that subsisting on forest resources is an integral part of their tradition and it is one way of connecting with their ancestral practice. Therefore, dependence on forest resources has sustained their culture for generations.

Community Empowerment through Forests

The participants expressed that the adjacent forest harbours business opportunities which could be explored for the benefit of the community. Among others, they mentioned trading in non-timber forest products (NTFPs) (e.g., wild food, construction materials, arts and crafts), agroforestry, and venturing into tourism. In regard to NTFPs trading, the production of oil from selected tree species was a frequently mentioned business opportunity by most participants. The oil which is extracted from *Schinziophyton rautenii* nuts was deemed a profitable entity. On this note, the participants stressed that the community should be helped to establish small business enterprises for production of *Schinziophyton rautenii* oil. The different stages of the value chain are likely to provide employment opportunities to the locals, e.g., harvesting of the *Schinziophyton rautenii* fruits. Moreover, this opportunity could benefit vulnerable segments of the community, e.g., women and the youth. During their separate focus group discussion, the women identified the production of *Schinziophyton rautenii* oil as a promising economic opportunity for improving their livelihoods. They called for support with infrastructure capital to pursue this business, e.g., machinery for crushing *Schinziophyton rautenii* nuts. Other potential NTFPs opportunities included weaving baskets and floor mats, jewellery

32 *Handbook on tourism and conservation*

made from tree seeds, and selling thatching grass. Thatching grass was identified as a lucrative business opportunity, given its demand in the construction sector. Besides, the participants also mentioned practising agroforestry. The participants are trying to establish an indigenous tree nursery, where they would also cultivate other cash crops for selling to complement their household income.

Potential of Forest-based Tourism

Beyond household subsistence on forest resources, the participants indicated that the forested area has the potential to be explored for tourism-related activities. On this note, ecotourism was the most recurring tourism-related activity mentioned by the participants across the focus groups. Through ecotourism, the participants endeavour to offer walking trails, where tourists are taken on a tour of the forests to conduct species identification and learn about the various uses of the species. Also mentioned were excursion trips for harvesting fruits. Cultural tourism was also identified to complement the forest-based tourism activities. Through cultural tourism, the participants endeavour to host annual cultural festivals, where they display and teach tourists about their culture. The Basarwa are still practising their traditional culture, which is rich in resources, e.g., attire, tools, lifestyle and skills, food, dance, arts and crafts. This cultural festival could also provide an avenue for selling forest by-products. The participants opined that the cultural events are important for generating income as well as stimulating the local economy. The participants were of the view that the ecotourism activities should be managed by the community. The ecotourism activities were envisaged to create employment opportunities and contribute to local economic development in the area. Besides, the adjacent forest is also endowed with aesthetic value, which offers a tranquil environment for relaxation and appreciating the beauty of the landscape. The locals can harness this opportunity by establishing eco-friendly infrastructure such as a campsites and mobile safaris.

SWOT Analysis for the Proposed Forest-based Tourism

A SWOT (strengths, weakness, opportunities, and threats) analysis was conducted to guide the development of the proposed forest-based tourism activities in Shaikarawe settlement. Table 3.2 summarises the results from the analysis.

DISCUSSION

The study findings demonstrated that forest products remain an integral part of the livelihoods of the marginalised communities of the Okavango Delta. The participants indicated that the majority of the community were heavily reliant on forests for sustaining livelihoods. For some, forest products remain the only source of dependable livelihoods. Based on the foregoing, forest resources are likely to continue to be an important source of livelihood for the foreseeable future in Shaikarawe settlement. This is likely to be the case since the settlement is not yet gazetted, hence there are limited opportunities available to diversify livelihoods. Forests provide food, medicine, household energy needs, and raw materials (arts and crafts, construction). Besides, forest products were also an important alternative income generating activity. For example, the oil extracted from *Schinziophyton rautenii* tree emerged as a viable

The prospects of forest-based tourism for marginalised communities   33

Table 3.2   Summary from SWOT analysis

Parameter	Attributes
Strengths	Rich traditional ecological knowledge
	Rich culture
	Community willingness to partake in forest-based tourism
	Community willingness to partake in forest conservation
	Abundance of natural resources
	Registered community trust
Weaknesses	Lack of entrepreneurial skills
	Lack of product diversification
	Lack of infrastructure
	Lack of financial capital
Opportunities	Availability of land for conducting forest-based tourism, pending allocation by land board
	Proximity to Namibia, could be a stopover for international tourists en route to Namibia and Angola
	Possibility to diversify wildlife-based tourism in the Okavango Delta
	Technical support from government, NGOs, BUAN, LEA, wildlife department, department of forestry and range resources
	Active community participation in tourism industry, especially the youth and women
	Active role of communities in forest conservation
	Community participation in fire management
Threats	Veldt fire – destroys the resources
	Unsustainable harvesting practices of resources, e.g., cutting live trees
	Habitat fragmentation due to wild animals, e.g., elephants that colonise the area and destroy the vegetation
	Climatic change which affects the vitality of the forest, e.g., fruit trees might take longer to produce fruits or possibly not bear fruits at all

business opportunity for the community. The findings on the dependence on forest resources are consistent with other local studies which demonstrated the significance of forests to rural livelihoods and well-being (Lepetu et al., 2009; Garekae et al., 2020). In these studies, over three-quarters of the sampled households were highly dependent on forest resources. Moreover, other studies from elsewhere demonstrate that the contribution of forests in the form of forest income could be at par and/or outweigh that of dominant activities such as agriculture (McElwee, 2010). On this note, the forest adjacent communities are likely to suffer should access to and right to forest resources be curtailed.

In addition, the forests are an important resource for community empowerment, especially through forest-based tourism. The forest harbours tourism potential, which can be further explored for improvement of livelihoods and community development. Following our discussions, ecotourism emerged as a viable business opportunity in the settlement. The identified ecotourism activities could be established and managed by the community. The findings are consistent with those from Chobe Enclave in northern Botswana, where the community stressed that Botswana's forest reserves have the potential to alleviate poverty through conducting pro-poor tourism activities in them (Manwa & Manwa, 2014). The latter study profiled the benefits of forest-based tourism in Chobe Enclave community and categorised them as follows: direct effects, secondary effects, and dynamic effects. As explained by Mitchell and Ashley (2010), direct effects entail the actual benefits accruing to the community through their involvement in the tourism industry/business, such as employment creation, setting up small–medium enterprises/business, etc. Secondary effects are benefits arising from the emerging linkages between the community and the tourism operators. A typical example could be the stimulation of local economy through sourcing supplies from the community. This is likely to boost other sectors such as agriculture. Lastly, the dynamic effects refer to structural changes which may arise from the tourism development. For example, the development of key infrastructure in the community such as roads, shops and other amenities.

Framing the development of forest-based tourism within the lens of pro-poor tourism framework is vital for 'unlocking the opportunities for the poor within the tourism, rather than expanding the overall size of the sector' (Roe & Khanya, 2001, p. 2). Pro-poor tourism is anchored on maximising economic and non-economic benefits and the production of a conducive policy environment. Therefore, pro-poor tourism can be a viable tool for facilitating inclusive tourism activities which aimed at improving the welfare of the poor (De Beer & De Beer, 2011). Moreover, pro-poor tourism is considered a vital strategy for enhancing livelihood security and poverty alleviation at the micro-level (Wen et al., 2021). Against this backdrop, pro-poor tourism is more relevant to marginalised communities who have suffered deprivation and exclusion from the mainstream economy. Therefore, pro-poor tourism emerges as an important source of emancipation for communities that are marginalised despite living in resource rich areas. Through pro-poor tourism, the communities will be able to come up with their own tourism initiatives to improve their livelihoods and well-being as well as expanding the local economy. Drawing on the above, forest-based tourism holds out promise for the implementation of pro-poor tourism in the marginalised communities of the Okavango Delta. This is pivotal for socio-economic development at the local level.

CONCLUSION

The study assessed the potential of forest-based tourism in contributing to sustainable livelihoods of the marginalised communities in the Okavango Delta, Botswana. Drawing from the participant's perspective, the forest harbours business opportunities which could be exploited for the benefit of the community. Trading in non-timber forest products and ecotourism were the most recurring business opportunities across the discussions. The community is enthusiastic about ecotourism, they have registered a Shaikarawe Community Trust which will be responsible for managing ecotourism projects in the proposed community forest. The ecotourism projects are likely to stimulate local economic development, improve livelihoods and promote pro-conservation behaviour. The local communities in this settlement are marginalised and beset with a high unemployment rate. Therefore, the creation of this kind of tourism enterprise will help boost their capacity to create employment and generate revenue for the community trust and local people while also achieving conservation objectives of the pristine Shaikarawe Forest Reserve.

REFERENCES

Bock, J. (2009). What makes a competent adult forager? In *Hunter-Gatherer Childhoods: Evolutionary, Developmental and Cultural Perspectives*, ed. B. Hewlett & M. Lamb. Piscataway, NJ: Transaction Publishers, pp. 109–120.

The prospects of forest-based tourism for marginalised communities   35

Chen, B., Qiu, Z., Usio, N., & Nakamura, K. (2018). Tourism's impacts on rural livelihood in the sustainability of an aging community in Japan. *Sustainability, 10*(8), 2896. https://doi.org/10.3390/su10082896.

Chifon, G. (2010). The role of sustainable tourism in poverty alleviation in South Africa: A case study of Spier Tourism Initiative. MA thesis, University of Western Cape, South Africa.

De Beer, A., & De Beer, F. (2011). Reflections on pro-poor tourism in South Africa: Challenges of poverty and policy in the search for a way forward. *Journal of Contemporary Management, 8*(1), 591–606.

Department of Tourism (2012). *Tourism Statistics: 2006–2010.* Gaborone: Botswana Government Printers.

Garekae, H., Lepetu, J., & Thakadu, O. T. (2020). Forest resource utilisation and rural livelihoods: Insights from Chobe enclave, Botswana. *South African Geographical Journal, 102*(1), 22–40.

Lebotse, T. (2009). Victims or actors of development: The case of the San people at D'Kar, Botswana. MA thesis. Norway: University of Tromso, Tromso.

Lenao, M. (2015). Challenges facing community-based cultural tourism development at Lekhubu Island, Botswana: A comparative analysis. *Current Issues in Tourism, 18*(6), 579–594.

Lepetu, J., Alavalapati, J., & Nair, P. K. (2009). Forest dependency and its implication for protected areas management: A case study from Kasane forest reserve, Botswana. *International Journal of Environmental Research, 3*(4), 525–536.

Le Roux, W. (1999). *Torn Apart: San Children as Change Agents in a Process of Acculturation. A Report on the Situation of Education of San Children over the Southern African Region.* Botswana: Kuru Development Trust and Working Group of Indigenous Minorities in Southern Africa.

Manwa, H., & Manwa, F. (2014). Poverty alleviation through pro-poor tourism: The role of Botswana forest reserves. *Sustainability, 6*(9), 5697–5713.

Manzoor, F., Wei, L., Asif, M., Haq, M. Z., & Rehman, H. (2019). The contribution of sustainable tourism to economic growth and employment in Pakistan. *International Journal of Environmental Research and Public Health, 16*(19), 3785.

Mbaiwa, J. E. (2017). Poverty or riches: Who benefits from the booming tourism industry in Botswana? *Journal of Contemporary African Studies, 35*(1), 93–112.

McElwee, P. D. (2010). Resource use among rural agricultural households near protected areas in Vietnam: The social costs of conservation and implications for enforcement. *Environmental Management, 45*, 113–131.

Mitchell, J., & Ashley, C. (2010). *Tourism and Poverty Reduction: Pathways to Prosperity.* London: Earthscan.

Mogende, E., & Moswete, N. (2018). Stakeholder perceptions on the environmental impacts of wildlife-based tourism at the Chobe National Park River Front, Botswana. *PULA: Botswana Journal of African Studies, 32*(1), 48–67.

Mogomotsi, P. K., Kolobe, M., Raboloko, M., & Mmopelwa, G. (2018). The economic contribution of tourism to local communities: The case of Khumaga and Moreomaoto villages, Botswana. *PULA: Botswana Journal of African Studies, 32*(1), 84–99.

Mugizi, F., Ayorekire, J., & Obua, J. (2018). Contribution of tourism to rural community livelihoods in the Murchison Falls Conservation Area, Uganda. *African Journal of Hospitality, Tourism and Leisure, 7*(1), 1–17.

NWDC (North West District Council) (2003). *Ngamiland District Development Plan 6: 2003–2009.* Ngamiland Development Committee, Ministry of Local Government, Republic of Botswana.

Nyumba, T. O., Wilson, K., Derrick, C. J., & Mukherjee, N. (2018). The use of focus group discussion methodology: Insights from two decades of application in conservation. *Methods in Ecology and Evolution, 9*(1), 20–32.

Otte, J., & Knips, V. (2005). *Livestock Development for Sub-Saharan Africa.* Rome: Food and Agriculture Organisation (FAO).

Porter, B. A., Orams, M. B., & Lück, M. (2018). Sustainable entrepreneurship tourism: An alternative development approach for remote coastal communities where awareness of tourism is low. *Tourism Planning & Development, 15*(2), 149–165.

Power, A. (2010). Ecosystem services and agriculture: tradeoffs and synergies. *Philosophical Transactions of the Royal Society B: Biological Sciences, 365*, 2959–2971.

Ramaano, A. I. (2021). Prospects of using tourism industry to advance community livelihoods in Musina Municipality, Limpopo, South Africa. *Transactions of the Royal Society of South Africa*, 76(2), 201–215.

Rizio, D., & Gios, G. (2014). A sustainable tourism paradigm: Opportunities and limits for forest landscape planning. *Sustainability*, 6(4), 2379–2391.

Roe, D., & Khanya, P. U. (2001). *Pro-Poor Tourism: Harnessing the World's Largest Industry for the World's Poor*. London: International Institute for Environment and Development.

Safa, R. P., Yasouri, M., & Hesam, M. (2021). Effects of tourism on sustainable rural livelihoods. Case study: Saravan, Rasht County, Iran. *Journal of Research and Rural Planning*, 3(34), 1–19.

Sgroi, F. (2020). Forest resources and sustainable tourism: A combination for the resilience of the landscape and development of mountain areas. *Science of the Total Environment*, 736. https://doi.org/10.1016/j.scitotenv.2020.139539.

Shackleton, C. M., Shackleton, S. E., Buiten, E., & Bird, N. (2007). The importance of dry woodlands and forests in rural livelihoods and poverty alleviation in South Africa, *Forest Policy and Economics*, 9(5), 558–577.

Sharpley, R. (2000). Tourism and sustainable development: Exploring the theoretical divide. *Journal of Sustainable Tourism*, 8(1), 1–19.

Statistics Botswana (2022). *2022 Population and Housing Census Preliminary Results v2*. Gaborone: Statistics Botswana.

Stone, L. S., Stone, M. T., & Mbaiwa, J. E. (2017). Tourism in Botswana in the last 50 years: A review. *Botswana Notes and Records*, 49, 57–72.

VanderPost, C. (2003). *Community Natural Resources of Bugakhwe and //Anikhwe in the Okavango Panhandle in Botswana*. Gaborone: The World Conservation Union/Netherlands Development Organization Community Based Natural Resource Management Support Programme in collaboration with the Teemacane Trust.

Wen, S., Cai, X., & Li, J. (2021). Pro-poor tourism and local practices: An empirical study of an autonomous county in China. *SAGE Open*, 11(2), 1–11.

WTTC (2015). *Travel & Tourism Economic Impact 2015: Botswana*. London: World Travel & Tourism Council.

WTTC (2021). *Travel & Tourism Economic Impact 2021*. London: World Travel & Tourism Council.

4. Human–elephant conflict: implications for rural livelihoods and wildlife conservation

Kenalekgosi Gontse, Joseph E. Mbaiwa and Olekae T. Thakadu[1]

INTRODUCTION

Human–wildlife conflict (HWC) is of great concern in agro-pastoral areas, as it negatively impacts on conservation efforts and livelihoods of people, particularly of rural communities that live in close proximity to wildlife-protected areas. The negative consequences of HWC include, among others, the destruction of food and cash crops, predation on livestock and subsequent loss of income, loss of livelihoods and sense of wellbeing (Hill, 2004; Hoare, 2000; Karanth et al., 2013; Linkie et al., 2007; Mc Guinness & Taylor, 2014). The negative consequences of HWC can sometimes be immense, such as loss of human life in human–elephant (*Loxodonta africana*) conflicts and exposure to diseases such as malaria, due to the need to guard fields day and night from wild animals (Gupta, 2013; Jadhav & Barua, 2012; Madden, 2004; Osborn & Parker, 2003; Treves, 2007).

According to Ogra (2009), protected areas are set aside to protect and sustain wildlife populations, but more than half of wildlife populations are found in lands outside protected areas. These areas are mostly agro-pastoral lands, and this results in raids by wildlife on agricultural lands, resulting in destruction of crops and farm implements (Ogra, 2009). This creates negative attitudes and perceptions in rural communities towards wildlife conservation, thus compromising conservation efforts as people begin to regard wildlife as a cost (Sifuna, 2010). Botswana is one of the countries endowed with a diverse range of wildlife resources, which contribute significantly to the country's economy, with elephants being the key tourism product. The elephant population in Botswana in 2018 is estimated to be 122,831 (Chase et al., 2018). Botswana's tourism industry is the second largest economic sector in the country, contributing 8.4 per cent to the gross domestic product (GDP) of Botswana (WTTC, 2014), which is second after mining which accounts for 22.4 per cent of Botswana's GDP (Central Statistics Office, 2015), while agriculture accounts for approximately 3 per cent of Botswana's GDP (Esterhuizen, 2015). Despite Botswana's tourism contribution to the GDP, local communities bear the brunt of conserving wildlife resources since the majority of rural communities in Northern Botswana live in and around areas with wildlife, therefore they are the first to deal with the negative impacts of wildlife (Gupta, 2013).

In Botswana a variety of species are reported to damage livestock and crops, including elephants (Darkoh & Mbaiwa, 2005; Department of Environmental Affairs, 2010; Graham, 2004; Mbaiwa & Rantsudu, 2003). Elephants are a key species that destroy crops and property and kill human beings in Botswana's wildlife areas. Wildlife induced damage is problematic for conservation because it has the potential to increase intolerance of wildlife by farmers in the HWC hotspots of Botswana (Graham, 2004; Gupta, 2013). Although there has been significant literature on HWC in northern Botswana (Gupta, 2013; Mosojane, 2004; Sifuna, 2010;

38   Handbook on tourism and conservation

Songhurst & Chase, 2008), the agricultural landscapes of Boteti sub-district such as Khumaga have not been adequately studied yet this area is a hotspot for HWC. Graham (2004) examined human–lion (*Panthera leo*) conflicts but did not focus on crop damage in the area. Monametsi (2008) studied the effects of electric fences in reducing HWC, but that was before the Boteti River flowed again and at the time the fence was still intact. We are of the view that there is need to critically understand the implications of crop damage to livelihoods and wildlife conservation in Khumaga because it will help in developing sustainable mitigation strategies for arable-land farmers in the Boteti sub-district (Ringrose et al., 1996).

SOCIAL EXCHANGE THEORY

This study is informed by social exchange theory (SET), which emerged in the early works of Malinowski (1922) and Mauss (1925). Social exchange is defined by Homans (1961) as the exchange of activity between at least two persons and their costs and benefits in the exchange. Costs are viewed in terms of alternative activities or opportunities forgone by the actors involved (Cook et al., 2013). In SET, people essentially take the benefits and subtract the costs to determine the value of a relationship (Blau, 1964; Homans, 1958). A positive relationship is one where the benefits outweigh the costs, while negative relationships occur when the costs are greater than the benefits (Homans, 1958).

Zafirovski (2005) has identified some weaknesses of SET. One of them is that the theory does not take into consideration the cultural context and differences of cultures. Social exchange theory is founded on the idea of reward or benefit, but cultures are different and some cultures may not necessarily seek a reward in a relationship. Miller (2005) also discussed the weakness of the social exchange theory of interpersonal relationships, noting that SET undervalues the impact of other factors besides costs and benefits, such as how one's evaluation of previous relationships affects one's assessment of the current relationship.

Even though Homans' theory was criticised for various reasons, today SET is used by a number of authors in different fields. For example, Andereck et al. (2005) used the theory for understanding residents' perceptions of community tourism; Brines (1994) addressed sociology, economic dependency, gender, and the division of labour at home; Choi and Murray (2010) studied resident attitudes towards sustainable community tourism; Hall (2003) applied SET in information science research; Kern and Willcocks (2000) used it in information technology; Pfouts (1978) used it for studying violent families and the coping responses of abused wives; and Son et al. (2003) used it to examine the effects of relational factors and channel climate on electronic data interchange (EDI) usage in customer–supplier relationships.

From a tourism and wildlife conservation perspective, SET proposes that individuals' attitudes towards wildlife are influenced by their evaluations of the outcomes for themselves and for their communities (Andereck et al., 2005). In the context of this study, we consider SET to be appropriate for a number of reasons. First, it will provide insights into whether farmers are willing or unwilling to promote conservation of wildlife resources irrespective of crop damage. Second, the theory will be used to determine a better land use option in the area between crop farming and wildlife conservation. Third, it will help to understand the relationship between arable farmers and the protected area, in this case the Makgadikgadi Pan National Park. Fourth, the theory will help to determine if people are prepared to coexist with wildlife species from Makgadikgadi Pan National Park even though the wildlife is a threat to their livelihoods. Luo (2002) argues that partners sometimes look beyond short-run inequities or risks and concentrate on long-run mutual gain. For example, people's participation in conservation in Nepal has been robust due to meaningful participation of local communities as they are able to derive benefits thus developing a positive attitude towards conservation. Fifth, SET can help to explain the costs and rewards of relationships and enables people to predict how to keep and sustain relationships. In most cases people participate in relationships that bear positive rewards or are profitable and reject those that have high costs. At times, parties involved in a problematic relationship seek to resolve it by resorting to violent action. In India, for example, more than 100 elephants are fatally injured every year through retributive actions by humans (Jadhav & Barua, 2012). In Kenya, pastoralists are reported to have poisoned and speared lions (Frank et al., 2006a, 2006b) because of the damage they cause to people's livelihoods. Figure 4.1 is a model showing SET in relation to human–wildlife interactions.

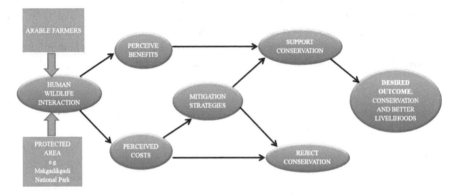

Source: Adapted from Gursoy and Kendall (2006, p. 607).

Figure 4.1   *The conceptual framework and application of SET in human–wildlife interactions, perceptions, attitudes and efforts towards wildlife conservation*

The conceptual framework explains that, when local communities directly benefit from Makgadikgadi Pan National Park through tourism development or Community Based Natural Resource Management (CBNRM) they will eventually support conservation. When there are perceived costs like crop damage, resulting from the interaction, people will look at available mitigation strategies, e.g. compensation. If mitigation strategies outweigh the costs, people will support conservation; they will reject it when the costs outweigh the benefits, even if the desired outcome is conservation of wildlife and better livelihoods.

METHODS

Study Area

This study was carried out in the Boteti sub-district in the north-central part of Botswana using Khumaga village as a case study (Figure 4.2). Khumaga village is one of the villages located around the Makgadikgadi Pans. Khumaga is located in Boteti, in the Central District of Botswana (Figure 4.2), with a population of 758 people (CSO, 2011). Most of the people of Khumaga are Banajwa. Other ethnic groups that have since come to settle in the village include Bayei, Basarwa, Basubiya, Bakalanga, Bakgalagadi, Baherero and Barotsi. The village is on the western side of Makgadikgadi Pan National Park near the Boteti River. The study was narrowed down to Khumaga village. Khumaga village is a classic case study because cases of HWC are higher and more common in the area, possibly due to its proximity to Makgadikgadi Pan National Park (Graham, 2004). The Makgadikgadi Pan National Park covers an area of 3,535 km^2 (DEA, 2010). There has been long-standing HWC in Khumaga due to wildlife crop damage (Gontse et al., 2018). Several farms and non-farm economic activities are sources of livelihood for the people living in Khumaga. Farm-based activities include livestock farming, dryland arable farming, and *molapo* or flood recession farming. Non-farm economic activities include informal employment, such as with the Ipelegeng (drought relief programme) and small businesses. Ipelegeng is a government initiative whose main objective is to provide short-term employment support and relief whilst carrying out essential development projects that have been identified and prioritised through the normal development planning process. It targets unskilled and semi-skilled labour for short-term assistance. The people of Khumaga

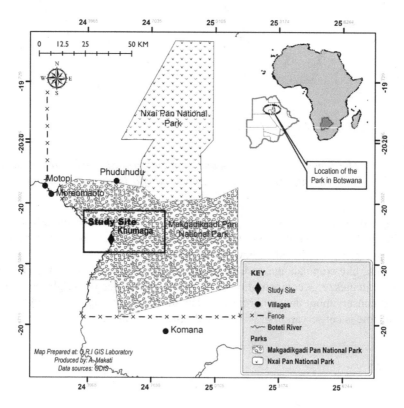

Source: Produced by Anastatia Makati, ORI GIS.

Figure 4.2   Map of the study area showing Khumaga

are mostly small-scale farmers who depend entirely on subsistence agriculture for their livelihoods.

CBNRM projects are weak in Khumaga. Mbaiwa (2015) claims that some CBNRM programmes in Botswana are failing due to a lack of availability of skilled personnel or lack of capacity building, reinvestment of CBNRM revenue or misappropriation of funds, and strong community cohesion or lack of it. Some of these factors may be affecting CBNRM in Khumaga. CBNRM is the management of resources such as land, forests, wildlife and water by collective and local institutions for local benefit. The CBNRM was introduced in Botswana to increase opportunities for local communities to benefit from wildlife and natural resources through tourism development. Communities participating in CBNRM in the Okavango Delta are believed to support wildlife conservation because they derive benefits from utilisation of natural resources through the programme (Mbaiwa, 2005).

Sampling, Data Collection, and Data Analysis

The study utilised both qualitative and quantitative data from primary and secondary data sources. Primary data collection involved face-to-face interviews with arable farmers. All interviews were conducted in Setswana (the national language spoken and understood throughout Botswana). A sampling frame consisting of 120 active arable farmers from Khumaga village was sourced from the Department of Crop Production in Khumaga. Active farmers in this study are defined as arable-land farmers who have been continuously ploughing for five years prior to the time of data collection. While the intention was to interview all active farmers, one of them did not consent to be interviewed.

Face-to-face interviews were conducted with farmers using an open- and closed-ended questionnaire. That is a questionnaire administered to farmers containing both open- and closed-ended questions. For closed-ended questions, farmers chose among the alternatives provided, such as Yes/No options. Even though an open-ended questionnaire was used, interviews progressed in a discussion style. The questionnaire also captured the socio-demographic characteristics of the respondents such as age, sex, ethnicity and educational level. To ensure independence of the data collected in this study, as well as to obtain as many widely representative views as possible, only an adult family head (man or woman) was interviewed from each household.

Data from arable farmers was supplemented by qualitative data obtained through unstructured interviews with six (6) key informants including a senior wildlife warden from the Department of Wildlife and National Parks (DWNP), community leaders (a village chief and three headmen of Khumaga village), and one agricultural demonstrator from the Department of Crop Production (DCP). Key informant interviews were conducted to triangulate the information collected from farmers and to pursue details on HWC.

Secondary sources included both published and unpublished work such as annual problem animal reports, crop production reports and related publications on HWC. The data from respondents were coded and descriptively analysed through the Statistical Package for Social Sciences (SPSS version 22). Content analysis was conducted by creating themes and patterns, and then the researcher made inferences about the messages within the text. Interviews with people were also recorded to analyse their perceptions on conservation.

42   *Handbook on tourism and conservation*

Table 4.1   Summary of respondents' demographics

Variable	Items	N (119)	M (SD)	%	n
Gender	Male			32	38
	Female			68	81
Age	25–30			3.4	4
	31–35			3.4	4
	36–40			10.9	13
	41–50			21.0	25
	51–55			10.1	12
	56–60			15.1	18
	61 and above		55.5 (14.5)	36.1	43
Education level	None			43	51
	Primary			34	41
	Secondary			23	27

RESULTS

Socio-demographic Characteristics

A total of 119 respondents (N=119) were interviewed in this study. The sample was predominantly female, accounting for 68 per cent of the total sample, with males accounting for only 32 per cent of the total sample. Multi-generational households are common at Khumaga. Approximately 82 per cent of the arable-land farmers were over 40 years old, illustrating that arable farming was mostly done by elderly people. The proportions of younger age groups (i.e., below 40 years) were very low. The mean age of the respondents was 55.5 (SD=14.5) years, and the mode of the age group was 61 and above. The majority of the arable-land farmers were therefore in old age (elderly people). The results indicate that 43 per cent of the farmers did not attend school at all, 34 per cent of the respondents went to primary school, and only 23 per cent were educated at the secondary school level. The least-represented age groups of arable farmers at Khumaga were those aged 25–30 and 31–35, accounting for only 3.4 per cent of the farmers interviewed (Table 4.1). The reason for the low participation of young people might be that they were employed elsewhere or were still pursuing education while the elderly people were unemployed and viewed agriculture as an important cultural aspect and as a legacy of their ancestors.

Factors Affecting Arable Farming at Khumaga

Interviews with arable farmers at Khumaga indicate that arable farming is largely affected by wildlife crop raiding. A majority (97 per cent) of respondents strongly agreed that wildlife crop raiding poses a threat to arable farming, while 94 per cent indicated that low rainfall is also a challenge to crop farming at Khumaga. A total of 60 per cent of respondents disagreed that shortage of machinery is a challenge to farming at Khumaga (Table 4.2). The agricultural demonstrator at Khumaga reported wildlife crop raiding, unreliable rainfall and pests to be problems for crop production. The demonstrator at Khumaga indicated that the reason why machinery is not a problem at Khumaga is because of the Integrated Support Programme for Arable Agriculture Development (ISPAAD). ISPAAD was introduced in 2008 by the Government of Botswana to assist arable farmers. The components of ISPAAD include the

Human–elephant conflict: implications for rural livelihoods and wildlife conservation 43

Table 4.2   *Factors affecting arable farming at Khumaga*

	Strongly agree	**Agree**	**Neutral**	**Disagree**	**Strongly disagree**
Wildlife crop raiding	115 (96.6%)	4 (3.4%)	0	0	0
Low rainfall	12 (10.1%)	100 (84.0%)	6 (5.0%)	0	1 (0.8%)
Shortage of machinery	7 (5.9%)	7 (5.9%)	34 (28.6%)	34 (28.6%)	37 (31.1%)

Table 4.3   *Mitigation strategies used against crop raiders*

	n	**%**
Making false human structure	109	91.6
Guarding crops	106	89.1
Loud sound	94	79.0
Use of chilli pepper	82	68.9
Firing with a gun	14	11.8
Traps	2	1.7
Beehive	1	0.8

provision of draught power for arable farmers, portable water and seeds. Subsistence farmers are assisted with 100 per cent subsidy for hybrid seeds to cover a maximum of five hectares and open pollinated seeds to cover a maximum of 16 hectares. One of the respondents (25 February 2015) sceptically remarked, "We are given seeds and machinery for free by government to plant for her elephants" when referring to the damage caused by elephants.

There are various types of passive or active mitigation measures that are being used by arable farmers of Khumaga to protect their crops from wildlife. Farmers have adopted locally available techniques to mitigate the crop depredation problems. Making false human structures was reported by 91.6 per cent (n=109) to be the most common techniques used by farmers to protect crops against wildlife, guarding crops reported by 89.1 per cent (n=106), and 68.9 per cent (n=82) used chilli pepper as indicated in Table 4.3.

Crop Damage and High Numbers of Elephants

The elephant population in Botswana has significantly increased from about 99,000 in 1999 to 205,545 in 2012 (DWNP, 2012) (Figure 4.3). The increase is partly attributed to the in-migration of elephants from Zimbabwe, Namibia and Zambia into Botswana since the 2014 hunting ban in Botswana. As a result, the lack of elephant hunting has resulted in elephants being found in areas where they were previously non-existent, e.g. Central Kalahari Game Reserve (CKGR), Ghanzi District, etc. The increase in elephant populations in Botswana has resulted in increased incidents of crop damage which has a negative effect on household income in rural areas such as Khumaga village. A 91-year-old man (25 February 2015) who was also a headman of the village noted that "since that devil called elephant came to our land no one has ever harvested here in Khumaga and we are dying of hunger because of elephant's crop damage, we have grown without that creature on our land. Since it came, we are always in fear and scared of walking on our own." This indicates that rural livelihoods of farmers at Khumaga are affected by elephants.

Wildlife crop damage was found to be severe in this study. A total of 85 per cent of farmers interviewed reported having experienced crop damage by elephants in the last five years (2010–2014). This is mainly attributed to the significant increase of the elephant population

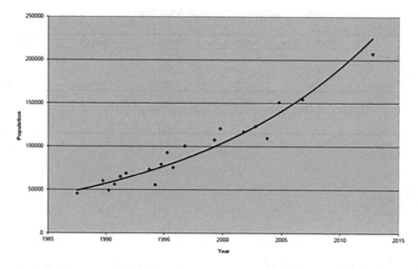

Source: DWNP (2012).

Figure 4.3 Elephant population in Botswana (1999–2012)

in the area as indicated by 19.3 per cent (n=23). Interviews with a DWNP officer at Rakops corroborated that there were many elephants in the study area and they were likely to cause much crop damage.

Ineffective Measures to Control Elephants from Makgadikgadi Pan National Park into Khumaga Village

The Government of Botswana has attempted to control the movement of elephants from Makgadikgadi Pan National Park into crop fields owned by farmers from Khumaga village through the erection of an electric fence along the Boteti River which serves as a boundary line between the village and the park. This electric fence was meant to keep elephants on the other side of the river within the park. Similarly, the Botswana Government erected a parallel standard cattle fence of 1.4 m in height which ran along the game-proof fence along the Boteti River to protect crop fields from associated wildlife damage. These measures were undertaken to reduce HWC as the Khumaga and Moreomaoto communities were complaining of crop loss and predation (DEA, 2010). However, the respondents indicated that the fence was effective when maintained, but due to disrepair the fence is currently ineffective (Figure 4.4).

The majority of residents in Khumaga (96 per cent) noted that the fence had been useful in minimising crop damage by wildlife. Data collected from the Department of Wildlife and National Park record books showed that there was a significant difference in elephant crop raids before the fence (2000–2004) and after the fence (2005–2009) (Figure 4.5). The majority of the respondents (87 per cent) indicated that crop damage by wildlife decreased after the construction of the fence (2005–2009), and 99 per cent of respondents noted that from 2010–2014, crop damage incidents had increased and were frequent. Monametsi (2008) also found that there was a reduction in elephant crop raids before and after the fence was built.

Human–elephant conflict: implications for rural livelihoods and wildlife conservation 45

Source: Picture by K. Gontse (2015).

Figure 4.4   *Makgadikgadi game-proof fence between Khumaga village and Makgadikgadi Pan National Park*

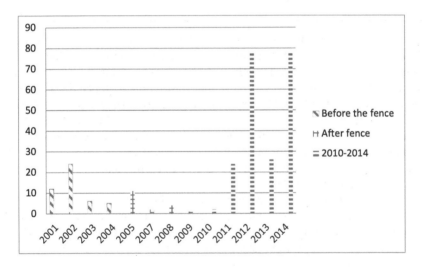

Figure 4.5   *Number of crop raids by elephants reported to the DWNP before the erection of the Makgadikgadi electric fence from 2000–2004 and after the erection of the Makgadikgadi electric fence in 2005–2009 and 2010–2014 in Khumaga*

The flooding of the Boteti River since 2009 is blamed for increasing incidents of crop damage in Khumaga. Approximately 90 per cent of the farmers indicated that, before the river started flowing again, crop damage was low and that the situation got worse after the river reflowed. This was because, before the river starting reflowing again (2004–2008), the electric fence was still intact and was still working properly.

In addition to the electric fence, there are many other mitigation measures that the government had once introduced at Khumaga village, including frequent ground patrols by the

46   Handbook on tourism and conservation

Table 4.4   Expected yield, actual yield and loss per crop in 2014

Crops planted	Expected yield per crop	Actual yield per crop	Total loss per crop
Millet	575.5	50	525.5
Maize	618	139.75	478.25
Sorghum	362.5	23.5	339
Watermelon	310	40.5	269.5
Beans	422.5	30.25	392.25
Groundnuts	20.25	0	20.25
Sweet reeds	248.5	18	230.5
Melon	8.5	0	8.5

Table 4.5   Data on crop production at Khumaga village from Department of Crop Production at Letlhakane (2013–2014)

Farmer's category	Amount obtained/production (kg)												
	Sorghum	Maize	Millet	Cowpeas	Groundnuts	Mung bean	Watermelons	Melons	Pumpkins	Sweet reeds	Sunflower	Lablab	Total
All farmers	375	3262	787.5	617.5	0	0	20,000	1000	1000	3500	0	500	33,042
Small scale	375	3262	787.5	617.5	0	0	20,000	1000	1000	3500	0	500	33,042
Commercial	-	-	-	-	-	-	-	-	-	-	-	-	0
Molapo	-	-	-	-	-	-	-	-	-	-	-	-	0
Youth	-	-	-	-	-	-	-	-	-	-	-	-	0

DWNP. However, 98 per cent of the farmers noted that the DWNP no longer conducts patrols in their area. According to a DWNP officer, before the construction of the Makgadikgadi game-proof fence, there was a DWNP Problem Animal Control Unit office (PAC) at Khumaga, whose responsibility was to assist in the protection of human life and the reduction of property damage caused by wildlife by patrolling the area. The unit also advised residents on mitigation strategies and ensured that compensation was paid to those who experienced losses due to HWC. Patrols and other responsibilities of PAC officers at Khumaga were neglected after the construction of the electric fence based on the assumption that the fence would completely mitigate conflicts. The officers were then transferred to other stations immediately after construction of the fence.

The DWNP officer at Rakops corroborated the position of the respondents by revealing that they lack resources (equipment and personnel), resulting in delayed or late attendance of reported cases. The officer revealed that there were approximately four DWNP officers at the Rakops office "who are supposed to take care of Boteti sub District with only one car since it is a vast area" (26 February 2015).

Implications of Elephant Crop Damage for Livelihoods

Crop damage affects livelihoods at Khumaga village. The results indicate that 100 per cent of the respondents alleged that their livelihoods had been negatively affected by wildlife crop damage during the previous five years. In the year 2014 arable farmers in Khumaga said they expected a maize harvest of 618 bags but only 139.75 bags were harvested. This represented a 77 per cent loss of maize harvest. During the same year, 362.5 bags of sorghum were expected by farmers, but only 23.5 bags were harvested, representing a 94 per cent loss (Table 4.4).

Data collected from the Department of Crop Production shows that in 2013–2014 farmers yielded 3,262 kg of maize, which is about 65.2 bags of maize, and 375 kg of sorghum, which is about 7.5 bags of sorghum (Table 4.5) at Khumaga. Production recorded by the Department of Crop Production was lower from the records collected from arable farmers on actual yield acquired by arable farmers in 2014.

Eighty-four per cent (n=100) of arable farmers claimed that they have abandoned flood-recession farming (*molapo* farming) in favour of dryland farming. The village administrator at Khumaga contended that the majority of the people in Khumaga have traditionally depended on flood-recession farming for their livelihoods but they have subsequently shifted to dryland farming.

Implications of Elephant Crop Damage for Wildlife Conservation

There are ramifications for wildlife conservation at Khumaga as a result of elephant crop damage. The perceptions of farmers towards elephants were found to be negative towards the conservation of elephants. Elephants were noted by farmers as the most destructive wildlife species in Khumaga compared to other crop raiders. Approximately 70 per cent of the farmers reported that they hate elephants, 8 per cent dislike them, and only 16 per cent said they like elephants. A total of 70 per cent of the farmers want elephants to be relocated from their area.

Respondents noted that they are generally not happy with the compensation for crop loss caused by elephants. One of the respondents said that "I once received a cheque of P10.00 from DWNP and for me to cash that cheque I was supposed to travel from Khumaga to Rakops which is about 70 kilometres, which means the cost of travel is higher than the claim" (26 February 2015), and this is consistent with SET. Timely payment can help victims to get over their anger and may reduce their incentives to retaliate against the animals that caused the damage (Nyhus et al., 2005). Compensation is a widely recommended and often used technique to reduce the economic impact of losses from wildlife behaviour and also to protect wildlife from being persecuted by farmers (Barua et al., 2013; Bulte & Rondeau, 2007; Hoare, 2000).

DISCUSSION

Though Botswana's tourism is based on wildlife and wilderness that need to be conserved, arable-land farmers at Khumaga incur costs from wildlife that are likely to undermine conservation efforts. Elephants destroy and trample crops of the arable-land farmers at Khumaga. The increased level of conflict was due to the expansion of the national elephant population into Boteti, making HWC an increasing issue of concern, with no signs of abating (DEA, 2010). Farmers of Khumaga experience persistent wildlife crop raiding that seriously affects their livelihoods and farmers have begun to hate elephants. For example, a 36-year-old woman at Khumaga said, "How can I like something that is not created by God? God cannot create something of that kind. Elephant was made by Satan" (23 February 2015), when referring to the body structure of elephants and the damage they caused in her fields. This indicates that farmers at Khumaga are afraid of elephants, and they tend to hate them as anticipated by SET. Elephants inspire fear and frustration and are deemed pests to be controlled or exterminated if their behaviour infringes on the livelihoods or security of the humans around them. These animals, with their great body size and strength, overpower humans when they interfere with their activity (Warner, 2008). Campbell-Smith et al. (2010), in their study on human–

48   Handbook on tourism and conservation

orangutan (*Pongo abelii*) conflicts, suggest that efforts to mitigate may not, per se, change negative perceptions of those who live with the species because these perceptions are often driven by fear, and this might also be the case at Khumaga.

To address this situation, the Government of Botswana has introduced different mitigation strategies through the DWNP such as the use of electric fences around the park to reduce HWC. Respondents indicated that the erection of the non-lethal electrified game-proof fence helped to reduce wildlife crop damage in Khumaga. The fence was, however, rendered ineffective by the river in 2009, and the solar panels that were producing electricity were stolen. The river that separates the park from the neighbouring villages dried up in the mid-1980s and started flowing again in 2009 (DEA, 2010). The respondents indicated that the fence was poorly maintained by DWNP, so elephants trespassed onto people's ploughed fields, resulting in an increasing number of crop damage incidents and raising concerns and complaints from farmers as they regard wildlife as a cost to their livelihoods (DWNP, 2012).

Fencing is one method used as a mitigation strategy to minimise HWC in Botswana. According to Mbaiwa and Mbaiwa (2006), Botswana has a history of fence use for mitigating conflict between the Ministry of Agricultural Development and Food Security and the Ministry of Environment, Natural Resources Conservation and Tourism. Mbaiwa and Mbaiwa (2006) reported that fences can succeed in separating wildlife and livestock, such as the buffalo fence that separates buffalo (*Syncerus caffer*) populations – which are known for transmitting foot and mouth disease – from cattle populations within the inner parts of the Okavango Delta.

Fences, however, are also known to disrupt wildlife migrations (Mbaiwa & Mbaiwa, 2006; Okello & D'Amour, 2008; Perkins & Ringrose, 1996). Fences are blamed for completely hindering the movement of wildlife species and for separating wildlife families (Mbaiwa & Mbaiwa, 2006). Others indicated that fences are responsible for the deaths of many migratory wildlife species (Perkins and Ringrose, 1996). In their study, Okello and D'Amour (2008) contend that the use of electric fences is not a solution to the problem of HWC, as wildlife are forced to change their migration routes, hence simply shifting HWC elsewhere rather than solving the problem. Even though fences have negative effects as highlighted by some authors, arable-land farmers in Khumaga concur with the claim made by Thouless and Sakwa (1995) that electric fences are effective in mitigating wildlife crop damage provided that the fences are well maintained. If the Government of Botswana can intervene and help local farmers with the suggested mitigation strategy, farmers may turn to support conservation of problem animals as a source of income to the country rather than as a cost to their livelihoods as suggested by SET.

Farmers of Khumaga have also abandoned flood-recession farming (*molapo* farming) because of wildlife crop damage. Woodroffe et al. (2005) argue that HWC can bring opportunity costs, where people forgo economic or lifestyle choices due to impositions placed upon them by the presence of wild animals. The shift from *molapo* farming to dryland farming is in line with Gupta's (2013) study, indicating that for some farmers, crop damage is one of the reasons why they have stopped farming their larger arable landholdings, intended for both commercial and subsistence purposes. Relying only on dryland farming is, however, problematic in times of no rain, as the crops of dryland farming rely on rainwater, as opposed to *molapo* farming, which relies on available river water, as already indicated. According to Magole and Thapelo (2005), *molapo* cropping is less risky as the residual flood water in the soil acts as a supply of moisture in seasons of either low or poorly distributed rainfall. *Molapo* farming is reported to produce more mature crops than dryland farming due to the unlimited

Human–elephant conflict: implications for rural livelihoods and wildlife conservation 49

moisture that makes it more sustainable and profitable (Bendsen & Meyer, 2002; Kashe et al., 2015).

Kashe et al. (2015) claim that maize (*Zea mays L.*) plant heights and grain yields were significantly higher in *molapo* fields at Lake Ngami than for the dryland fields at Shorobe. They reported grain yields to be 3.40 t ha-1 at Lake Ngami compared to 2.58 t ha-1 at Shorobe. In this regard, wildlife crop damage is causing either a decline in or an abandonment of a profitable traditional livelihood option for the respondents. According to the village chief, Khumaga farmers have since shifted to dryland farming which has become a predominant farming system and is dependent on the rains. Dryland farming is practised only during the rainy season (Magole & Thapelo, 2005). Even though the farmers of Khumaga have shifted to dryland farming, their crops are still destroyed by wildlife, leading to food insecurity. An economic loss to the local people as a result of crop damage is one of the major issues that triggers HWC and impedes the achievement of long-term conservation goals (DEA, 2010).

Social exchange theory postulates that if the benefits of wildlife conservation outweigh their costs, a community is more likely to support conservation initiatives, but if the costs of conservation outweigh the benefits, the local communities will not support conservation. Thus, if residents benefit from wildlife conservation through CBNRM, tourism, and high compensation, they are more likely to ignore the negative effects of wildlife and continue to support wildlife conservation because of the benefits they receive from it (Andereck et al., 2005). Farmers at Khumaga do not want elephants in their area because they do not benefit from them, hence suggesting relocation of elephants as a way of transferring the costs elsewhere since the laws of Botswana prohibit killing problematic animals. Arable farmers do not want to see elephants within or in the vicinity of their arable lands, and they even call them satanic creatures. They want the elephants to be removed from their area because living with elephants is a cost to them, not a benefit. Only few of the respondents reported positive associations towards elephants; this might be because they understand the importance of elephants as a key tourism product of the country.

CONCLUSION: IMPLICATIONS FOR CONSERVATION

This chapter concludes that communities generally suffer the consequences of living with wildlife due to crop loss, and hence have negative attitudes towards wildlife such as elephants. The government should, therefore, recognise the potential and capabilities of local communities in sustainably managing wildlife by promoting community participation in the formulation of wildlife management institutions. CBNRM projects are highly recommended at Khumaga. In CBNRM, rural communities are granted rights to utilise wildlife, but these rights are part of a larger conservation plan that focuses on the sustainable use of natural resources. Based on the National Parks and Game Reserves Regulations of 2000, the management plan for a national park or game reserve may designate an area as a community-use zone, and community-use zones shall be for the use of designated communities living in or immediately adjacent to a national park or game reserve to conduct commercial tourism activities. This is important in giving communities a sense of ownership toward wildlife resources and will probably shift their attitudes from negative to positive.

This can reduce hatred and the negative perceptions of farmers if they are allowed to become custodians of wildlife and make decisions about wildlife use (Metcalfe, 1994).

50   *Handbook on tourism and conservation*

Mwenya et al. (1991) contend that successful wildlife conservation is an issue of who owns and who should manage wildlife; if people see wildlife as theirs and derive benefits of owning wildlife resources, there is higher potential for them to conserve wildlife in their area. This can increase the benefits from living with elephants and other wildlife. The CBNRM should therefore be in sync with Botswana's original community-based natural resources management model. According to Cassidy (2021) CBNRM in Botswana has now lost its original meaning; the current structure of CBNRM cannot be said to be community based because decisions are imposed on communities by higher-level authorities.

Patrols by DWNP officers are an essential requirement in Khumaga. PAC offices should be located at Khumaga as they were before the construction of the fence around the park; this will enable the officers to attend to reported cases on time and to carry out patrols around the village. Well-constructed electric fences will deny the access of migratory elephants into people's ploughing fields. However, the cooperation and participation of local people for such an activity through regular monitoring of the fence and its maintenance are essential. Farmers of Khumaga seek innovative methods to keep animals away from their fields, and they still have a feeling that electric fences can help in mitigating HWC in their area. There is a need for long-term research and monitoring: This study found that there is a lack of research at Khumaga on human–wildlife interaction. Further studies should be carried out to design the site-specific appropriate strategy to tackle HWC and to improve local livelihoods.

ACKNOWLEDGEMENTS

We thank two anonymous reviewers who provided insightful comments. We would also like to express our gratitude to the people of Khumaga and government officials for providing us with the information used in this study. We also wish to express our gratitude to the Southern African Science Service Centre for Climate Change and Adaptive Land Management (SASSCAL) Project for funding this study. Finally, we wish to thank our colleagues (ORI staff) for their support.

NOTE

1.   The authors have declared no potential conflicts of interest with respect to the research, authorship, and/or publication of this chapter. The authors received no financial support for the research, authorship, and/or publication of this chapter.

REFERENCES

Andereck, K. L., Valentine, K. M., Knopf, R. C., & Vogt, C. A. (2005). Residents' perceptions of community tourism impacts. *Annals of Tourism Research, 32*(4), 1056–1076.

Barua, M., Bhagwat, S. A., & Jadhav, S. (2013). The hidden dimensions of human–wildlife conflict: Health impacts, opportunity and transaction costs. *Biological Conservation, 157,* 309–316.

Bendsen, H., & Meyer, T. (2002). The dynamics of land use systems in Ngamiland: Changing livelihood options and strategies. Paper presented at the Environmental Monitoring of Tropical Wetlands, Proceedings of a Wetland Conference, Maun, Botswana.

Blau, P. (1964). *Exchange and Power in Social Life*. New York: Wiley.

Human–elephant conflict: implications for rural livelihoods and wildlife conservation 51

Brines, J. (1994). Economic dependency, gender, and the division of labor at home. *American Journal of Sociology, 100*(3), 652–688.

Bulte, E., & Rondeau, D. (2007). Compensation for wildlife damages: Habitat conversion, species preservation and local welfare. *Journal of Environmental Economics and Management, 54*(3), 311–322.

Campbell-Smith, G., Simanjorang, H. V., Leader-Williams, N., & Linkie, M. (2010). Local attitudes and perceptions toward crop-raiding by orangutans (*Pongo abelii*) and other nonhuman primates in northern Sumatra, Indonesia. *American Journal of Primatology, 72*(10), 866–876.

Cassidy, L. (2021). Power dynamics and new directions in the recent evolution of CBNRM in Botswana. *Conservation Science and Practice, 3*(1), e205.

Chase, M., Schlossberg, S., Sutcliffe, R., & Seonyatseng, E. (2018). *Dry Season Aerial Survey of Elephants and Wildlife in Northern Botswana: July–October*. Elephants Without Borders.

Choi, H. C., & Murray, I. (2010). Resident attitudes toward sustainable community tourism. *Journal of Sustainable Tourism, 18*(4), 575–594.

Cook, K. S., Cheshire, C., Rice, E. R., & Nakagawa, S. (2013). Social exchange theory. In J. DeLamater & A. Ward (eds), *Handbook of Social Psychology*. Dordrecht: Springer, pp. 61–88.

Central Statistics Office (2011). *Population and Housing Census, 2011*. Gaborone: Statistics Botswana.

Central Statistics Office (2015). *Indices of the Physical Volume of Second Quarter 2015, Mining Production*. Gaborone: Statistics Botswana.

Darkoh, M., & Mbaiwa, J. E. (2005). *Natural Resource Utilization and Land Use Conflicts in the Okavango Delta, Botswana*. Department of Environmental Science and Harry Oppenheimer Okavango Research Centre, University of Botswana.

Department of Environmental Affairs (2010). *The Makgadikgadi Framework Management Plan*. Gaborone: Government of Botswana.

Department of Wildlife and National Parks (2012). *Community Support & Outreach Division*. Annual Report Gaborone.

Esterhuizen, D. (2015). *Global Agriculture Information Network*. Botswana Agricultural Economic Fact Sheet.

Frank, L., Hemson, G., Kushnir, H., & Packer, C. (2006a). Lions, conflict and conservation in Eastern and Southern Africa. Eastern and Southern African Lion Conservation Workshop. Johannesburg.

Frank, L., Maclennan, S., Hazzah, L., Bonham, R., & Hill, T. (2006b). Lion killing in the Amboseli-Tsavo ecosystem, 2001–2006, and its implications for Kenya's lion population. Living with Lions, Nairobi, Kenya.

Gontse, K., Mbaiwa, J. E., & Thakadu, O. T. (2018). Effects of wildlife crop raiding on the livelihoods of arable farmers in Khumaga, Boteti sub-district, Botswana. *Development Southern Africa, 35*(6), 791–802.

Graham, H. (2004). The ecology of conservation of lions: Human–wildlife conflict in semi-arid Botswana. DPhil thesis, University of Oxford.

Gupta, A. C. (2013). Elephants, safety nets and agrarian culture: Understanding human–wildlife conflict and rural livelihoods around Chobe National Park, Botswana. *Journal of Political Ecology, 20*(1), 238–254.

Gursoy, D., & Kendall, K. W. (2006). Hosting mega events: Modeling locals' support. *Annals of Tourism Research, 33*(3), 603–623.

Hall, H. (2003). Borrowed theory: Applying exchange theories in information science research. *Library & Information Science Research, 25*(3), 287–306.

Hill, C. M. (2004). Farmers' perspectives of conflict at the wildlife–agriculture boundary: Some lessons learned from African subsistence farmers. *Human Dimensions of Wildlife, 9*(4), 279–286.

Hoare, R. (2000). African elephants and humans in conflict: The outlook for co-existence. *Oryx, 34*(01), 34–38.

Homans, G. C. (1958). Social behavior as exchange. *American Journal of Sociology, 63*(6), 597–606.

Homans, G. C. (1961). *Social Behavior*. New York: Harcourt Brace and World.

Jadhav, S., & Barua, M. (2012). The elephant vanishes: Impact of human–elephant conflict on people's wellbeing. *Health & Place, 18*(6), 1356–1365.

Karanth, K., Naughton-Treves, L., DeFries, R., & Gopalaswamy, A. (2013). Living with wildlife and mitigating conflicts around three Indian protected areas. *Environmental Management, 52*(6), 1320–1332.

52   *Handbook on tourism and conservation*

Kashe, K., Mogobe, O., Moroke, T., & Murray-Hudson, M. (2015). Evaluation of maize yield in flood recession farming in the Okavango Delta, Botswana. *African Journal of Agricultural Research, 10*(16), 1874–1879.

Kern, T., & Willcocks, L. (2000). Exploring information technology outsourcing relationships: Theory and practice. *The Journal of Strategic Information Systems, 9*(4), 321–350.

Linkie, M., Dinata, Y., Nofrianto, A., & Leader-Williams, N. (2007). Patterns and perceptions of wildlife crop raiding in and around Kerinci Seblat National Park, Sumatra. *Animal Conservation, 10*(1), 127–135.

Luo, X. (2002). Trust production and privacy concerns on the Internet: A framework based on relationship marketing and social exchange theory. *Industrial Marketing Management, 31*(2), 111–118.

Madden, F. (2004). Creating coexistence between humans and wildlife: Global perspectives on local efforts to address human–wildlife conflict. *Human Dimensions of Wildlife, 9*(4), 247–257.

Magole, L., & Thapelo, K. (2005). The impact of extreme flooding of the Okavango river on the livelihood of the Molopo farming community of Tubu village, Ngamiland Sub-district, Botswana. *Botswana Notes and Records, 37*, 125–137.

Malinowski, B. (1922). *Argonauts of the Western Pacific*. London: Routledge & Kegan Paul.

Mauss, M. (1925). *The Gift: Forms and Functions of Exchange in Archaic Society*. New York: Free Press.

Mbaiwa, J. E. (2005). Wildlife resource utilisation at Moremi Game Reserve and Khwai community area in the Okavango Delta, Botswana. *Journal of Environmental Management, 77*(2), 144–156.

Mbaiwa, J. E. (2015). Community-based natural resource management in Botswana. In R. van der Duim, M. Lamers, & J. van Wijk (eds), *Institutional Arrangements for Conservation, Development and Tourism in Eastern and Southern Africa: A Dynamic Perspective*. Dordrecht: Springer, pp. 59–80.

Mbaiwa, J. E., & Mbaiwa, O. I. (2006). The effects of veterinary fences on wildlife populations in Okavango Delta, Botswana. *International Journal of Wilderness, 12*(3), 17–23.

Mbaiwa, J. E., & Rantsudu, M. (2003). *Socio-Economic Baseline Survey of the Bukakhwe Cultural Conservation Trust of Gudigwa, Botswana*. Maun: Conservation International.

Mc Guinness, S., & Taylor, D. (2014). Farmers' perceptions and actions to decrease crop raiding by forest-dwelling primates around a Rwandan forest fragment. *Human Dimensions of Wildlife, 19*(2), 179–190.

Metcalfe, S. (1994). The Zimbabwe communal areas management programme for indigenous resources (CAMPFIRE). In D. Western & R. M. Wright (eds), *Natural Connections: Perspectives in Community-Based Conservation*. Washington, DC: Island Press, pp. 161–192.

Miller, K. (2005). *Communication Theories*. New York: McGraw-Hill.

Monametsi, N. F. (2008). The effect of the electric fence in reducing human–wildlife conflict in the Western part of Makgadikgadi Pan National Park, Botswana. MSc thesis, Norwegian University of Life Sciences.

Mosojane, S. (2004). Human–elephant conflict in the eastern Okavango Panhandle. MSc thesis, University of Pretoria.

Mwenya, A., Lewis, D., & Kaweche, G. (1991). *Policy, Background and Future: National Parks and Wildlife Services, New Administrative Management Design for Game Management Areas*. United States Agency for International Development, Lusaka, Zambia.

Nyhus, P. J., Osofsky, S. A., Ferraro, P., Madden, F., & Fischer, H. (2005). Bearing the costs of human–wildlife conflict: The challenges of compensation schemes. In R. Woodroffe, S. Thirgood, & A. Rabinowitz (eds), *People and Wildlife, Conflict or Co-existence?* Cambridge: Cambridge University Press, pp. 107–121.

Ogra, M. (2009). Attitudes toward resolution of human–wildlife conflict among forest-dependent agriculturalists near Rajaji National Park, India. *Human Ecology, 37*(2), 161–177.

Okello, M., & D'Amour, D. (2008). Agricultural expansion within Kimana electric fences and implications for natural resource conservation around Amboseli National Park, Kenya. *Journal of Arid Environments, 72*(12), 2179–2192.

Osborn, F. V., & Parker, G. E. (2003). Towards an integrated approach for reducing the conflict between elephants and people: A review of current research. *Oryx, 37*(1), 80–84.

Perkins, J., & Ringrose, S. (1996). Development co-operation's objectives and the Beef Protocol: The case of Botswana. A study of livestock/wildlife/tourism/degradation linkages. Department of Environmental Science, University of Botswana, Gaborone.

Pfouts, J. H. (1978). Violent families: Coping responses of abused wives. *Child Welfare: Journal of Policy, Practice, and Program*, *57*(2), 101–111.

Ringrose, S., Chanda, R., Nkambwe, M., & Sefe, F. (1996). Environmental change in the mid-Boteti area of north-central Botswana: Biophysical processes and human perceptions. *Environmental Management*, *20*(3), 397–410.

Sifuna, N. (2010). Wildlife damage and its impact on public attitudes towards conservation: A comparative study of Kenya and Botswana, with particular reference to Kenya's Laikipia Region and Botswana's Okavango Delta Region. *Journal of Asian and African Studies*, *45*(3), 274–296.

Son, J.-Y., Narasimhan, S., & Riggins, F. J. (2003). Effects of relational factors and channel climate on EDI usage in the customer-supplier relationship. *Journal of Management Information Systems*, *22*(1), 321–353.

Songhurst, M. A., & Chase, M. (2008). *Elephant Survey August 2008 Eastern Okavango Panhandle, Botswana (NG11, NG12 and NG13)*. Elephants Without Borders.

Thouless, C., & Sakwa, J. (1995). Shocking elephants: Fences and crop raiders in Laikipia District, Kenya. *Biological Conservation*, *72*(1), 99–107.

Treves, A. (2007). *Balancing the Needs of People and Wildlife: When Wildlife Damage Crops and Prey on Livestock*. Land Tenure Center, Nelson Institute of Environmental Studies, University of Wisconsin.

Warner, M. Z. (2008). Examining human–elephant conflict in southern Africa: Causes and options for coexistence. Master's thesis, University of Pennsylvania.

Woodroffe, R., Thirgood, S., & Rabinowitz, A. (eds) (2005). *People and Wildlife, Conflict or Co-existence?* Cambridge: Cambridge University Press.

World Travel and Tourism Council (2014). *Travel & Tourism Economic Impact Botswana*. London: WTTC.

Zafirovski, M. (2005). Social exchange theory under scrutiny: A positive critique of its economic-behaviorist formulations. *Electronic Journal of Sociology*, *2*(2), 1–40.

5. Residents' perception of ecotourism development at Tachila Nature Reserve, North-East District, Botswana

Unabo Tafa and Joseph E. Mbaiwa

INTRODUCTION

In recent years communities that experience shortage of land question the development of ecotourism as a suitable approach to poverty alleviation and rural development in their local areas. In the North-East District (NED) of Botswana, there is a gap in knowledge on how local people perceive or benefit from ecotourism development at the Tachila Nature Reserve (TNR). In Botswana, more especially in rural areas, rural livelihoods are dependent on natural resources. Livelihood activities are those that together determine the living gained by the individual or household (Ellis, 2000). Livelihood activities in the NED include arable farming, pastoral farming, collection of veldt products and mining. In the NED the most significant competition and conflicts are those between arable farming, livestock farming and the desire to reserve massive chunks of land for ecotourism purposes. The development of ecotourism at TNR creates conflicts over natural resources such as land and forest use. Buckles and Rusnak (1999) lament that local communities everywhere have to compete for the natural resources they need to sustain their livelihoods. The objective of this chapter is to analyse the performance of ecotourism as a viable model designed to achieve improved livelihoods and conservation in the NED, Botswana.

ECOTOURISM IN BOTSWANA: UNDERSTANDING ITS CONTEXT

The concept of ecotourism has been adopted in developing countries as a tool for conservation and alleviating poverty in rural areas. Ecotourism can succeed in some areas but it can also fail to take off in other areas. Mbaiwa (2015) noted that in areas where ecotourism is a success, it has the potential to generate economic benefits such as employment opportunities and household income, resulting in support for ecotourism and conservation among communities. In areas where ecotourism has failed, however, lack of managerial skills, entreprencurship and marketing skills of local communities have been cited as key factors contributing to failure of projects. In tourism planning, Botswana as a country has popularized ecotourism as a tool that can meet the needs of local communities, development, and improved livelihoods. Since the 1990s, Botswana adopted the ecotourism approach to address natural resources depletion and how communities living in such areas with natural resources could benefit from them (Mbaiwa, 2008). However, the contribution of ecotourism to improved rural livelihoods in the NED has not been adequately researched and analysed. An essential goal of ecotourism is that the framework recognizes the local communities as active participants and direct beneficiaries that should enjoy a tourism product in their host destinations. Honey (2008, p. 39) defines eco-tourism as 'responsible travel to natural areas that conserve the environment and improves the well-being of local people'. Researchers (Mbaiwa, 2005a; Rozemeijer & Van der Jagt, 2000) point out that ecotourism projects have the goal of reducing rural poverty, promoting conservation and empowering communities to manage natural resources for their long-term benefits. Murphree (1999) noted that local people's participation and involvement in natural resources management is associated with accrued benefits at household level. Without access to those benefits, or where no benefits are gained or the distribution of such benefits is inequitable, the local people are unlikely to participate. Local people cannot be expected to provide support for protected areas if the costs of doing so outweigh the benefits (Kiss, 1990; Western et al., 1994).

A study in Colorado conducted by Perdue et al. (1990) showed that ecotourism development was positively related to the perceived positive or negative impacts of tourism. Getz (1994), in a study conducted in Scotland, found that increased negative attitudes towards ecotourism development suggested that residents believed that benefits have declined or not matched their expectations. A study conducted in Trentino, Italy (Brida et al., 2011) demonstrated that new conservation strategies contributed to negative conservation attitudes among residents. As a result, these strategies usually lead people to harbour negative perceptions towards ecotourism and conservation initiatives within the boundaries of the protected areas (Weladji et al., 2003; Vodouhê et al., 2010). Communities evaluate ecotourism in terms of social exchange, that is, in terms of expected benefits or costs obtained in return for the services they supply (Ap, 1992). A study conducted in Myanmar (Burma) indicates that protected areas are essential for the long-term conservation of biodiversity (Dasmann, 1984; Machlis & Tichnell, 1985; Zube 1986; Brandon & Wells, 1992; Newmark et al., 1993; Fiallo & Jacobson, 1995; Furze et al., 1996) and a positive public attitude is a key indicator of the success of a protected area (Struhsaker et al., 2005). However, the relationship between people and the protected area is often contentious, as protected area establishment often entails resettling or depriving people of access to resources upon which they have depended for generations (Western & Pearl, 1989; West & Brechin, 1991). For example, in Africa, 500 people were forcibly relocated from the Nechasar National Park, Ethiopia in 2004 (Adams & Hutton, 2007), while in the early 1960s, San/Basarwa were removed from the Moremi Game Reserve, Botswana for natural resource conservation (Bolaane, 2004). During the late nineteenth century and much of the twentieth century, efforts to protect biodiversity in Africa emphasized the designation of protected areas (Adams & McShane, 1992).

The introduction of protected areas and conservation areas in Africa came with an increasing interest from foreign lords who had the passion to rule and govern African states. Rodney (1972) and Darkoh (1996) further explain that colonialism and modernization in Africa alienated African societies from the natural resources upon which they had previously based their livelihood under a system of collective rights. Since many indigenous communities were evicted by colonial governments from their ancestral lands when protected areas were proclaimed, indigenous people generally viewed wildlife as a threat (Magome & Murombedzi, 2003). This scenario led to the development of negative perceptions and attitudes towards wildlife conservation by the local people. This was the beginning of human–wildlife conflict, a common theme discussed by Collett (1987), Grove (1987), Lindsay (1987), and Marekia (1991) about Kenya; Moganane and Walker (1995) about Botswana; and Chenje and Johnson (1994) about Southern Africa. The indigenous people and the government clashed when wild-

56 *Handbook on tourism and conservation*

life resources were declared state property under the colonial legislation, making it illegal for rural people to make any use of the resources in their areas (Mbaiwa, 1999).

Most of the local people around protected areas have negative attitudes about state policies and conservation programmes. The alienation of grazing land for the exclusive use of wildlife and tourists has had a direct impact on the pastoralist communities and prompted them to raise questions about African wildlife policy as it leads to human–wildlife conflict. According to Irandu (2004), the local communities living near and around the national parks and game reserves are the first to pay the price for wildlife conservation through the destruction of their property and death or injuries caused by wild animals. A study by Mutanga et al. (2015), indicates that the creation of protected areas forced the relocation of local communities from their original areas of residency, depriving them of access to resources in the protected areas such as meat, grazing areas and firewood. This deprivation seems to have disconnected local communities from the adjacent protected areas. Many African conservation protected areas have a long history of being dominated by coercive conservation policies that have later become known as fortress conservation (Mutanga et al., 2015). These policies exclude local communities' participation and have often caused negative relationships between protected areas and local communities, resulting in conflicts and problems such as increased illegal hunting, habitat encroachment and destruction, violence and poverty among indigenous communities. Local people adjacent to four protected areas in Zimbabwe have developed negative attitudes and this background continues to influence the communities' perception on wildlife conservation and tourism to date (Mutanga et al., 2015). Studies in Zimbabwe by Murphree (1993) and Mwenya et al. (1991), in South Africa by Prosser (1996), and in Namibia by Ashley (1995) and Rihoy (1995) have shown that local people tend to support the wise use of natural resources such as wildlife in their local environment if they derive socio-economic benefits from them.

The creation of protected areas alienated or distanced communities living closer to national parks from access to resources (Boyd & Timothy, 2001; Mayoral-Phillips, 2002). In some instances, the local people have been forcibly relocated outside parks for conservation purposes (Bolaane, 2004; Child, 2009). The history of negative attitudes of local people towards conservation especially wildlife in Botswana began during the British colonial rule of the country (1885–1966). The centralization of wildlife resources and the establishment of protected areas resulted in the displacement of local communities from their homelands and denied them access to resource use in parks (Adams & McShane, 1992; Bolaane, 2004). For example, when Moremi Game Reserve was established in 1963, several San (Basarwa) communities in Khwai, Mababe, and Gudigwa villages were relocated from their homeland (Bolaane, 2004; Mbaiwa, 2005b). Mbaiwa (1999) notes that tourism development is often found to be conflicting with other land-use activities leading to land-use conflicts between tourism and residents of the neighbouring villages.

SOCIAL EXCHANGE THEORY

This study is informed by social exchange theory (SET). The theory has its origins in the work of a sociologist called George Homans in 1958. The SET model assumes that rewards and costs drive relationships (Homans, 1958). Both parties in a social exchange take responsibility for each other and depend on each other, thus forming a system of relational life (Thibaut & Kelley, 1959).

Blau (1964) indicated that human beings evaluate relationships in terms of costs and benefits. People calculate the overall worth of a particular relationship by subtracting its costs from the rewards it provides (Jones, 1976). The worth of a relationship influences its outcome, or whether people will continue with a relationship or terminate it. Positive relationships are expected to endure, whereas negative relationships will probably terminate.

The principle of the SET has been adopted widely by tourism researchers since the 1990s (Perdue et al., 1987; Madrigal, 1993; Getz, 1994; Hernandez et al., 1996; McGehee & Andereck, 2004; Andereck et al., 2005). The SET helps to understand the costs and rewards of relationships and also helps people to predict how to keep and sustain relationships. In the tourism perspective, the SET helps to understand that if there is need for a development then it should benefit the local people. If ecotourism is to be regarded a viable model of development it should bring rewards for local communities to appreciate and support it. Human beings appreciate when something is done for them and they in turn want to do something because of their need to reciprocate.

Although the SET has been widely used, Zafirovski (2005) has identified the following as the weaknesses of the SET. The theory neglects culture context and variations of cultures. The social exchange theory is based on a reward concept, but all cultures are different and in some cultures people may not seek a reward from a relationship. Despite these weaknesses, the SET has been adopted in many tourism studies. The theory can accommodate explanation of both positive and negative perceptions and can examine relationships at the individual or collective level as compared to other models which lack a common comparability and a theoretical framework that is able to give a clear common explanation (Stolte et al., 2001). The theory is used in this study to demonstrate why residents would perceive tourism impacts to be negative or positive and influence their position in the support for tourism development.

STUDY AREA

The study was conducted in north-eastern Botswana, using Tachila Nature Reserve and its neighbouring villages as case studies (Figure 5.1). The nature reserve was established in 2007 and officially launched in 2008.The reserve is situated 5 km east of the city of Francistown and offers tourism activities which include horse riding, snake viewing, education, camping, picnic sites and wildlife viewing. The TNR includes Lady Mary Farm and part of the Sam Estates Farm, situated between the Tati River to the east and the main A1 Road to the west (Francistown–Gaborone road). Recently developed waterworks (Mambo) provide a permanent flow of water into the Tati River bordering the land, which not only provides an essential supply of water for wildlife, but also a permanent wetland habitat that attracts many additional wildlife species. In the past, people from the villages of Matshelagabedi, Ditladi, and Patayamatebele used to fetch water from the Tati River, harvest forest products, and hunt and graze their livestock in the area currently occupied by TNR. TNR is a new tourism resort in the NED established to serve ecotourism purposes and it is located on land owned by Tati Company (TC), and managed by 'local communities' and TC. This makes TNR an interesting study since the NED has few tourism destinations as compared to the Chobe region.

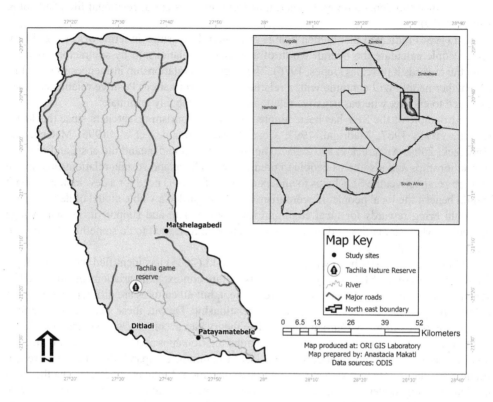

Source: Okavango Research Institute GIS Lab, map prepared by Anastacia Makati.

Figure 5.1   Map of Tachila Nature Reserve and study villages

To gather information on residents' perceptions and attitudes towards tourism at TNR, three communities living around the reserve were sampled: the villages of Patayamatebele, Matshelagabedi and Ditladi.

Patayamatebele

Patayamatebele is a small village of about 349 people (Statistics Botswana, 2001). The village is 18 km south-east of TNR. The people belong to various ethnic groups such as the Bangwato, Basarwa, Bakhurutshe, Bakalaka and Babirwa. Patayamatebele lies between the Shashe and Tati Rivers. The village was established in 1989 and officially gazetted in 2007. The main economic activities in Patayamatebele are farming and gathering wild berries. Residents of Patayamatebele used the area currently occupied by TNR as their hunting grounds. They also fetched water from the Tati River which passes through the reserve. Their livestock also used the area for grazing. Patayamatebele is a middle-aged village. It is a rural poor village with little development.

Matshelagabedi

The village of Matshelagabedi is about 12 km north-east of TNR (Figure 5.1). There are about 2,871 people in the village (Statistics Botswana, 2011). The people of Matshelagabedi belong to various ethnic groups such as Bakhurutshe, Bakalaka, Basarwa and Bangwato. The people in this village are employed in various economic sectors such as agriculture, mining and quarrying, manufacturing, water and electricity, construction, wholesale, transport and communication, business services, health services, education and domestic services (Statistics Botswana, 2011). The village used the area now occupied by TNR as grazing land, a place where they used to harvest thatch grass and other veld products; it was also their hunting ground. Some of the founders of Matshelagabedi first settled where it is now TNR before they were relocated by the Tati Company during the colonial period (1885–1966). Matshelagabedi is a relatively very old village. In a personal communication with one elder of this village, he indicated that the village was established in 1971.

Ditladi

Ditladi is situated 10km south of TNR (Figure 5.1). The population of Ditladi is about 1,344 (Statistics Botswana, 2011) and the main activities are arable farming and pastoral farming whilst some people are temporarily employed in the Ipelegeng projects. The village has long used the area currently occupied by TNR as grazing land and for hunting. Ditladi was established in the early 1970s. A village elder indicated that Ditladi was established as a small rural settlement in the early 1970s and people regarded the settlement as a cattle-post until in 1990 when the village was officially gazetted. Up to now, Ditladi is still a very small village with a low level of development.

TNR was used in the study because it is the first development project in NED to offer ecotourism services. The objective of the study is to assess how nature reserves developed on private land can contribute to improved livelihoods, land-use conflict resolution and socio-economic development for the local people. In NED, local people argue that they have been displaced from their land and therefore rendered landless (Manatsha, 2014). The land where TNR is established is characterized by land politics.

METHODS

Data Collection

Data for this study were collected between December 2014 and August 2015. Qualitative and quantitative data collection methods were used in the study. Primary data collection involved household interview surveys using open- and closed-ended questions, face-to-face interviews and focus group discussions. The researcher developed a questionnaire that was used as an interview guide. The researcher conducted interviews with household heads guided by the questionnaire which had open- and closed-ended questions. Respondents were asked to select an answer from among a list provided by the researcher. The questions provided uniformity of responses, which allowed for easy processing of data.

60 *Handbook on tourism and conservation*

Table 5.1 *Total number of key informants*

Village	Village chief, chief advisers, ward head men	Village development committee members	Village elders	Total
Matshelagabedi	1	3	3	7
Patayamatebele	1	1	3	5
Ditladi	2	2	2	6
Total	4	6	8	18

Open-ended questions were also used in household survey interviews to collect data. Here respondents were asked to provide their own answer to the questions. Questions were asked such as 'What is your level of agreement to ecotourism development at TNR?' 'Do you get benefits from Tachila?' and then progressed based, primarily, upon the initial response. Then a Likert scale was developed to summarize the attitudes in a fairly brief statement. Babbie (2016) notes that a Likert scale is a format in which respondents are asked to strongly agree, agree, disagree or strongly disagree or strongly approve, approve and so forth. The question-naires were mainly used to collect information on the perceptions and attitudes of people towards the development of tourism at TNR. These mainly collected qualitative data. The collection of data was conducted in Setswana (a national language spoken widely throughout Botswana). Access to households was easy and permissible during the household interviews. Face-to-face interviews were also conducted with the reserve staff to get their views about the contribution of the reserve to livelihoods and the involvement of the people in the reserve planning and decision-making.

Focus group discussions (FGDs) were used to check the validity of results obtained from household interviews and key informants. This discussion was guided, monitored and recorded by a researcher. Focus groups are used for generating information on collective views and the meanings that lie behind those views. A total of two (2) FGDs were conducted in each village. The discussants were categorized according to their ages; the youth (people aged 16 to 39) and elders (people aged 40 to 85). Elderly participants' education level ranged from non-formal education to high school level, while the youths' ranged from lower primary to senior high school. A total of 16 females and 14 males participated in the focus groups. Each focus group had 5 participants and was composed of people of similar socio-economic backgrounds so as to limit biases and to ensure free deliberations of the discussants. Groups of FGDs were identified by means of purposive sampling. Focus groups were used mainly to collect only qualitative data. The interviewer suggested the general theme of the discussion and posed further questions as these came up in the spontaneous development of the interaction between the interviewer and the research participant (Welman & Kruger, 2001). The data obtained from focus groups were analysed by means of systematic coding through content analysis.

Key informants were also used to provide information for this study. A key informant is an expert source of information. Burgess (1982) calls such individuals 'natural observers', Sjoberg and Nett (1968) describe such individuals as 'strategic informants'. The principal advantage of the key informant technique relates to the quality of data that can be obtained in a relatively short period of time. Purposive sampling was used to select these key inform-ants (Table 5.1). These were the community leaders such as village chief, village committee members, village councillors, members of parliament for the district, village advisers (elders) and the local communities. A chief from each of the three villages was interviewed.

Residents' perception of ecotourism development at Tachila Nature Reserve 61

Table 5.2   Household sample size

Matshelagabedi (N = 267)	Patayamatebele (N = 51)	Ditladi (N = 112)
267/1+267(0.05)²= 160	51/1+51(0.05)²= 45	112/1+112(0.05)²= 87

The research also obtained data from documentary materials, which include district tourism departments, TNR project management plan, policy documents, journal articles, theses and dissertations regarding the development of tourism and its contribution to livelihoods. Documented board meeting minutes were also used to gather information on how decision-making processes are made, the achievements of the project, the activities in place and how the board related to the local communities. The TNR management plan also provided information on the dates the project was established, when it was started and the benefits distribution plan and actual benefits to the local people. The advantage of using existing data is that data collection is quick, easy and inexpensive.

Sampling Design

Taro Yamane's formula (1967) was used to determine each sample in each of the villages:

$$*n = N/1 + N(e)2$$

where: n = Sample size; N = Population size; e = Level of precision or Sampling Error.

There was a total of 267 households in Matshelagabedi, 51 households in Patayamatebele and 112 households in Ditladi. Therefore, the sample size of households was calculated as shown in Table 5.2, where N = Total population of households.

The application of Taro Yamane's formula resulted in 292 households used in the study. Household surveys were undertaken with 292 household heads in the three villages that neighbour the nature reserve. From Matshelagabedi village 160 household surveys were collected, 45 in Patayamatebele and 87 in Ditladi.

All the households in each study village were listed prior to household sampling. Each household was given a consecutive number (e.g. 001, 002, 003 and so on) for identification. Simple random sampling was then used to select the desired number of household units in each study village. In simple random sampling, each member of the population has the same chance of being included in the sample (Babbie and Mouton, 2001). After counting all the units of analysis in the sampling frame and giving them consecutive numbers, a table of random numbers was developed. Thereafter a pencil was used to make a mark on the table in order to select the number closest to the mark. From the pencil mark a decision was made whether to select numbers on a row or column. Numbers were noted down until a collection of numbers was made equal to the size of the desired sample in each village.

Data Analysis

Data collected were finally analysed both qualitatively and quantitatively. Thematic analysis was used to analyse all the qualitative data. Thematic analysis involves data reduction into themes and patterns to be reported. Themes are identified by bringing together components or fragments of ideas or experiences, which often are meaningless when viewed alone

62   *Handbook on tourism and conservation*

(Leininger, 1985). In thematic analysis, themes that emerge from the informants' stories are pieced together to form a comprehensive picture of their collective experience (Aronson, 1994). Therefore, qualitative data from households, key informants' interviews and focus group discussions were summarized into specific themes and patterns of resident perceptions and attitudes towards the development of tourism at TNR.

Responses collected through open-ended questions were then categorized and coded into specific themes. Quotations of some key informants and focus group discussions were used to give the final report a deep and well supported analysis. These were presented in a descriptive form. Quantitative data were analysed using computer packages such as the Special Package for Social Sciences (IBM SPSS) and Excel. Statistical tests such as t-test and one-way ANOVA were performed to compare the mean averages between groups. T-test was used to evaluate the differences in means between two groups. The *p*-value reported with a t-test represents the probability of error involved in accepting the research hypothesis about the existence of a difference in the means. Assumptions were tested and the t-statistic assumes normality of the group distributions or variance (Welman & Kruger, 2001). A one-way ANOVA was also used to determine whether groups or treatments differed statistically with regard to the group mean scores from one dependent variable. Age (independent variable) was grouped into three categories and the influence of age on perception assessed.

RESULTS AND DISCUSSION

Local Communities' Perceptions of the Development of Tourism at TNR

This study sought to answer the question on how local communities perceive tourism development at TNR. That is, the status of support local people placed on the development of tourism at TNR was used as an indicator to measure attitudes of residents in study villages towards ecotourism development. Respondents were asked 'Do you support ecotourism development at TNR' and indicated by 'Yes or No' and then supported their response. Those who chose 'No' represented negative attitudes and those who chose 'Yes' had positive attitude. Results in Table 5.3 indicate that the majority (87%, n = 234) of the households at the villages of Ditladi, Patayamatebele and Matshelagabedi have strong negative perceptions of the development of tourism at TNR. They do not support tourism development because of the following: the reserve has deprived them of access to forest harvest; reduced their grazing land by fencing a huge area; there are no employment opportunities; the people want the land they have lost to the Tati Company; they also felt they were not consulted during the initial stages of the project and they also do not know how the nature reserve will benefit them. Only 11 per cent (29 respondents) indicated that they had positive perceptions of TNR. Even though the number of residents who perceive positive benefits from tourism is small, the local people indicated that TNR could bring the following: conservation education to the people and school children; employment benefits; and conservation of forests and wild animals. Researchers on tourism have identified that local residents perceive tourism positively due to its propensity to create jobs, generate income and provide social services and infrastructure in local communities (Andereck et al., 2005; Jurowski et al., 1997; Kuvan & Kuvan, 2005; Murphy, 1985; Teye et al., 2002; McGehee & Andereck, 2004). TNR can benefit local people and if people are involved in the planning and management of the nature reserve, they will support the develop-

Residents' perception of ecotourism development at Tachila Nature Reserve   63

Table 5.3   Perceptions of households regarding the development of TNR

Village	Total no. of respondents	Negative perception		Positive perception		Neutral	
		Number of respondents	% of respondents	Number of respondents	% of respondents	Number of respondents	% of respondents
Ditladi	80	72	90	7	8.7	1	1
Patayamatebele	40	31	68.8	6	13	3	6.6
Matshelagabedi	150	131	81.8	16	10	3	1
Total	270	234	87	29	11	7	3

ment and also perceive positive benefits. Studies (Cordes & Ibrahim, 1999; Cole, 2006; Dyer et al., 2007; Lepp, 2004; Walpole & Goodwin, 2001) have indicated that when residents were involved in the tourism industry or recreation activity, they tended to show more support for additional tourism development. Also, residents who showed positive environmental behaviours expressed support for tourism (Jurowski et al., 1997; Kuvan & Kuvan, 2005; Perdue et al., 1990). However, local people with negative perceptions and attitudes about tourism showed less support for its development (Andereck & Jurowski, 2006; Banks, 2003; Kuvan & Kuvan, 2005; Teye et al., 2002; Wilson & Hart, 2001). The local people want to be left free to access the resources such as forest harvest, land for cultivation and grazing and fuel wood from the reserve. A study (Stone, 2012) has indicated that the establishment of a bird sanctuary at Nata pushed out all the livestock that used to graze on the communal land demarcated for bird rearing. Ecotourism projects may bring resource use competition like land use as shown by the TNR case.

Key informants (chiefs) in all the three villages, Ditladi, Patayamatebele and Matshelagabedi, indicated that the development of TNR came as a surprise to them when most of the residents wanted land which was taken by the Tati Company. The key informants also indicated that there were no proper consultative meetings held in their villages. A 32-year-old man from Matshelagabedi village highlighted that they were simply told about a nature reserve that was to be developed. The results from household interviews suggest that the idea of establishing this reserve was conceived by just a few people, and the local communities were informed about the decision later. Manatsha (2014) noted that TNR is an elite project.

Household interviews (90%, n = 243) indicated that TNR is for the rich people who will surely benefit from the project. Some households (87%, n = 234) stated that the nature reserve serves the interest of foreigners since the board members are not local people. Results indicate that the local people do not know how they relate to the reserve. In ecotourism projects the equitable distribution of benefits is one of the challenges confronting such nature-based projects (Mbaiwa and Darkoh, 2006). TNR Board of Trustees members are influential individuals. An interview with the manager at TNR indicated that the criteria used to be a member of the board of trustees was by voluntarism and most of the people who volunteered are business people. This gives some local communities the impression that it is 'owned' by a few business people hence the failure to distribute benefits to the local communities. These results, therefore, indicate that the people in the study have negative attitudes towards TNR. If people are deprived of access to resources because of the development of ecotourism, it is unlikely that they will support the conservation of the natural resources upon which ecotourism is based (Okech & Bob, 2009).

Local People's Perceptions of Ecotourism

Residents were asked about how they perceive ecotourism at TNR. Results indicated that people (87%, n = 234) have negative perceptions towards the development of ecotourism at TNR. This is because communities do not benefit from the development of TNR as it is supposed to be an ecotourism project. Ecotourism development should improve the livelihoods of host communities and be beneficial to them. The development of TNR for ecotourism purposes is something else as there is no community involvement and participation. The communities in the three villages of Matshelagabedi, Patayamatebele and Ditladi do not know who owns TNR and how they relate to it. If ecotourism projects are poorly communicated they bring confusion and frustration among the members of the community (Chenier et al., 1999; Suliman, 1999). Stone (2012) noted that a lack of awareness and understanding about the workings of community-based tourism projects are indicative of a lack of communal ownership of the project. In Nata, respondents in a study of what is registered as a community project indicated that they did not know who owns the project, 'their belief was the project is owned by white people who drive 4 x 4 trucks and always hide their eyes with dark glasses' (Stone, 2012). This is the same understanding of TNR where the local communities do not know who owns TNR.

Chen (2006), in a study done in Chinghai Biosphere Mountain reserve in China, noted that local people held negative attitudes toward forest reserves. The study shows that 60.4 per cent of the people had negative attitudes and attributed their dislike of the reserve to income loss due to strict forest use rules, crop damage by wild pigs, the restriction on killing wild animals viewed as pests, inequitable distribution of mountain resources and their potential benefits and inadequate attention to community development after the ban on collecting mountain resources was established.

The negative perceptions that people have date back to the colonial era. The interest and concern of the local African people were not considered in the establishment of these protected areas. Mackenzie (1998) further rightly argues that foreign interest and not the interest of the African people influenced the legislation of wildlife management and protected areas in particular. In many cases, creation of these protected areas deprived local people of a resource that they had been accessing for a long time for both their cultural and economic values (Barrow & Murphree, 2001). The increasing human population and the resulting pressure on land resources increase the conflict between protected areas, managers and the neighbouring communities, hence the local people develop negative attitudes.

A one-way between group analysis of variance was conducted to further explore the impact of age levels on support for ecotourism development at TNR. Participants were divided into three groups according to their age (group 1: 18–35; group 2: 36–55 years; group 3: 56 years and above). There was no statistical significance at $p < .05$ level in the scores for the three age groups. $F (2,267) = .276$, $p = .759$. The mean score for group 1 (m = 1.52, SD = .975) did not differ with group 2 (m = 1.56, SD = 1.079) and also for group 3 (m = 1.42, SD = .988). The results indicate that support for ecotourism development at TNR is not associated with age. The results also imply that the study population ages do not influence perceptions towards TNR; the people hold negative perceptions regardless of their age.

Local People's Perceptions of Conservation

In relation to local community attitudes towards conservation, results indicate that residents of Ditladi, Matshelagabedi and Patayamatebele hold negative perceptions of conservation at TNR. This is because people do not differentiate between TNR and Tati Company. The negative perception they hold towards TC translates to negative perceptions towards TNR. Understandably though, it is hard to erase the sad memories of the brutalities of the colonial Tati Company in the minds of those who witnessed them. For instance, in a study on the creation of a landless state Manatsha (2020) states that in Matshelagabedi an elderly man accused the government of being lenient with Tati Company, which oppressed and stole their land in the first place. He was quoted as follows:

> TC stole our land and rendered our people destitute in the process because the people's fertile land for ploughing and cattle grazing was forcibly taken away from them. [Tati Company] would simply command you to take your belongings from your house and then they would set the whole homestead on fire as a form of eviction. They had no mercy, so why is the government being so patient and lenient with them at our expense. [We] were treated badly and lived in fear of this company which is of foreign origins ... [TC] turned the whole land, which was previously tribal into private property. (Manatsha, 2020, p. 114)

The negative views about Tati Company make it impossible for residents to support conservation at TNR. A manager at TNR said that he was aware that some people still have the 'colonial stigma' about the company (Manatsha, 2010). Communities said that their land for grazing that they used for so many years has now been fenced by TNR resulting in loss of grazing land and important natural water sources. The area where TNR is being developed used to attract a lot of wild animals and provided an opportunity for hunting. The local communities of the villages of Ditladi, Patayamatebele and Matshelagabedi have also lost their traditional hunting grounds. The household interviews shows that 87% (n = 234) would prefer using land for growing crops and cattle rearing as compared to ecotourism. In the NED local people prefer cattle rearing for the economic and social value cattle have. Cattle are used as a sign of wealth; they provide milk and they can also be sold for money. Ecotourism at TNR is unable to match the benefits of cattle rearing.

The board of trustees of TNR does not have any members from the villages of Ditladi, Patayamatebele and Matshelagabedi. This makes the accountability of the board of trustees questionable since it does not include representatives from other villages surrounding TNR while TNR is an ecotourism project. The board of trustees lacks community representatives. It consists of wealthy individuals who are influential and are guarding their business interests using the nature reserve. In Ditladi a 57-year-old man indicated that they do not even know the management of the nature reserve. This is a clear sign that the nature reserve is not a community project. Some private individuals owning ranches within the Tati Company land have identified themselves to local people and have been helping in the village developments, whereas TNR is secretive and very exploitative according to one of the respondents. TNR, as an ecotourism project, lacks the guiding programme of Community Based Natural Resources Management (CBNRM). In Botswana, CBNRM is seen as a development approach that supports natural resource conservation and the alleviation of poverty through community empowerment and the management of resources for long-term social, economic and ecological benefits.

66 *Handbook on tourism and conservation*

The study used the SET to evaluate perceptions of residents. The SET postulates that residents evaluate tourism in terms of social exchange, that is, in terms of expected benefits or costs obtained in return for the services they supply (Ap, 1992). Hence, it is assumed that host residents seek tourism development for their community to satisfy their economic, social and psychological needs and to improve the community's well-being.

The perceptions of local people in the villages of Ditladi, Patayamatebele and Matshelagabedi need to be changed through public workshops and seminars. Failure to do that means TNR will remain unjust in the eyes of some villagers. The TNR Trust should genuinely include the local communities in this project for it to be meaningful to them. It should engage in public relations and education campaigns to fully explain itself to the villagers in the simplest language they can understand. There is urgent need for the management of TNR and the government to show a clear documentation and commitment to implement a plan that can also benefit the local people as TNR is supposed to be an ecotourism project.

If the local people are not involved in the decision-making for tourism development and conservation, they are likely to develop negative attitudes towards natural resources management as shown by the case at TNR. Since the 1980s, tourism literature calls for the inclusion and involvement of local communities in tourism development (Mbaiwa, 2008). Failure to include the local people means that residents have the potential to disown the tourism product (Hardy et al., 2002). Residents of Patayamatebele, Ditladi and Matshelagabedi perceive tourism to have created costs that impinge on them adversely.

An independent sample t-test was conducted to compare the perception scores for males and females. The results indicated that there was a significant statistical difference in scores for males (M = 1.22, SD = .490) and females (M = .059, SD = .368; t (268) = 2.073, p = .039). This implies that males and females perceive the development of ecotourism at TNR differently. Depending on a number of factors, individuals may tend to favour or disfavour an innovation or development. The results indicate that most men in the studied villages strongly disagree with the development of tourism at TNR. In most of the interviews they indicated that the development of TNR has created costs to them, and these costs included restrictions to use their previous grazing land and hunting land, and restrictions to forest product harvest and this has affected their livelihoods negatively. These results are also supported by research scholars on tourism (e.g. Haley et al., 2005; Haralambopoulos & Pizam, 1996) who also found that people do not support tourism development if it brings more costs to them than expected benefits.

Moreover, a one-way ANOVA was performed to explore the mean difference of perceptions for the three villages of Matshelagabedi, Ditladi and Patayamatebele. The results have shown that the mean difference is the same for all the three villages: F (2,267) = 1.5, p = .217. The mean score for the village of Matshelagabedi was found to be (M = 1.47, SD = .946), Ditladi (M = 1.49, SD = 1.006) and Patayamatebele (M = 1.78, SD = 1.209). The people of the studied villages have negative perceptions and attitudes towards the development of TNR.

CONCLUSIONS

Perceptions of tourism development and natural resources management can determine whether communities can accommodate tourism development in protected areas. The overall findings from the study villages of Matshelagabedi, Ditladi and Patayamatebele indicate that the development of TNR as an ecotourism project has 'robbed' the communities' resources such as land, forests and important traditional water sources. Such natural resources are important for communities to sustain their livelihoods (Buckles & Rusnak, 1999). The communities have accrued more costs than benefits due to TNR and ecotourism. The communities were not consulted about such a development and this has caused frustration over land-use change from cattle grazing to conservation and ecotourism. The Social Exchange Theory argues that individuals' attitudes towards tourism and their subsequent level of support for its development will be influenced by their evaluations of the outcomes of tourism for themselves and their communities (Andereck et al., 2005). According to the SET, residents should derive economic benefits that exceed costs for them to be obliged to support tourism development and conservation in their local area (Blackie, 2006; Swatuk, 2005; Twyman, 2000; Tsing et al., 1999; Leach et al., 1999). The study has shown that people of Ditladi, Patayamatebele and Matshelagabedi have negative attitudes (87%, n = 234) towards the development of tourism at TNR. This is mainly because the development of tourism at TNR does not address the urgent needs of the people. Related studies (e.g. Hernandez et al., 1996; Schroeder, 1996; Ryan & Montgomery, 1994; Nicholas, 2007; Williams & Lawson, 2001) found that the SET significantly predicts support for ecotourism and conservation. If people do not derive significant benefits, they are likely not to support the development of ecotourism projects. For community-based tourism to bring more benefits for locals, more interaction is needed between local people and the trust management. The development of tourism at TNR is not welcomed by many local people therefore rendering it a failure. The study concludes that members of the villages of Matshelagabedi, Ditladi and Patayamatebele perceived that they received few benefits but incurred increased costs from this ecotourism project.

ACKNOWLEDGEMENTS

I would like to express my gratitude to Matshelagabedi, Patayamatebele and Ditladi residents, Tachila Management and Tati Company for providing me with information used for this study. I also wish to express my gratitude to my supervisors Professor J. E. Mbaiwa and Dr O. T. Thakadu for the discussions we had. Finally, I wish to thank Mrs Makati for the production of the map used in the chapter.

REFERENCES

Adams, J. S., & McShane, T. O. (1992). *The Myth of Wild Africa: Conservation without Illusion.* Berkeley: University of California Press.

Adams, W. M., & Hutton, J. (2007). People, parks and poverty: Political ecology and biodiversity conservation. *Conservation and Society, 5*(2), 147–183.

Andereck, K., & Jurowski, C. (2006). Tourism and quality of life. In G. Jennings & N. Nickerson (eds), *Quality Tourism Experiences.* Amsterdam: Elsevier, pp. 136–154.

Andereck, K. L., Valentine, K. M., Knopf, R. C., & Vogt, C. A. (2005). Residents' perceptions of community tourism impacts. *Annals of Tourism Research, 32*(4), 1056–1076.

Ap, J. (1992). Residents' perceptions on tourism impacts. *Annals of Tourism Research, 19*(4), 665–690.

Aronson, J. (1994). Pragmatic view of thematic analysis. *The Qualitative Report, 2*(1), 1–3.

Ashley, C. (1995). *Tourism, Communities, and the Potential Impacts on Local Incomes and Conservation* (No. 10). Directorate of Environmental Affairs, Ministry of Environment and Tourism.

68   Handbook on tourism and conservation

Babbie, E. R. (2016). *The Practice of Social Research* (14th edn). Boston: Cengage Learning.

Babbie, E., & Mouton, J. (2001). *The Practice of Social Science Research*. Belmont, CA: Wadsworth.

Banks, S. K. (2003). Tourism related as perceived by three resident typology groups in San Pedro, Belize. PhD dissertation, North Carolina State University, USA.

Barrow, E., & Murphree, M. (2001). Community conservation: From concept to practice. In D. Hulme & M. Murphree (eds), *African Wildlife and Livelihoods: The Promise and Performance of Community Conservation*. Oxford: James Currey, pp. 24–37.

Blackie, P. (2006). Is small really beautiful? Community-based natural resource management in Malawi and Botswana. *World Development, 34*(11), 1942–1957.

Blau, P. M. (1964). *Exchange and Power in Social Life*. New Brunswick, NJ: Transaction Books.

Bolaane, M. (2004). The impact of game reserve policy on the River BaSarwa/Bushmen of Botswana. *Social Policy & Administration, 38*(4), 399–417.

Boyd, S. W., & Timothy, D. J. (2001). Development partnerships: Tools for interpretation and management. *Tourism Recreation Research, 26*(1), 47–53.

Brandon, K. E., & Wells, M. (1992). Planning for people and parks: Design dilemmas. *World Development, 20*(4), 557–570.

Brida, J. G., Osti, L., & Faccioli, M. (2011). Residents' perception and attitudes towards tourism impacts: A case study of the small rural community of Folgaria (Trentino-Italy). *Benchmarking: An International Journal, 18*(3), 359–385.

Buckles, D., & Rusnak, G. (1999). Conflict and collaboration in natural resource management. In D. Buckles (ed.), *Cultivating Peace: Conflict and Collaboration in Natural Resource Management*. Ottawa: IDRC/World Bank, pp. 1–12.

Burgess, R. G. (1982). *Field Research: A Sourcebook and Manual*. London and New York: Routledge.

Chen, F. J. (2006). Interplay between forward and backward transfer in L2 and L1 writing: The case of Chinese ESL learners in the US. *Concentric: Studies in Linguistics, 32*(1), 147–196.

Chenier, J., Sherwood, S., & Robertson, T. (1999). Copan, Honduras: Collaboration for identity, equity, and sustainability. In D. Buckles (ed.), *Cultivating Peace: Conflict and Collaboration in Natural Resource Management*. Ottawa: IDRC/World Bank, pp. 221–235.

Chenje, M., & Johnson, P. (1994). *State of the Environment in Southern Africa*. Maseru, Lusaka & Harare: SADC.

Child, G. (2009). The growth of park conservation in Botswana. In H. Suich, B. Child, & A. Spenceley (eds), *Evolution and Innovation in Wildlife Conservation: Parks and Game Ranches to Transfrontier Conservation Areas*. London: Earthscan, pp. 187–200.

Cole, S. (2006). Information and empowerment: The keys to achieving sustainable tourism. *Journal of Sustainable Tourism, 14*(6), 629–644.

Collett, D. (1987). Pastoralists and wildlife: The Maasai. In D. Anderson & R. Grove (eds), *Conservation in Africa: People, Policies and Practice*. Cambridge: Cambridge University Press, pp. 129–148.

Cordes, K. A., & Ibrahim, H. M. (1999). *Applications in Recreation and Leisure: For Today and the Future* (2nd edn). New York: McGraw-Hill.

Darkoh, M. B. K. (1996). Environmental problems in Kenya's arid and semi-arid lands. In M. J. Eden & J. T. Parry (eds), *Land Degradation in the Tropics: Environmental and Policy Issues*. London: Commonwealth Foundation, pp. 126–143.

Dasmann, R. F. (1984). The relationship between protected areas and indigenous peoples. In J. A. McNeely & K. R. Miller (eds), *National Parks, Conservation, and Development: The Role of Protected Areas in Sustaining Society*. Washington, DC: Smithsonian Institution, pp. 118–123.

Dyer, P., Gursoy, D., Sharma, B., & Carter, J. (2007). Structural modeling of resident perceptions of tourism and associated development on the Sunshine Coast, Australia. *Tourism Management, 28*(2), 409–422.

Ellis, F. (2000). *Rural Livelihoods and Diversity in Developing Countries*. Oxford: Oxford University Press.

Fiallo, E. A., & Jacobson, S. K. (1995). Local communities and protected areas: Attitudes of rural residents towards conservation and Machalilla National Park, Ecuador. *Environmental Conservation, 22*(03), 241–249.

Furze, B., Lacy, T. D., & Birckhead, J. (1996). *Culture, Conservation and Biodiversity: The Social Dimension of Linking Local Level Development and Conservation through Protected Areas.* Chichester: John Wiley & Sons.

Getz, D. (1994). Residents' attitudes towards tourism: A longitudinal study in Spey Valley, Scotland. *Tourism Management, 15*(4), 247–258.

Grove, R. (1987). Early themes in Africa conservation: The Cape in the nineteenth century. In D. Anderson & R. Grove (eds), *Conservation in Africa: People, Policies and Practice.* Cambridge: Cambridge University Press, pp. 21–39.

Haley, A. J., Snaith, T., & Miller, G. (2005). The social impacts of tourism: A case study of Bath, UK. *Annals of Tourism Research, 32*(3), 647–668.

Haralambopoulos, N., & Pizam, A. (1996). Perceived impacts of tourism: The case of Samos. *Annals of Tourism Research, 23*(3), 503–526.

Hardy, A., Beeton, R. J., & Pearson, L. (2002). Sustainable tourism: An overview of the concept and its position in relation to conceptualisations of tourism. *Journal of Sustainable Tourism, 10*(6), 475–496.

Hernandez, S. A., Cohen, J., & Garcia, H. L. (1996). Residents' attitudes towards an instant resort enclave. *Annals of Tourism Research, 23*(4), 755–779.

Homans, G. C. (1958). Social behavior as exchange. *American Journal of Sociology, 63*(6), 597–606.

Honey, M. 2008. *Ecotourism and sustainable development: Who owns paradise?* (2nd edn). New York: IslandPress.

Irandu, E. M. (2004). The role of tourism in the conservation of cultural heritage in Kenya. *Asia Pacific Journal of Tourism Research, 9*(2), 133–150.

Jones, M. R. (1976). Time, our lost dimension: Toward a new theory of perception, attention, and memory. *Psychological Review, 83*(5), 323–355.

Jurowski, C., Uysal, M., & Williams, D. R. (1997). A theoretical analysis of host community resident reactions to tourism. *Journal of Travel Research, 36*(2), 3–11.

Kiss, A. (ed.) (1990). *Living with Wildlife: Wildlife Resource Management with Local Participation in Africa.* Washington, DC: The World Bank.

Kuvan, Y., & Kuvan, P. (2005). Residents' attitudes towards general and forest-related impacts of tourism: The case of Belek, Antalya. *Tourism Management, 26*(5), 691–706.

Leach, M., Mearns, R., & Scoones, I. (1999). Environmental entitlements: Dynamics and institutions in community-based natural resources management. *Wildlife Development, 27*(2), 225–247.

Leininger, M. M. (1985). Ethnography and ethnonursing: Models and modes of qualitative data analysis. In M. M. Leininger (ed.), *Qualitative Research Methods in Nursing.* Orlando, FL: Grune & Stratton, pp. 33–72.

Lepp, A. (2004). Tourism in rural Ugandan village: Impacts, local meaning and implications for development. PhD dissertation, University of Florida, USA.

Lindsay, W. K. (1987). Integrating parks and pastoralists: Some lessons from Amboseli. In D. Anderson & R. Grove (eds), *Conservation in Africa: People, Policies and Practice.* Cambridge: Cambridge University Press, pp. 149–167.

Machlis, G. E., & Tichnell, D. L. (1985). *The State of the World's Parks.* Boulder, CO: Westview Press.

Mackenzie, A. F. D. (1998). *Land, Ecology and Resistance in Kenya, 1880–1952.* Edinburgh: Edinburgh University Press.

Madrigal, R. (1993). A tale of tourism in two cities. *Annals of Tourism Research, 20*(2), 336–353.

Magome, H., & Murombedzi, J. (2003). Sharing South African national parks: Community land and conservation in a democratic South Africa. In W. Adams & M. Mulligan (eds), *Decolonizing Nature: Strategies for Conservation in a Post-Colonial Era.* London: Earthscan, pp. 108–135.

Manatsha, B. T. (2010). Land reform in the north east district of Botswana: An elite-hijacked project? *Botswana Notes and Records, 42*, 90–99.

Manatsha, B. T. (2014). The politics of Tachila Nature Reserve in the north east district, Botswana: A historical perspective. *South African Historical Journal, 66*(3), 521–545.

Manatsha, B. T. (2020). Chiefs and the Politics of Land Reform in the North East District, Botswana, 2005–2008. *Journal of Asian and African Studies, 55*(1), 111–127.

Marekia, E. N. (1991). Managing wildlife in Kenya. In A. Kiriro & C. Juma (eds), *Gaining Ground: Institutional Innovations in Land-Use Management in Kenya.* Nairobi: ACTS Press, pp. 155–179.

70   *Handbook on tourism and conservation*

Mayoral-Phillips, A. J. (2002). Trans-boundary areas in Southern Africa: Meeting the needs of Conservation or Development? The Commons in an Age of Globalisation, 9th Conference of the International Association for the Study of Common Property.

Mbaiwa, J. E. (1999). Prospects for sustainable wildlife resource utilisation and management in Botswana: A case study of East Ngamiland District. MSc thesis, Department of Environmental Science, University of Botswana, Gaborone.

Mbaiwa, J. E. (2005a). Enclave tourism and its socio-economic impacts in the Okavango Delta, Botswana. *Tourism Management, 26*(2), 157–172.

Mbaiwa, J. E. (2005b). Wildlife resource utilisation at Moremi Game Reserve and Khwai community area in the Okavango Delta, Botswana. *Journal of Environmental Management, 77*(2), 144–156.

Mbaiwa, J. E. (2008). *Tourism Development, Rural Livelihoods, and Conservation in the Okavango Delta, Botswana*. College Station, TX: Texas A&M University Press.

Mbaiwa, J. E. (2015). Ecotourism in Botswana: 30 years later. *Journal of Ecotourism, 14*(2–3), 204–222.

Mbaiwa, J. E., & Darkoh, M. K. (2006). *Tourism and Environment in the Okavango Delta, Botswana*. Gaborone: Pula Press.

McGehee, N. G., & Andereck, K. L. (2004). Factors predicting rural residents' support of tourism. *Journal of Travel Research, 43*(2), 131–140.

Moganane, B. O., & Walker, K. P. (1995). *The Role of Local Knowledge in the Management of Natural Resources with Emphasis on Woodland, Veld Products, and Wildlife. Botswana Case Study: Final Report*. Forestry Association of Botswana.

Murphree, M. W. (1993). *Communities as Resource Management Institutions*. London: IIED.

Murphree, M. W. (1999). *Enhancing Sustainable Use: Incentives, Politics, and Science*. Berkeley Workshop on Environmental Politics. Working Paper 2/99.

Murphy, P. E. (1985). *Tourism: A Community Approach*. London: Methuen.

Mutanga, C. N., Vengesayi, S., Muboko, N., & Gandiwa, E. (2015). Towards harmonious conservation relationships: A framework for understanding protected area staff–local community relationships in developing countries. *Journal for Nature Conservation, 25*, 8–16.

Mwenya, A. N., Lewis, D. M., & Kaweche, G. B. (1991). *Policy, Background and Future: National Parks and Wildlife Services, New Administrative Management Design for Game Management Areas*. Lusaka: United States Agency for International Development.

Newmark, W. D., Leonard, N. L., Sariko, H. I., & Gamassa, D.-G. M. (1993). Conservation attitudes of local people living adjacent to five protected areas in Tanzania. *Biological Conservation, 63*(2), 177–183.

Nicholas, L. N. (2007). Stakeholder perspectives on the Pitons management area in St. Lucia: Potential for sustainable tourism development. DPhil dissertation, University of Florida, USA.

Okech, R. N., & Bob, U. (2009). Sustainable ecotourism management in Kenya. *Ethiopian Journal of Environmental Studies and Management, 2*(1), 57–65.

Perdue, R. R., Long, P. T., & Allen, L. (1987). Rural resident tourism perceptions and attitudes. *Annals of Tourism Research, 14*(3), 420–429.

Perdue, R. R., Long, P. T., & Allen, L. (1990). Resident support for tourism development. *Annals of Tourism Research, 17*(4), 586–599.

Prosser, R. (1996). *Managing Environmental Systems*. London: Nelson.

Rihoy, E. (1995). From state control of wildlife to co-management of natural resources: The evolution of community management in southern Africa. In E. Rihoy (ed.), *The Commons Without Tragedy? Strategies for Community-based Natural Resources Management in Southern Africa*. Proceedings of the Regional Natural Resources Management Programme Annual Conference: SADC Wildlife Technical Co-ordinating Unit, Kasane, pp. 1–36.

Rodney, W. (1972). How Europe underdeveloped Africa. In P. Rothenberg (ed.), *Beyond Borders: Thinking Critically about Global Issues*. New York: Worth Publishers, pp. 107–125.

Rozemeijer, N., & Van der Jagt, C. (2000). *Community-Based Natural Resource Management in Botswana: How Community-Based Is Community-Based Natural Resource Management in Botswana?* Occasional Paper Series, IUCN/SNV CBNRM Support Programme, Gaborone.

Ryan, C., & Montgomery, D. (1994). The attitudes of Bakewell residents to tourism and issues in community responsive tourism. *Tourism Management, 15*(5), 358–369.

Schroeder, T. (1996). The relationship of residents' image of their state as a tourist destination and their support for tourism. *Journal of Travel Research*, *34*(4), 71–73.

Sjoberg, G., & Nett, R. (1968). *A Methodology for Social Research*. New York: Harper & Row.

Statistics Botswana (2011). *Botswana Populations and Housing Census*. Ministry of Finance and Development Planning, Gaborone, Botswana.

Stolte, J. F., Alan, G., & Cook, K. S. (2001). Sociological miniaturism: Seeing the big through the small in social psychology. *Annual Review of Sociology*, *27*(1), 387–413.

Stone, P. R. (2012). Dark tourism and significant other death: Towards a model of mortality mediation. *Annals of Tourism Research*, *39*(3), 1565–1587.

Struhsaker, T. T., Struhsaker, P. J., & Siex, K. S. (2005). Conserving Africa's rain forests: Problems in protected areas and possible solutions. *Biological Conservation*, *123*(1), 45–54.

Suliman, M. (1999). Nuba Mountains of Sudan: Resource access, violent conflict, and identity. In D. Buckles (ed.), *Cultivating Peace: Conflict and Collaboration in Natural Resource Management*. Ottawa: IDRC/World Bank.

Swatuk, L. A. (2005). From "project" to "context": Community based natural resource management in Botswana. *Global Environmental Politics*, *5*(3), 95–124.

Teye, V., Sonmez, S. F., & Sikaraya, E. (2002). Residents' attitudes toward tourism development. *Annals of Tourism Research*, *29*(3), 668–688.

Thibaut, J. W., & Kelley, H. H. (1959). *The Social Psychology of Groups*. New York: John Wiley & Sons.

Tsing, A. L., Brosius, J. P., & Zerner, C. (1999). Assessing community-based natural resource management. *Ambio*, *28*(2), 197–198.

Twyman, C. (2000). Participatory conservation? Community-based natural resource management in Botswana. *The Geographical Journal*, *166*(4), 323–335.

Vodouhê, F. G., Coulibaly, O., Adégbidi, A., & Sinsin, B. (2010). Community perception of biodiversity conservation within protected areas in Benin. *Forest Policy and Economics*, *12*(7), 505–512.

Walpole, M. J., & Goodwin, H. J. (2001). Local attitudes towards conservation and tourism around Komodo national park, Indonesia. *Environmental Conservation*, *28*(2), 160–166.

Weladji, R. B., Moe, S. R., & Vedeld, P. (2003). Stakeholder attitudes towards wildlife policy and the Benoue Wildlife Conservation area, North Cameroon. *Environmental Conservation*, *30*(04), 334–343.

Welman, J. C., & Kruger, S. J. (2001). *Research Methodology for the Business and Administrative Sciences*. New York: Oxford University Press.

West, P. C., & Brechin, S. R. (1991). *Resident Peoples and National Parks: Social Dilemmas and Strategies in International Conservation*. Tucson: University of Arizona Press.

Western, D., & Pearl, M. (eds) (1989). *Conservation for the 21st Century*. New York: Oxford University Press.

Western, D., Wright, M., & Strum, S. (eds) (1994). *Natural Connections: Perspectives in Community-Based Conservation*. Washington, DC: Island Press.

Williams, J., & Lawson, R. (2001). Community issues and resident opinions of tourism. *Annals of Tourism Research*, *28*(2), 269–290.

Wilson, G. A., & Hart, K. (2001). Farmer participation in agri-environmental schemes: Towards conservation-oriented thinking? *Sociologia Ruralis*, *41*(2), 254–274.

Yamane, T. (1967). *Statistics: An Introductory Analysis* (2nd edn). New York: Harper and Row.

Zafirovski, M. (2005). Social exchange theory under scrutiny: A positive critique of its economic-behaviorist formulations. *Electronic Journal of Sociology*, *2*(2), 1–40.

Zube, E. H. (1986). Local and extra-local perceptions of national parks and protected areas. *Landscape and Urban Planning*, *13*, 11–17.

PART II

TOURISM AND CLIMATE CHANGE

6. Tourism and climate change adaptation in protected areas

Kaarina Tervo-Kankare

INTRODUCTION

Tourism research has addressed climate change since the mid-1980s, when Geoffrey Wall and his team published the article "The implications of climatic change for camping in Ontario" (1986). Since then, protected areas, such as national parks, marine parks, and heritage sites have continued to serve as research arenas for climate change related studies in tourism. Already in 2004, there were over 100,000 protected areas globally, with a coverage of over 12 per cent of the Earth's terrestrial surface (Chape et al., 2005). More current estimations by the United Nations Environment Programme, World Conservation Monitoring Centre (UNEP-WCMC) and International Union for Conservation of Nature (IUCN) suggest that about 15 per cent of land areas and about 4 per cent of the Earth's oceans are protected (UNEP-WCMC & IUCN, 2016). These areas have an important role in conserving biodiversity and cultural heritage, but they also offer attractive visitation locations for tourists and recreationists all over the world. Visitors are attracted to these aesthetically pleasing, healthy and historical environments, and any changes in the cultural or natural landscapes can negatively impact their amenity value (UN World Tourism Organization (UNWTO) & UNEP, 2008).

Changing climate will alter these cultural and natural landscapes, and negatively influence biodiversity conservation in protected areas (Aloia et al., 2019). This has implications for tourism in protected areas as well, making them susceptible to climate change. This vulnerability leads to a situation where action is required from the stakeholders involved. This action is called adaptation. In general, the term adaptation refers to the adjustments that are needed in natural or human systems as a response to changing environments (Yohe & Tol, 2002). These adjustments can refer to ways to cope with actual or expected climatic stimuli or their impacts, and to moderate harm and/or to exploit beneficial opportunities (IPCC, 2007).

In 2008, UNWTO and UNEP provided a broad description of the implications of climate change:

> Climate affects a wide range of the environmental resources that are critical attractions for tourism, such as snow conditions, wildlife productivity and biodiversity, water levels and quality. Climate also has an important influence on environmental conditions that can deter tourists, including infectious disease, wildfires, insect or water-borne pests (e.g., jellyfish, algae blooms), and extreme events such as tropical cyclones. (UNWTO & UNEP, 2008, p. 28)

The UNWTO-UNEP report also provided examples of the potential implications of climate change for protected areas. Regardless of the type of the protected area (i.e., whether it conserves natural or cultural heritage), the consequences mostly meant reduced attractiveness, decreased visitation, and increased risks to visitor safety, together with increased management costs. In addition, the UNWTO-UNEP report (2008) overviewed climate change adaptation

74 *Handbook on tourism and conservation*

Table 6.1 *A list of potential adaptation strategies for natural and cultural heritage sites and nature-based destinations in general according to the World Tourism Organization and United Nations Environment Programme*

Natural and cultural heritage destinations	Nature-based tourism destinations in general
Master plans and response plans: e.g., water supply planning (in drought susceptible destinations), risk assessment and preparedness strategies, and implementation of early warning systems (e.g., flooding).	Improve adaptive capacity of authorities and managers of protected areas through capacity building initiatives, especially in biodiversity hotspots of least developed countries and developing countries.
Scientific monitoring survey programmes to assess (ecosystem) changes and necessary protection (monitoring activities could especially focus on species and habitats most vulnerable to climate change impacts and most important for tourism activities: e.g., levels of endemic species; flood protection; glacial lake levels to prevent outburst flooding).	Promote the application of integrated tourism carrying capacity assessment techniques (considering physical, economic, environmental, socio-cultural and managerial aspects) in protected areas as a tool for tourism planning.
Reconstruction and stabilization of historic assets such as architecturally rich buildings and archaeological sites using a combination of traditional materials and skills (to preserve their historic aesthetics and attraction), and modern engineering techniques to enhance their longevity.	Improve visitor and congestion management to prevent overuse of sites and physical impacts of visitation.
Product diversification; for example: opening up new 'micro' destinations and attractions within and adjacent to an already popular national park or heritage site.	Promote mitigation options amongst environmentally conscious eco-tourists, e.g., through offsetting their trips to nature-based tourism destinations.
Translocation; a final strategy for species such as flowering plants that will not survive in their current location involving safer wild habitats or storing the genetic resources in gene or seed banks.	Ensure active participation of local communities living within or near protected areas, in policy making and management processes.
Protected area re-design/redefinition; i.e., zoning certain areas, protecting a larger area, creation of migratory corridors to allow threatened species to more easily find new geographic ranges and alleviate the effects of climate change.	Take into consideration local and traditional knowledge to develop coping and adaptation strategies.
Combining traditional materials and skills with modern engineering when reinforcing, stabilizing and renovating historic sites.	Develop replicable methodologies and share knowledge across nature-based destinations.
Education and awareness raising on minimizing external stresses; increasing the profile and knowledge base of users and stakeholders of the undermining nature of external stresses to a destination struggling to deal with the impacts of climate change.	Reduction or removal of external stresses such as overuse, pollution and in the case of marine resources, agricultural run-off.

Source: UNWTO & UNEP, 2008, pp. 8–9 and Table 905 (modified by the author).

in protected areas. Several suggestions were made for adapting in nature-based destinations and especially in natural or cultural heritage sites. These potential strategies are presented in Table 6.1.

In climate change and tourism studies, there continue to exist knowledge gaps, especially from the geographical perspective; not much published research has been available in the context of less developed countries (Scott et al., 2019). Human behaviour in changing climate is another knowledge gap, identified, among others, by Gössling et al. (2012), and echoed by Groulx et al. (2017; see also Hall, 2013). In addition, tourism adaptation in protected

Tourism and climate change adaptation in protected areas   75

areas has not raised much interest previously, even though tourism adaptation to climate change in general has been reviewed by many (e.g., Hoogendoorn & Fitchett, 2018; Kaján & Saarinen, 2013; Njoroge, 2015). A quick review of literature on this topic indicates that the UNWTO-UNEP overview from 2008 may be the most extensive one published so far.

This chapter reviews studies on climate change, adaptation, and tourism in protected areas. The aim is to create understanding of the future of tourism in protected areas in changing climate, based on a scoping review on scientific articles on tourism adaptation in this context. First, the chapter provides a short overview on studies focusing on this topic in different parts of the world and on the implications of changing climate in diverse protected environments. Understanding the direct and indirect impacts of the change on the resources, infrastructure and tourism demand is crucial for describing the potential adaptation methods and strategies (see Tervo-Kankare, 2012), which will be discussed in the latter sections of the chapter. Rather than drawing conclusions and making suggestions on the ways that tourism could and should adapt to climate change, the chapter aims at comparing the current knowledge and progress in adaptation to the situation in 2008, when the previous overview and suggestions on adaptation in protected areas were published. Thus, the diverse adaptation methods, and their applicability in diverse protected areas will be examined. Interesting topics relating to adaptation are maladaptation and the phenomenon of 'last chance tourism', both of which will also be briefly touched upon. The concluding section draws the main findings of the review together and discusses the challenges and some future research needs concerning climate change adaptation in protected areas.

Scoping Review on Climate Change and Tourism and Recreation in Protected Areas

The method chosen for reviewing literature in this chapter is a scoping study, or scoping review. The aim of the review is to examine the extent and nature of research activity on the chosen topic (Arksey & Malley, 2005). However, the idea is not to analyse the development and phase of protected areas' climate change research in numerical terms or form a policy for successful adaptation in protected areas as could be the objective of a systematic literature review model (Moher et al., 2015). Rather, the review aims at gaining an understanding about the general issues affecting tourism and climate change adaptation in and around the protected areas, and the development that has happened since 2008. Thus, the general research question was: What can we tell about climate change adaptation in tourism in the context of protected areas?

The studies chosen for review were identified by a Scopus document search conducted in February 2021. The terms 'protected areas/national parks' and 'tourism' and 'climate change' were used to search the abstract, keyword, and title sections in Scopus. This search generated 131 hits. The search was refined with the term 'adaptation', which narrowed the number to 58. The number of articles was further narrowed down by reviewing the abstracts and estimating their relevance to tourism in protected areas, climate change and adaptation. For example, articles where 'tourism' was only mentioned as one potential ecosystem service and as a minor issue were excluded. Finally, the list of articles was complemented with a Google Scholar search, and a few full-text articles that could not be retrieved were removed. In conclusion, 40 articles were chosen for review, the earliest one published in 2007 and the latest ones in 2020. The chosen articles were mostly published in journals, but a few book chapters and one report were included as well.

76 *Handbook on tourism and conservation*

The literature collected was analysed qualitatively in three stages: First, an overall picture about the geographical coverage and topics discussed in literature was drawn. Then, the review focused on the impacts and implications of changing climate to tourism in protected areas. The third stage focused on adaptation strategies (the spectrum of potential adaptation methods, considers the overarching goals of adaptation), methods (actions, practical level measures), and the challenges relating to them.

OVERVIEW ON RESEARCH ON TOURISM AND CLIMATE CHANGE IN PROTECTED AREAS

The reviewed articles covered all continents except South America and Antarctica. This may be due to the language used for the document search, as the aim was to cover only English literature. This can, of course, cause bias in the review. Antarctica was mentioned in some articles, but none of them focused on it or presented empirical research conducted there. In addition, it is worthwhile to mention that the Alpine region, where plenty of climate change research has been conducted, was not well represented in the Scopus document search. This, most probably, was due to the term 'protected areas/national parks' utilized in the literature search. Referring to protected areas or national parks is not very common in this context (Alps).

Tourism in protected areas is in general vulnerable to changes in climatic conditions if they alter the ecological stage and biodiversity of these areas, which has further implications. Environmental changes affect the attractiveness of the areas and thus influence visitor numbers, accessibility, and seasonality (of visitation). The climate change studies in protected areas tend to focus on analysing the impact of changing climatic conditions (mostly temperature and precipitation) on recreational opportunities, visitor behaviour patterns and demand (e.g., Coldrey & Turpie, 2020; Groulx et al., 2017; Liu et al., 2020; Pongkijvorasin & Chotiyaputta, 2013; Scott et al., 2008; Welling et al., 2020), occasionally combined with reactions to adaptation methods (Purdie et al., 2020; Welling et al., 2020). Another central research topic is the perceptions of the stakeholders, mostly tourism operators (Mushawemhuka et al., 2018; Welling & Abegg, 2019). In several studies, the value of tourism (and recreation) in conservation areas is emphasized, and the implications of change are described in monetary terms or as visitor numbers.

However, as Tolvanen and Kangas (2016) suggest, it is important to pay attention also to the ecological changes that derive from the integrated impacts of changing tourism and recreation patterns and climate change. In addition, one should acknowledge the role of the protected areas in mitigating global environmental change and its manifestations (Carter et al., 2014; Hoffmann & Beierkuhnlein, 2020). It is somewhat surprising that even though protected areas have an important role in conserving ecological and cultural values, this aspect and the conservation areas' futures under warming climate and increasing tourism pressure have not received much attention in research. Tolvanen and Kangas (2016) reviewed research literature on the impacts of tourism and recreation on terrestrial ecosystems in Northern Fennoscandia and concluded that the impacts of changing climate, recreation and tourism on wildlife and vegetation were not among researched topics. The same tendency seems to prevail in this review: of the 40 articles reviewed for this chapter, only five papers discussed the future of protected areas from an ecological perspective. In addition to the above-mentioned study by

Tolvanen and Kangas, there were three studies from Australia (Scherrer & Pickering, 2001; Dutra et al., 2017; Jacobs et al., 2019) and one from the USA (Halofsky et al., 2018), all of them focusing on terrestrial areas.

However, the studies reviewed did acknowledge the ecological aspects to some extent. For example, the protected areas are considered important for educating people about climate change, and increasing environmental consciousness (Choe & Schuett, 2020; Purdie et al., 2020; Welling et al., 2020). Another example is the study by Dutra et al. (2017), where consideration of ecological issues arose from shared risk perception. Dutra et al. found that climate change induced sea level rise, when perceived as a common threat by the locals, was an opportunity to bring together Indigenous and non-Indigenous knowledge and governance systems to enhance ecological sustainability.

IMPACTS OF CLIMATE CHANGE IN PROTECTED AREAS

Climate change will have direct and indirect impacts on conservation areas. They are dependent on the geographical location, as it defines the magnitude and speed of the change(s). Rising temperatures, changing precipitation patterns and increasing occurrence of extreme weather events such as storms, peak winds, heavy rain, and longer and more severe droughts together with sea level rise (Hoegh-Guldberg et al., 2018) will influence the flora and fauna as well as the landscapes of these areas. These further affect the local communities, their livelihoods and coexistence with the protected areas' nature and, of course, the visitors' behaviour. Estimating these impacts and their magnitude is rather difficult, since the conservation areas, and nature-based areas in general, have high diversity of resources (UNWTO & UNEP, 2008).

Especially in regions where the local communities are poor and livelihoods dependent on nature, the indirect impacts of climate change may exacerbate the situation further. This seems to be the case for example in Zimbabwe, where Musakwa et al. (2020) found that the food insecurity caused by droughts increased the unsustainable use of the protected area. This unsustainable use included poaching of wildlife for subsistence purposes or for extra income (participating in commercial poaching networks to hunt big game), and unsustainable harvesting of vegetation and biodiversity resources. This kind of behaviour escalates the influence of climate change and decreases the potential of adaptation and mitigation to be successful. Liu (2016) approached low rainfall from the perspective of the challenges it causes as it leads to water shortages and potentially to increasing environmental pressure.

However, most studies examining the future of protected areas in changing climate focus on the impacts on park visitation rather than on local communities, their behaviour, and the induced impacts on them. Temperature, rainfall and weather in general seem to be the most important elements affecting the visitation in protected areas (Coldrey & Turpie, 2020; Fisichelli et al., 2015; Liu, 2016; Liu et al., 2020). In some locations, temperature is the dominating condition (Fisichelli et al., 2015; Liu et al., 2020), while in other locations, rainfall has major influence on visitation (Liu, 2016). Location of the parks in relation to main visitor home regions seems to affect visitation as well, like in all nature-based tourism: when the visitors live close by, they are able to monitor the weather conditions in real-time and cancel at short notice, while those who travel longer distances make fewer last-minute changes in their visitation plans (e.g., Liu, 2016).

78   *Handbook on tourism and conservation*

In the USA, future projections based on historical monthly temperature and visitation data suggest increasing visitation numbers and lengthening visitation season (Fisichelli et al., 2015). These findings are supported by Urioste-Stone et al. (2015) whose study focused on the perceptions of visitors: weather seemed to be very important in the visitors' decision to travel to the region (to visit the national park). However, the respondents of this study also emphasized the natural assets as the top motivators for them to visit, suggesting that the changes in flora and fauna will eventually define the future of the parks. This perspective has been discussed also in the African context, where changing migration habits, starvations, and drownings of the iconic wildlife have raised concern (Kilungu et al., 2017).

In areas where visitation patterns follow animal or bird migration, plant flowering, etc., changing precipitation and seasons may lead to mismatches between visitors and the attraction. In Serengeti National Park, for example, the variability of rainfall seasons has caused chaos and extra costs (time- and money-wise) for both the visitors and tourism actors when the visitors have had to chase the migrating animals outside the pre-planned viewing locations (Kilungu et al., 2017). A study in Balearic Islands' wetland in Mallorca suggested that climate change will increase salinization and stresses over migratory bird species, which will lead to reductions in abundance and diversity of bird species (Faccioli & Riera, 2015). According to Faccioli and Riera (2015), the latter affects also the visitors' 'welfare' negatively. The findings from Australia's wetlands present similar issues: sea level rise will lead to saltwater inundation of freshwater ecosystems, and to widespread impacts on biodiversity, Indigenous cultural values, and built infrastructure (Dutra et al., 2017).

Infrastructure raises concern in many locations, as especially the weather extremes put a strain on basic infrastructure and cause safety issues such as flooding of sewers. However, also the gradual changes may have unpredictable impacts on roads, foundations of buildings, bridges, etc. Heavy rainfalls in national parks in Taiwan, for example, have damaged infrastructure, and washed away roads (Liu, 2016) and caused landslides (Cheablam & Shrestha, 2015). Heavy rainfalls have been challenging also in Zimbabwe, where visitors' (and locals') access can be seriously affected (Mushawemhuka et al., 2018).

The visitors' comfort and willingness to visit the protected areas in future seem to differ between regions, and there continues to be uncertainty in the projections. Even though the changing visitation patterns and levels may not directly link to the ecological state of the protected areas, it is important to remark, as Tolvanen and Kangas (2016, p. 59) remind us, that the "increasing tourism pressure together with climate warming may enhance the ecological changes through, for example, the increase of alien species under warmer conditions, changed species interactions and enhanced plant regeneration rates through the increase of tolerant species".

One important perspective of climate change impacts that is not yet covered is the appearance of so-called last chance tourism (LCT), as discussed, among others, by Lemieux et al. (2018). The presence of climate change and awareness of it has stimulated LCT, where concern over vanishing destinations increases the tourists' and recreationists' willingness to visit them before it is too late (Lemelin et al., 2010; Wang et al., 2017). This phenomenon poses an additional threat to the future of protected areas and their tourism and recreation. The unique and rare species, or the conditions which the sites aim to preserve, are also the attractions that allure visitors, while travelling to the sites and activities offered contribute heavily to carbon emissions that influence climate change and thus may promote the extinction of these species. This is the case especially in the more remote protected areas, such as Churchill in

Canada, where Dawson et al. (2010) estimated the greenhouse gas emissions to be 6–34 times higher per visitor than the average tourism experience. However, according to Lemieux et al. (2018) LCT can also provide an opportunity for enhancing climate change awareness, especially if the motive to participate in LCT is "coupled with a desire to learn about environmental change" (p. 667).

ADAPTATION IN PROTECTED AREAS

Adaptation in the reviewed literature related to the impacts presented in the previous section. The discussed and proposed adaptation mostly dealt with anticipating visitor behaviour and responding to the geographical, seasonal, and attraction-related changes. The level of precision varied considerably; both practical and general suggestions were made, thus both adaptation methods and strategies were presented (see Table 6.2 and compare the suggestions by Dube & Nhamo and Wang et al.).

According to Fisichelli et al. (2015), adaptation strategies should be developed for both protected areas and their neighbouring communities in order to be able to capitalize on opportunities and minimize the negative impacts from changing visitation – mostly increasing visitor numbers. However, they do not suggest any practical methods for adaptation, but indicate that more research is required to enable the protected area managers to "remain effective resource stewards while promoting visitor experience" (Fisichelli et al., 2015, p. 11). Urioste-Stone et al. (2015, p. 41) go a bit further, by discussing the educational opportunities of managers to increase visitors' awareness and knowledge about the ecological changes in destinations, visitors' role in reducing their carbon footprint, "climate-friendly services offered by the park" and the adaptation strategies taking place.

In the case of Balearic wetlands, the adaptation plans were examined from the perspective of visitor satisfaction. The concluding remarks state that increasing environmental awareness and tourist demand for high environmental quality will lead to a positive contribution to wetland and environment protection via tourism (Faccioli & Riera, 2015). Also Welling et al. (2020) approach adaptation from the visitors' perspective: they divide the visitors into three groups and suggest different adaptation strategies that will appeal to visitors with different motivations. They conclude that the adaptation process may require trade-offs between sustainability and visitor satisfaction. Disclosure of these trade-offs is needed to choose the most suitable adaptation strategies – this requires that the managers of the protected areas have a clear understanding about the objectives of adaptation.

Jacobs et al. (2019) emphasize that anticipation is important for the management side of the protected areas and suggest wide participation of managers to support adaptation. They do not, however, discuss the participation of other stakeholders and interest groups (e.g., private service providers, communities, NGOs) whose engagement is important for the successful execution of any adaptation strategy. In Nyongesa's and Vacik's (2019) study about the use of multi-criteria analysis for evaluating and choosing the best management strategy, an extensive group of stakeholders was involved in the planning of the management strategies to reduce fire danger. The participation of diverse stakeholders had positive implications for the structuring of the problem, and the transparency and the quality of the decision-making processes. However, the study also revealed some weaknesses or issues that need to be taken into consideration when applying this kind of participatory method: the participants had

80 *Handbook on tourism and conservation*

Table 6.2 *Examples of adaptation listed in the studies examined*

Authors	Suggested adaptation
Dube & Nhamo, 2020	– effective weather communication needs to be at the heart of tourism management, especially in the light of increased incidences of extreme weather events.
	– improved wildlife management in the park, to be applied in such a manner to ensure access to food reserves and natural water during years of drought.
	– reviewing conservation fees to cater for the impact of climate change on national parks is imperative in order to reduce the burden on the national fiscus.
	– conducting perception studies on how the visitors will respond to changes imposed by climate change.
	– revising land use planning in the park to reduce the disasters associated with extreme weather events such as fire and flooding to minimize loss and damage. This includes controlled fire burning trials.
	– retrofitting of old camp buildings and green building design into new infrastructure so as to align the park operations in line with Sustainable Development Goals.
	– revising adequacy of insurance cover by all business operating in the national parks.
	– exploring how various animal and plant species are responding to climate change.
	– financing to ensure that research and innovation address sustainability.
Wang et al., 2017	– promoting travel in new seasons (e.g., spring, autumn).
	– enhancing the quality of the tourism service for the growing tourist market.
	– establishing an environmental security monitoring system especially in glacier tourist areas, to safeguard against the frequent geological disasters caused by climate change.
	– establishing a dedicated website for providing accurate forecasts of risk assessment of tourism-related climate.
	– providing financial support from the government in the form of subsidies and interest-free loans to climate-change-sensitive small businesses.
	– delivering prioritized support for research projects focusing on tourism vulnerability on the protected areas.
	– advocating responsible tourism and low-carbon tourism activities.
	– making sure that tourism enterprises fully consider all relevant meteorological disasters and geological hazards when designing tourism products and tourist routes or when building tourism infrastructure.
	– planning and implementing public activities that promote rare animal protection and climate change adaptation.
	– doing empirical research based on first-hand fieldwork of climate change adaptation in the tourism sector.

difficulties in expressing their preferences, the terminology used was not familiar to all, and there were diverse perceptions about the development process itself. These findings highlight well the challenges that wide participation of stakeholders may bring – reaching a satisfactory consensus is time and energy consuming and may require compromises that are against the conservation principles of protected areas. In the study by Dutra et al. (2017), these kinds of challenges were overcome with a special learning approach that enabled the use of local and traditional knowledge and accumulative learning in the development of adaptation strategies.

In Taiwan, Liu (2016, p. 274) suggests that economic change should be taken into account when planning the future of parks:

Because tourism at each park reacts differently to an overall economic change, there must be parks with comparatively few visitors during every period. During such periods, parks should actively engage in ecological recovery and/or public infrastructure construction and maintenance. Making necessary management decisions and taking measures to adjust the offerings of national parks in

Tourism and climate change adaptation in protected areas 81

accordance with overall economic changes will increase the elasticity of Taiwan's national park system in response to climate change.

In the studies analysed, some reported adaptation methods related to the concept of maladaptation. Maladaptation is considered adaptation that "worsens the existing and/or future conditions for individuals/civil society/corporations/governments, or environmental values" (Glover & Granberg, 2021, p. 69). In the context of this chapter, it can, for example, refer to occasions where the adaptation method has negative ecological impact, it increases greenhouse gas emissions, or it influences the visitor experience in ways that decrease attractiveness.

In the study on Canadian glaciers, the visitors perceived the naturalness of the site to diminish because of adaptation methods (e.g., construction of paths, helicopter tours, snowmobiles, etc.; Groulx et al., 2017). Similarly, in New Zealand, the "visitors had a strong awareness of climate change, but somewhat ironically, one of the key adaptive strategies to maintaining mountain access has been an increase in the use of aircraft" (Purdie et al., 2020, p. 1). In Thailand, in Mu Ko Surin National Park, the bleaching of coral reefs led to the opening of new sites to satisfy the tourist demand (Cheablam & Shrestha, 2015), which does not necessarily support the ecological wellbeing of the park. However, their suggested adaptation also includes the development of new forms of tourism which are less dependent on the most climate vulnerable elements of the park. In Gambia's Tanbi Wetland National Park (a mangrove estuary) the adaptation methods presented by the local tourism workers included purchasing bigger boats and expanding their sightseeing zones, which leads to higher costs (and possibly increased pollution) (Ceesay et al., 2016). Adaptation methods that increase energy use, such as installing air conditioning or improving indoor facilities in an attempt to create a more comfortable environment for the visitors (Mushawemhuka et al., 2018), can easily fall in the category of maladaptation as well.

CONCLUSIONS: FROM ADAPTATION MEASURES TO ADAPTATION STRATEGIES

The reviewed climate change impact and adaptation studies provide diverse scenarios about the future of tourism in protected areas. In particular, forecasts of changing visitor patterns and visitor numbers are popular. However, there is great uncertainty if, and when, these forecasts will materialize. In addition, the estimations seem not to – at least yet – contribute to adaptation practices or to the development of comprehensive adaptation strategies. Rather, the studies call for better management without explaining what this management implies or entails in real life. From this perspective, it seems that not much has happened in protected areas' adaptation since the publication of the UNWTO-UNEP report in 2008. The adaptation covered in the studies reviewed provided little new knowledge in addition to the adaptation strategies gathered in the mentioned report some fifteen years ago.

However, this ostensible incremental development may be needed to lay a solid foundation for protected areas' futures in a changing climate. Since the areas differ considerably and tailored solutions are needed, it is of the utmost importance to understand the local perspectives concerning the direct and indirect impacts of the change and the potential of diverse adaptation methods before developing an adaptation strategy. In addition, since in many areas, as indicated by Dube and Nhamo (2020), tourism is the driving force and funder for most conservation activities in the parks, it is understandable to focus on this aspect.

There seems to be two major issues in protected areas' adaptation that need more thorough examination: agency and the avoidance of maladaptation. Agency refers to the roles, responsibilities and motivations of the different stakeholders participating in the protected areas' management and utilization. For example, a topic that was not reflected upon was the responsibilities in adaptation, and the ways to motivate adaptation among the various stakeholders. A study conducted in Northern Finland, in the vicinity of and within a national park, revealed that in some locations, the local stakeholders have not yet faced any significant negative impacts (i.e., costs) because of climate change (Tervo-Kankare et al., 2018), and thus are not developing or engaging in anticipatory adaptation. If the key stakeholders in other protected areas have experienced that they currently mostly benefit from the change, then persuading them to anticipatory action can be difficult. And if the stakeholders feel overwhelmed with the changes, what kind of support is needed to encourage them to take even small steps in adaptation (and conservation of the areas for future generations)? This also raises the question about responsibility. Who are, or should be, responsible for adapting, and whose responsibility is it to make sure that maladaptation does not happen? What kind of mandates should be given to protected areas' managers to secure 'good' adaptation?

The development of successful adaptation strategies requires continuous involvement and engagement of different stakeholders from scientists and managers to local communities and visitors. As the examples from USA (Choe & Schuett, 2020), Zimbabwe (Musakwa et al., 2020), and Australia (Jacobs et al., 2019) emphasize, understanding the potential and importance of the traditional and Indigenous knowledge is crucial. The learning approach proposed by Dutra et al. (2017, p. I-J) supports this: "natural and cultural resource management necessitates that stakeholders learn from each other and from past and current management actions and processes." In addition, collaboration that supports sustainable decision-making among the users of the protected areas is essential. For example, the case of Iceland's glacier tourism shows that the operators cannot be left alone in planning and realizing their adaptation, but land-use management and the scientific community need to be involved to improve decision-making (Welling & Abegg, 2019). This will secure more sustainable future(s) and leads back to the theme of maladaptation.

The avoidance of maladaptation relates to the central question of securing the conservation principle in protected areas. How to make sure that adaptation is in accordance with the ultimate objective of the parks/sites? This seems to be an issue where the protected areas' managers have a key role: They are the ones who should evaluate the terms according to which the use of protected areas supports their fundamental objectives – conserving biodiversity and natural and cultural heritage.

Identification of maladaptation is important, as reviewed studies have indicated that some adaptation methods may impact in ways that increase carbon emissions and thus speed up climate change. In protected areas, the appropriate adaptation toolbox may be more limited than in environments where conservation policies are non-existent or weaker. In addition, a successful adaptation strategy in one location may not work in a similar way in other locations. This underlines that the park managers must be educated to understand the special characteristics of each protected area: They must possess the skills to avoid the development of unsustainable tourism or other activities that may negatively influence the protected areas. They need to acknowledge the different users of the areas. Finally, they need to understand and be aware of the perspectives and needs of the various stakeholders, balance them, and create trust between the stakeholders. The stakeholders to be considered include also the visitors, as Urioste-Stone et al. (2015, p. 42) indicate: "research on visitors' perceptions on mitigation strategies and resource stewardship in parks and protected areas is highly needed, especially those that reflect visitors' potential support for management strategies that address climate change." Acknowledging the visitor perspective includes preservation of 'naturalness' (which is of high importance for the visitors) and education of visitors so that they understand the meaning of adaptation and their own roles in sustaining protected areas.

Balancing the stakeholders' needs and the conservation ideology is not an easy task in a changing climate. Groulx et al. (2017, p. 1030) aptly conclude that in order to avoid failures in designing and implementing adaptation policies, the managers of the protected areas should "pay careful attention to the fit between the unique identity of protected landscapes and visitors' desire for a natural experience". Their understanding, which is easy to support, is that this kind of approach will most likely contribute to adaptations that are more sensitive to factors that motivate visitation and may also "reinforce the mandate of PA agencies to prioritize ecological integrity over competing development pressures". It also forces the managers to take into consideration "the social and cultural values that are supported (or not supported) through adaptation". It is easy to agree with them, but at the same time it shows how multifaceted task adaptation in protected areas can be.

REFERENCES

Aloia, C. C. D., Naujokaitis-Lewis, I., Blackford, C., Chu, C., Martensen, A. C., Rayfield, B., Sunday, J. M., & Xuereb, A. (2019). Coupled networks of permanent protected areas and dynamic conservation areas for biodiversity conservation under climate change. *Frontiers in Ecology and Evolution*, 7, 1–8. https://doi.org/10.3389/fevo.2019.00027.

Arksey, H., & Malley, L. O. (2005). Scoping studies: Towards a methodological framework. *International Journal of Social Research Methodology*, 8(1), 19–32.

Carter, R. W. B., Walsh, S. J., Jacobson, C., & Miller, M. L. (2014). Global change and human impact challenges in managing iconic national parks. *The George Wright Forum*, 31(3), 245–255.

Ceesay, A., Wolff, M., Njie, E., Kah, M., & Koné, T. (2016). Adapting to the inevitable: The case of Tanbi Wetland National Park, the Gambia. In W. L. Filho, H. Musa, G. Cavan, P. O'Hare, & J. Seixas (eds), *Climate Change, Adaptation, Resilience and Hazards*. Cham: Springer, pp. 257–274.

Chape, S., Harrison, J., Spalding, M., & Lysenko, I. (2005). Measuring the extent and effectiveness of protected areas as an indicator for meeting global biodiversity targets. *Philosophical Transactions of the Royal Society of London B: Biological Sciences*, 360(1454), 443–455.

Cheablam, O., & Shrestha, R. P. (2015). Climate change trends and its impact on tourism resources in Mu Ko Surin Marine National Park, Thailand. *Asia Pacific Journal of Tourism Research*, 20(4), 435–454.

Choe, Y., & Schuett, M. A. (2020). Stakeholders' perceptions of social and environmental changes affecting Everglades National Park in South Florida. *Environmental Development*, 35, 100524. https://doi.org/10.1016/j.envdev.2020.100524.

Coldrey, K., & Turpie, J. K. (2020). Potential impacts of changing climate on nature-based tourism: A case study of South Africa's national parks. *Koedoe*, 62(1), 1–12.

Dawson, J., Stewart, E. J., Lemelin, H., & Scott, D. (2010). The carbon cost of polar bear viewing tourism in Churchill, Canada. *Journal of Sustainable Tourism*, 18(3), 319–336.

Dube, K., & Nhamo, G. (2020). Evidence and impact of climate change on South African national parks: Potential implications for tourism in the Kruger National. *Environmental Development*, 33, 100485. https://doi.org/10.1016/j.envdev.2019.100485.

84 *Handbook on tourism and conservation*

Dutra, L., Bayliss, P., McGregor, S., & Christophersen, P. (2017). Understanding climate-change adaptation on Kakadu National Park, using a combined diagnostic and modelling framework: A case study at Yellow Water wetland. *Marine and Freshwater Research*, 69(7), 1146–1158.

Faccioli, M., & Riera, A. (2015). Valuing the recreational benefits of wetland adaptation to climate change: A trade-off between species' abundance and diversity. *Environmental Management*, 55, 550–563. https://doi.org/10.1007/s00267-014-0407-7.

Fisichelli, N. A., Schuurman, G. W., Monahan, W. B., & Ziesler, P. S. (2015). Protected area tourism in a changing climate: Will visitation at US national parks warm up or overheat? *PLOS ONE*, 10(6), 1–13. https://doi.org/10.1371/journal.pone.0128226.

Glover, L., & Granberg, M. (2021). The politics of maladaptation. *Climate*, 9(5), 69.

Gössling, S., Scott, D., Hall, C. M., Ceron, J. P., & Dubois, G. (2012). Consumer behaviour and demand response of tourists to climate change. *Annals of Tourism Research*, 39(1), 36–58.

Groulx, M., Lemieux, C. J., Lewis, J. L., Brown, S., Groulx, M., Lemieux, C. J., Lewis, J. L., Brown, S., & Groulx, M. (2017). Understanding consumer behaviour and adaptation planning responses to climate-driven environmental change in Canada's parks and protected areas: A climate futurescapes approach. *Journal of Environmental Planning and Management*, 60(6), 1016–1035.

Hall, C. M. (2013). Framing tourism geography: Notes from the underground. *Annals of Tourism Research*, 43, 601–623. https://doi.org/10.1016/j.annals.2013.06.007.

Halofsky, J. E., Peterson, D. L., Dante-Wood, S. K., Hoang, L., Ho, J. J., & Joyce, L. A. (2018). *Climate Change Vulnerability and Adaptation in the Northern Rocky Mountains Part 2*. United States Department of Agriculture.

Hoegh-Guldberg, O., Jacob, D., Taylor, M., Bindi, M., Brown, S., et al. (2018). Impacts of 1.5°C of global warming on natural and human systems. In V. Masson-Delmotte, P. Zhai, H.-O. Pörtner, D. Roberts, J. Skea, et al. (eds), *Global Warming of 1.5°C*. An IPCC Special Report on the impacts of global warming of 1.5°C above pre-industrial levels and related global greenhouse gas emission pathways, in the context of strengthening the global response to the threat of climate change. Geneva: IPCC, pp. 175–311.

Hoffmann, S., & Beierkuhnlein, C. (2020). Climate change exposure and vulnerability of the global protected area estate from an international perspective. *Diversity and Distributions*, 26(11), 1496–1509.

Hoogendoorn, G., & Fitchett, J. M. (2018). Tourism and climate change: A review of threats and adaptation strategies for Africa. *Current Issues in Tourism*, 21(7), 742–759.

IPCC (2007). *Climate Change 2007: Impacts, Adaptation and Vulnerability*. Contribution of working group II to the fourth assessment report of the Intergovernmental Panel on Climate Change (M. L. Parry, O. F. Canziani, J. P. Palutikof, van der L. P. J., & C. E. Hanson (eds)). Cambridge: Cambridge University Press.

Jacobs, B., Boronyak, L., & Mitchell, P. (2019). Application of risk-based, adaptive pathways to climate adaptation planning for public conservation areas in NSW, Australia. *Climate*, 7(4), 58.

Kaján, E., & Saarinen, J. (2013). Tourism, climate change and adaptation: A review. *Current Issues in Tourism*, 16(2), 167–195.

Kilungu, H., Leemans, R., Munishi, P. K. T., & Amelung, B. (2017). Climate change threatens major tourist attractions and tourism in Serengeti National Park, Tanzania. In W. Leal Filho, S. Belay, J. Kalangu, W. Menas, P. Munishi, & K. Musiyiwa (eds), *Climate Change Adaptation in Africa*. Cham: Springer, pp. 375–392.

Lemelin, H., Dawson, J., Stewart, E. J., Maher, P., & Lueck, M. (2010). Last-chance tourism: The boom, doom, and gloom of visiting vanishing destinations. *Current Issues in Tourism*, 13(5), 477–493.

Lemieux, C. J., Groulx, M., Halpenny, E., Stager, H., Dawson, J., Stewart, E. J., & Hvenegaard, G. T. (2018). The end of the Ice Age? Disappearing world heritage and the climate change communication imperative. *Environmental Communication*, 12(5), 653–671.

Liu, T. (2016). The influence of climate change on tourism demand in Taiwan national parks. *Tourism Management Perspectives*, 20, 269–275. http://dx.doi.org/10.1016/j.tmp.2016.10.006.

Liu, W., Huang, Y., & Hsieh, C. (2020). The impacts of different climate change scenarios on visits toward the National Forest Park in Taiwan. *Forests*, 11(11), 1203. http://dx.doi.org/10.3390/f11111203.

Moher, D., Stewart, L., & Shekelle, P. (2015). All in the family: Systematic reviews, rapid reviews, scoping reviews, realist reviews, and more. *Systematic Reviews*, 4. https://doi.org/10.1186/s13643-015-0163-7.

Musakwa, W., Gumbo, T., Parazda, G., Mpofu, E., Nyathi, N. A., & Selamolela, N. B. (2020). Partnerships and stakeholder participation in the management of national parks: Experiences of the Gonarezhou National Park in Zimbabwe. *Land*, 9(11), 399. https://doi.org/10.3390/land9110399.

Mushawemhuka, W., Rogerson, J. M., & Saarinen, J. (2018). Nature-based tourism operators' perceptions and adaptation to climate change in Hwange National Park, Zimbabwe. *Bulletin of Geography*, 42, 115–127. http://doi.org/10.2478/bog-2018-0034.

Njoroge, J. M. (2015). Climate change and tourism. *Tourism and Hospitality Management*, 21(1), 98–105.

Nyongesa, K. W., & Vacik, H. (2019). Evaluating management strategies for Mount Kenya Forest Reserve and National Park to reduce fire danger and address interests of various stakeholders. *Forests*, 10(5), 426.

Pongkijvorasin, S., & Chotiyaputta, V. (2013). Climate change and tourism: Impacts and responses. A case study of Khaoyai National Park. *Tourism Management Perspectives*, 5, 10–17. http://dx.doi.org/10.1016/j.tmp.2012.10.002.

Purdie, H., Hughes, J., Stewart, E., & Espiner, S. (2020). Implications of a changing alpine environment for geotourism: A case study from Aoraki / Mount Cook, New Zealand. *Journal of Outdoor Recreation and Tourism*, 29, 100235. https://doi.org/10.1016/j.jort.2019.100235.

Scherrer, P., & Pickering, C. M. (2001). Effects of grazing, tourism and climate change on the alpine vegetation of Kosciuszko National Park. *Victorian Naturalist*, 118, 93–99.

Scott, D., Dawson, J., & Jones, B. (2008). Climate change vulnerability of the US Northeast winter recreation-tourism sector. *Mitigation and Adaptation Strategies for Global Change*, 13(5–6), 577–596.

Scott, D., Hall, C. M., & Gössling, S. (2019). Global tourism vulnerability to climate change. *Annals of Tourism Research*, 77, 49–61.

Tervo-Kankare, K. (2012). Climate change awareness and adaptation in nature-based winter tourism: Regional and operational vulnerabilities in Finland. *Nordia Geographical Publications*, 41(2).

Tervo-Kankare, K., Kaján, E., & Saarinen, J. (2018). Costs and benefits of environmental change: Tourism industry's responses in Arctic Finland. *Tourism Geographies*, 20(2), 202–223. https://doi.org/10.1080/14616688.2017.1375973.

Tolvanen, A., & Kangas, K. (2016). Tourism, biodiversity and protected areas: Review from northern Fennoscandia. *Journal of Environmental Management*, 169, 58–66. http://dx.doi.org/10.1016/j.jenvman.2015.12.011.

UNEP-WCMC, & IUCN (2016). *Protected Planet Report 2016: How Protected Areas Contribute to Achieving Global Targets for Biodiversity*. Cambridge and Gland: UNEP-WCMC and IUCN.

UN World Tourism Organization and United Nations Environment Programme (2008). *Climate Change and Tourism: Responding to Global Challenges*. Madrid. https://doi.org/10.18111/9789284412341.

Urioste-Stone, S. M. De, Scaccia, M. D., & Howe-Poteet, D. (2015). Exploring visitor perceptions of the influence of climate change on tourism at Acadia National Park, Maine. *Journal of Outdoor Recreation and Tourism*, 11, 34–43. http://dx.doi.org/10.1016/j.jort.2015.07.001.

Wall, G., Harrison, R., Kinnaird, V., McBoyle, G., & Quinlan (1986). The implications of climatic change for camping in Ontario. *Recreation Research Review*, 13(1), 50–60.

Wang, L., Zeng, Y., & Zhong, L. (2017). Impact of climate change on tourism on the Qinghai-Tibetan plateau: Research based on a literature review. *Sustainability*, 9(9), 1539. https://doi.org/10.3390/su9091539.

Welling, J., & Abegg, B. (2019). Following the ice: Adaptation processes of glacier tour operators in Southeast Iceland. *International Journal of Biometeorology*, 65, 703–715. https://doi.org/10.1007/s00484-019-01779-x.

Welling, J., Þorvarður, Á., & Rannveig, Ó. (2020). Implications of climate change on nature-based tourism demand: A segmentation analysis of glacier site visitors in Southeast Iceland. *Sustainability*, 12(13), 5338.

Yohe, G., & Tol, R. S. J. (2002). Indicators for social and economic coping capacity: Moving towards a working definition of adaptive capacity. *Global Environmental Change*, 12(1), 25–40.

7. Tourism and climate change vulnerabilities: a focus on African destinations
Kaitano Dube

INTRODUCTION

The year 2020 was dominated by the impact of Covid-19 on global economies, particularly on the tourism industry. Due to the Covid-19 pandemic, the tourism industry contracted to significantly low levels, such as those seen in the early 1990s. While the impact of Covid-19 on the global tourism industry has been deep and widespread (Nhamo et al., 2020a), the pandemic did not put brakes on the severe impact of climate change-induced extreme weather events, which have become one of the most significant threats to tourism across the world (Scott et al., 2012, 2019; Fang et al., 2018).

The evidence suggests that the year 2020 was characterised by a record number of extremes attributed to climate change (Dube, 2022). However, some of these went unnoticed or under-reported as the world was focused on Covid-19. According to the World Meteorological Organization (2021), the year 2020 was one of the warmest years on record, with the global temperature averaging 14.9°C, 1.2° (± 0.1°C) above the pre-industrial (1850–1900) level. The warmest years on record have been recorded between 2015 and 2020, with 2016, 2019 and 2020 being the top three warmest years on record. Global warming, one of the main drivers of climate change, has also been reflected in other related climate change challenge markers. These include increasing sea surface temperature, sea level rise, increased ocean acidity, increased heatwaves, increased tornadoes, intense tropical cyclones, extreme droughts, fires, and melting polar ice, to mention but a few. It is now widely accepted that such extremes pose a danger and threat to tourism resorts, and the tourism economy as the foundation resources for tourism are often compromised. Nowhere else is this threat of extreme weather events more apparent and substantial than in Africa and Small Island Developing States (SIDS) (Intergovernmental Panel on Climate Change [IPCC], 2018).

In such a context, this study examines the vulnerability of tourism destinations to climate change-induced extreme weather events caused by global warming. The critical question of this study is understanding how global warming is threatening iconic tourism resorts in Africa. The study is a response to tourism climate change knowledge which several academics have well documented. There is a general concern that tourism and climate change remain under-researched in Africa, with vast knowledge gaps (Hall, 2008; Hoogendoorn & Fitchett, 2018; Scott et al., 2019). Over the years, however, there has been some progress in documenting the tourism–climate change nexus, particularly in some parts of South Africa, Zimbabwe, Botswana, and Tanzania.

Globally the relationship between tourism and climate has been of interest since the 1970s. However, the rise in the reported impact of extreme weather events reinvigorated interest in this field of study starting in the 1990s (Fang et al., 2018). Several tourism activities depend on ideal climatic conditions, with seasonality and climatic conditions being crucial determinants of the quality of tourism experience for travellers. This understanding led to tourism geographers developing various tourism climate change indexes to understand the relationship between tourism and climate, one of the first being the tourism climate change index (TCI) produced by Mieczkowski (1985). The TCI acknowledged the role of various weather elements across multiple tourism activities. Efforts at a regional level in Africa have been made to customise and determine tourism risks from a TCI perspective at many tourism destinations (Fitchett & Hoogendoorn, 2018; Noome & Fitchett, 2019; Mushawemhuka et al., 2020).

Tourist activities such as beachgoing are often shaped by certain weather conditions that pull or push people off the beaches. Beachgoers often favour ordinarily sunny days where wind conditions are calm. There is increasing awareness and fear that most tourism destinations worldwide are vulnerable to climate change. Before the advent of the Covid-19 pandemic, the tourism industry was one of the largest economic sectors that promised economic development, foreign currency generation, employment creation, and sustainability for many communities worldwide (Nhamo et al., 2020b). After the pandemic, there have been renewed hopes that tourism development will continue to trigger global economic development. Nevertheless, the threat of climate change remains one of the biggest challenges to many destinations, particularly in Africa, where the tourism sector is primarily based on nature-based tourism, which is vulnerable to the vagaries of climate change-induced extreme weather events.

RESEARCH DESIGN

Using secondary data and field observations, this study seeks to highlight some of the threats and challenges caused by climate change for some of the continent's iconic tourism resorts. A multiple case study approach is utilised, where multiple cases are investigated to address the study objectives. The study utilises historical events to track potential risk exposure of destinations based on previous extreme weather events and potential risks posed by the trend of various weather events. In this regard, primary and secondary data are utilised. The latest technological instruments utilised in tourism geography studies, such as remote sensing and archival data sources, are utilised in this study. In some instances, primary data from field observations are utilised. Customer experiences are derived from customer review comments from TripAdvisor. Using platforms such as TripAdvisor is an acceptable practice in tourism studies as they are a rich source of customer feedback. Several studies have previously used data from this platform, such as Taecharungroj and Mathayomchan (2019) and Chang et al. (2019). The study also used processed remotely sensed images to analyse glaciers at mountainous destinations, archival data from meteorological stations and the National Oceanic and Atmospheric Administration (NOAA), and published literature to arrive at conclusions and triangulation. Google Scholar was the main source of published data between 2010 and 2020. Only material that responds to the research questions and fits within the set geographic parameters was considered for this research.

Both content and thematic analysis approaches were utilised in this research, with themes derived from the study objectives. Content analysis was used to analyse data as dictated by Erlingsson and Brysiewicz (2017). Consequently, the analysis started with less abstraction of relevant material to obtain unit meaning. This was followed by the condensation of meaning units of data, which led to code development, categorisation, and thematic development using the research questions as guiding instruments. Data from field observations followed the same

88 *Handbook on tourism and conservation*

approach as only notes from field observations and imagery responding to various question parameters were used. Data from remotely sensed imagery from Google Earth was analysed and pre-processed for use and analysed using various Geographic Information System (GIS) tools to conduct change detection in the study area.

RESULTS AND DISCUSSION

The following section looks at various extreme weather elements experienced in Africa to identify vulnerabilities at crucial destinations across Africa with a strong bias on destinations in Southern and Eastern Africa familiar to the researcher.

Global Warming

Temperature increase is one of the biggest threats to biodiversity and nature that forms the core of Africa's tourism product. An increase in temperature impacts various tourism products and resorts across the continent both directly and indirectly. An increase in temperature has implications for most national parks on the continent, as some plants and animals might not withstand the new temperature thresholds resulting from global warming. New temperature anomalies are occurring because of global warming.

Some of the most populous national parks are in southern and eastern Africa, namely, Kruger, Hwange, Serengeti, Chobe, Masai Mara, Etosha, and South Luangwa Bwindi Impenetrable, Amboseli National Park and the Ngorongoro Conservation area, all of which are famous for wildlife. The southern hemisphere, where most of the national parks are located, experienced the highest temperature anomalies over the past period (Figure 7.1), a trend that can significantly impact flora and fauna. It is important to note that some southern hemisphere areas have had higher temperature anomalies, a feature masked in the regional picture. Van Wilgen et al. (2016) established that some temperature anomalies observed over the past decades in national parks in the southern Cape, South Africa, exhibited a temperature increase projected for 2035. An increase in temperature in parks is believed to significantly impact the bird population, which is feared to decline with temperature increases. This can drastically impact the area's ecological balance as birds act as pollinators. According to Cunningham et al. (2013), an increase in heatwaves in some South African national parks was to blame for bird mortality. A decline in bird species will have a serious impact on tourism for those particularly interested in bird wildlife (birders). It is anticipated that national parks with a similar climate regime across the continent might be experiencing a similar challenge.

One big concern arising from the increase in temperature is that it increases the vulnerability of tourism resorts and national parks to fire. While fire is generally good for some ecosystems, the frequency and intensity of fires that have occurred recently are a threat to particular plant and animal species that could go extinct. Of concern to Miller et al. (2019) is the impact of such fires on population and plant communities' composition, a major drawcard for many tourists to national parks across the continent.

A study of Kruger National Park revealed growing concern over the impact of rising temperatures, including increased demands from tourists for better cooling systems in their accommodation establishments (Dube & Nhamo, 2020). Tourists and employees alike face challenges in sleeping due to heat fatigue in the absence of air conditioning systems. In some

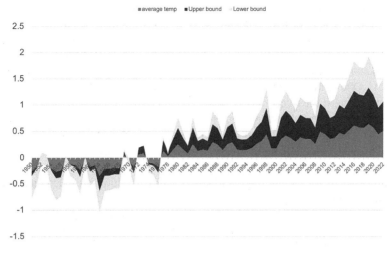

Source: Author.

Figure 7.1   Southern hemisphere temperature anomalies 1850–2019

instances, there was a demand for cooling systems in swimming pools in the hot summer months, particularly when temperatures ranged between 30°C and 40°C. In Victoria Falls, tourists avoided the heat by altering their daily activities during the summer months when temperatures rise to high levels (Dube & Nhamo, 2019).

Impact of Global Warming on Mountain Tourism in Africa

Perhaps even more significant is the impact of global warming on tourist resorts that are snow and ice dependent. Africa has three well-known resorts dependent on ice and snow: Mount Kenya, Mount Kilimanjaro, and Lesotho/South Africa's Maluti Mountains, which host the Afriski and the Tiffindell Ski resorts. Mount Kenya was proclaimed a World Heritage Site in 1997 and with an altitude of 5,199m is the second-highest mountain in Africa after Mount Kilimanjaro, another World Heritage Site 5,895m above sea level. There has been growing concern about the effect of increasing temperatures on glaciers, one of these mountains' dominant features. Due to increased temperature, the number of glaciers has fallen from 18 to 12. This is a serious problem as many tourists feel that the glacial ice on the top of the mountain is an important aesthetic value. As one tourist wrote on TripAdvisor: "I see it arise from farmlands to the thick bamboo jungle, zones of moors & glaciers to the peaks that offer stunning views! Snowy pinnacles & forested slopes appear as if in glorious technicolour on a Panavision screen etched against a dazzling blue sky!"

A study by Chen et al. (2018) that used remote sensing to reconstruct the glaciers revealed that between 87 and 88 per cent of the glaciers have been lost since the Last Glacial Maximum increased by between 5.2°C and 6.5°C. An earlier study by Ouma and Tateishi (2005) observed a glacial decline that changed from 456.3 × 10^3 m² to 227.7 × 10^3 m² between 1987 and 2000. The increase in glacial loss threatens the aesthetic value and tourism activities in and around the destination. Mount Kenya, for example, receives an annual average of more

90   *Handbook on tourism and conservation*

than 30,000 tourists (Evaristus, 2014). A combination of temperature increases is likely to reduce the glacier cover further and have a long-lasting impact on the water supply at the basin, affecting tour operators and other economic activities. One tourist expressed this fear on TripAdvisor: "You don't have to be a climber, just fit enough, and you'll make it to an amazing mountain, unique tropical vegetation at 4000m, snow and glaciers (but sadly not for a long time)." One of the waterfalls that are fed by the glacier, the Nithi Falls, might also be at risk of disappearing if the rivers dry up.

The drying up of rivers, lakes, and other water points on the mountain is also feared to be compounded by increased chances of fire on the mountain. A combination of fire and reduced water could prove problematic for this tropical alpine area. According to Downing et al. (2017), small fire incidence usually occurs, but big fire occurrences have been more frequent recently. The fires are adversely affecting moorland and other vegetation. Fires also threaten hikers on the mountain, who might require evacuation and disrupt tourists' itineraries as a result. With several tourism and other supporting facilities such as camps and lodges, fires are challenging on the mountain. The risk of fires for tourist activities and infrastructure on Mount Kenya is equally concerning and a significant threat, as noted by Nyongesa and Vacik (2018).

An analysis of eight Google images from Google Earth Pro of Mount Kilimanjaro shows that the glacier significantly declines over time. For the month of August in the years between 2005, 2016 and 2017, where three snapshots were analysed, there seems to be strong evidence of glacier variability and decline over the study period (Figure 7.2). The glaciers on the mountain are believed to have declined by as much as 80 per cent since 1912. There are fears that these might disappear in the next 20 years (World Wildlife Fund for Nature, 2010). Analysis done by NASA (2012) shows that as of October 2012, the mountain had lost about 85 per cent of its glaciers between 1912 and 2012, anticipating that at current rates, glaciers will have disappeared by 2060. Helama (2015) also attested that the glaciers were disappearing, arguing that there were conflicting views on the causes of such a phenomenon, amongst other theories being the issue of global warming. A study by Kilungu et al. (2019) established that the temperature in the area had increased by $1.3 \pm 0.06°C$ (p-value <0.05) between 1973 and 2013 at the nearest Kilimanjaro airport. The disappearance of glaciers on Mount Kilimanjaro was also observed by Hemp (2009), who noted that apart from disappearing glaciers, there was also a risk of climate change-induced fires on the mountain with far-reaching effects on vegetation, tourism infrastructure, and other supporting services.

From the analysis of tourist reviews on TripAdvisor, it is clear that tourists consider the glaciers a vital component of the resort. Tourists are equally worried about the decline of the glaciers on the mountain. Many tourists also appreciate their experience of the five climatic zones one must go through when hiking on the mountain. These climatic zones include montane forest, heath and moorland, alpine desert and summit. These zones are rich in endemic and endangered species sensitive to climate change. Climate change is an inherent threat to the aesthetic values that form part of the hikers' experience on Mount Kilimanjaro.

Kilimanjaro, a resort that attracts about 50,000 climbers every year, is also prone to fires and this has had an effect on tourist activities when hiking trails have been forced to close. One of the worst fires recently reported resulted in the burning of a significant portion of the park and disrupted tourists' activities, with many tour agencies issuing travel warnings. A huge contingent of resources, including more than 400 personnel and fire helicopters, were deployed to deal with the fire that caused extensive damage to vegetation. Apart from the immediate effects of fire, burnt vegetation is also an eyesore to tourists and endangers the lives of many

Tourism and climate change vulnerabilities: a focus on African destinations 91

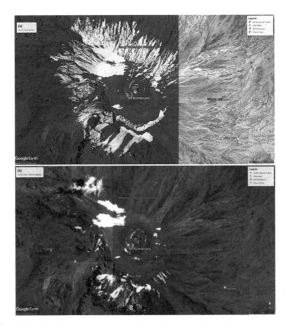

Source: Google Earth.

Figure 7.2 Impact of global warming on Mount Kilimanjaro Glacier: a snap survey of August 2005 and August 2017

animal species. Fire disruptions are costly for tour operators and tourists as the latter fork out vast sums of money to participate, which can be as high as $3,000 for an eight-day hiking trail tour. The increased fire frequency on Mount Kilimanjaro has been raised as a concern in previous studies, such as by Kilungu et al. (2019).

In the case of Afriski, with regard to climatic elements observed on the Climate Information Platform from the Climate Systems Analysis Group at the University of Cape Town, there have not been significant annual average changes to minimum temperature and precipitation at some of the stations near the resort and maximum temperatures have only slightly increased over the years between 1971 and 2000. However, looking at Google Earth images for the winter months shows variability between the years, suggesting that the resort could be vulnerable to global warming. While the snow at Afriski is mostly the result of human engineering because it requires a specific temperature range, warming in the winter months might render snow- and ice-making more expensive as a result of changes to the climate. A similar sentiment was echoed by Hoogendoorn et al. (2021), who pointed out that the company is looking into new equipment for snow-making as a climate change adaptation plan.

Impact of Global Warming on Coastal and Ocean Tourism

The Special Report on Ocean and Cryosphere indicated that one of the most vulnerable places on Earth due to global warming is the oceans which suffer multiple challenges arising from global warming, such as an increase in sea surface temperature, sea level rise and ocean acidification, to mention just a few (IPCC, 2019). Africa has a long coastline stretching about

92 *Handbook on tourism and conservation*

30,500 km and the ocean shores on the mainland and the various pristine islands such as Mauritius and Seychelles are popular tourism destinations. Evidence shows that sea surface temperature anomalies have affected Africa's coastline and ocean space (Mgadle et al., 2022). The increase in sea surface temperature is blamed for the death of coral reefs, a crucial part of Africa's tourism industry, particularly in the Mozambique channel. Sibitane et al. (2022) blame the increased heat and temperature for the significant decline in turtle populations, which could adversely affect some tourist activities in coastal areas which act as breeding spaces for turtles.

Evidence suggests that some areas are hard hit by coral bleaching within the Indian Ocean Basin, as 893 observations reported coral bleaching (Figure 7.3). In Figure 7.3, the areas around the Islands of Mayotte, the east coast of Tanzania, Mozambique, South Africa, and Madagascar seem to be some of the worst affected areas by coral bleaching. Areas around other island states, such as Mauritius, Seychelles, and off the coast of Equatorial Guinea, also face coral bleaching threats. With the increase in temperature, it is anticipated that there will be a lot more observed mortality of coral reefs south-west of the Indian Ocean. Figure 7.3 shows that in 2016, 2019, and 2020, there was a significant amount of bleaching. The results are not surprising, given that those years coincided with the period when the global temperature was at its highest. According to the World Meteorological Organization (2021), 2016 and 2019 are two of the three warmest years on record, with 2016 coinciding with one of the strongest El Niño years. There are visible markers of climate variability on coral mortality. With the expectation that climate change will worsen the severity and impact of El Niño events globally (Wang et al., 2017), it is anticipated that there could be even more bleaching in the future.

Tourism actors have been raising concerns over the issue of coral bleaching in the Indian Ocean for some time. Consequently, some tour operators took steps to consolidate and protect the corals in the Indian Ocean (Dube et al., 2020). According to Hoogendoorn and Fitchett (2018), coral bleaching threatens scuba diving and snorkelling tourism. This confirmed earlier findings by Ngazy et al. (2005), who reported the value of corals to the Zanzibar tourism economy. There is ample evidence to show that the SIDS are equally at the mercy of increased sea surface temperatures (SST) given several recent die-offs and which are on the increase (Figure 7.3). Other countries affected include Kenya, Madagascar, and South Africa. Coral bleaching poses a severe threat to the blue ocean economy, which is anticipated by the Sustainable Development Goals (SDGs) to be a means of strengthening economies and fighting poverty (Karani & Failler, 2020).

In North Africa, reefs are essential to Egyptian tourism as the clear water and the red reefs form a valuable part of the tourism experience. Some concerns are being raised over coral bleaching threatening the tourism economy. Hilmi et al. (2012) highlighted the centrality of coral reefs to tourism in the Hurghada coastal region, which is a pristine area with a significant volume of corals, making it attractive to tourists. El-Raey (2010) indicated that the Red Sea was experiencing coral bleaching, which could threaten the tourism experience and livelihoods dependent on tourism.

Sea Level Rise and Coastal Tourism in Africa

Apart from coral bleaching, one of the biggest threats to coastal tourism is the effect of global warming on sea levels and, as a result, the knock-on effects on tidal activity. As a result of polar ice melting and thermal expansion, sea levels have risen by 3.3 ± 0.3 mm/yr since 1993

Tourism and climate change vulnerabilities: a focus on African destinations 93

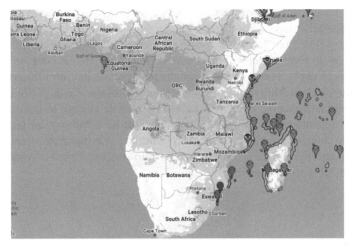

Source: Adapted from NOAA (2021).

Figure 7.3   Indian Ocean coral bleaching 2016–2020

(World Meteorological Organization, 2021). Regarding regional sea level rise, while increases in the African coastal areas are not as high as this, sea level rises in other parts of the world, such as the south coast of North America, southern parts of South Asia, and the Mediterranean region, will have a marked effect on tourism activity. But while much of the African coastline does not have as many permanent sea level markers, the general picture is that across the continent, Africa will experience an increase in sea level overall, with several hot spots (Figure 7.4). The most significant sea rise level is recorded in the continent's SIDS, such as Mauritius, Réunion Islands and Seychelles.

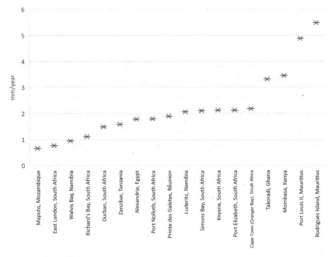

Source: Author data from NOAA (2021).

Figure 7.4   Sea level trends along the African coastline

The rising sea level will mainly affect the coastal beach resorts. A recent study by Dube et al. (2021) revealed that significant damage and erosion to beaches threatens beach tourism in and around many Blue Flag beaches across the Western Cape due to a combination of storms and sea level rise. In another study, Mather and Stretch (2012) pointed out the vulnerability of Durban beaches to sea level rise. Beaches are the major attraction for millions of people who flock to coastal resorts worldwide annually and hundreds of thousands of South Africans who take to the beaches annually. The threat caused by climate change to coastal infrastructure is disturbing. Even though the Western Cape does not have the highest sea level rise, the damage already witnessed in that place is concerning. In some instances, coastal erosion from increased tidal activity threatens coastal tourism infrastructures such as roads (Figure 7.5), railways, and other tourism amenities.

Source: Author 2020 fieldwork.

Figure 7.5   *Cut-off section of road to Monwabisi beach due to coastal erosion in Cape Town, South Africa*

A coastal vulnerability assessment conducted by Boateng et al. (2017) established that about 50 per cent of the 540 km shoreline of Ghana is vulnerable to rising sea levels. This threat is significant to heritage sites, some of which lie across the coast. According to Twumasi-Ampofo et al. (2020), Ghana's architectural and cultural heritage is already under significant threat. Sagoe-Addy and Addo (2013) also attested to the climate change risk for Ghana's coastal tourism that manifests in coastal inundation and flooding. In Mombasa, Kenya, there is growing concern about the impacts of sea level rise on its coastline and infrastructure (Kebede et al., 2012), which may damage some tourism products, resorts, and heritage.

In Mauritius and Seychelles, a recognised tourism threat is rising sea levels as the tourism industry's beaches are increasingly threatened by rising sea levels. Etongo (2019) argued that coastal flooding and sea level rise are threats that the Seychelles tourism economy has to battle with. Mauritius, the third most popular tourist destination in sub-Saharan Africa, and whose signature statement is "sea, sun and sand", acknowledges that one of the greatest threats to the destination is climate change, which manifests itself in increased tidal waves, surges, beach erosion, and deterioration of coral reefs due to increased sea surface temperature (Republic of Mauritius Ministry of Tourism, 2021). The threat poses a challenge to the destination's competitive advantage as it can potentially affect its attractiveness.

Tourism Vulnerability to Tropical Cyclones in Africa

One of the most under-researched – and contentious – issues, apart from climate change's impact on coastal areas and heritage, is the impact of tropical cyclones on tourism. Studying the impact of global warming on tropical cyclones, several scholars and models show that the frequency of high-intensity tropical cyclones will increase in a warming world (Knutson et al., 2010; Murakami et al., 2013; Fitchett, 2018). Tropical cyclone rainfall is believed to increase in a warmer climate, as is the cost and impact of tropical cyclones on human lives (Lin et al., 2015). Dube (2022) found that several tropical cyclones have occurred over the past five years within the South-West Indian Ocean, with several of them making landfall with some substantial socio-economic damage resulting in some instances (Nhamo & Dube, 2021). Severe storms and tropical cyclones thus already adversely affect tourism operations in such tourist attractions and destinations.

Given that tourists are generally disaster- and risk-averse individuals, the impacts of tropical cyclones and associated rainfall systems have been both a blessing and a curse for Africa's tourism industry. A tropical cyclone is characterised as an extreme weather event and warnings are often issued days in advance for people to avoid the storm tracks. Given the high impact of tropical cyclones, such events also significantly disrupt ferries and cruise ship operations used by tourists. This potentially disrupts SIDS in the Indian Ocean Basin and other areas where hurricanes or tropical cyclones occur.

When tropical cyclones occur on land, they are often accompanied by high tides, resulting in coastal erosion and flooding. In many instances, the coastal areas are lined with tourism infrastructure, so coastal tourism is the bedrock of these areas' economies. Tropical cyclones recently have resulted in the closure of critical tourism infrastructure. When Tropical Cyclone Idai hit Mozambique in 2019, Beira airport was temporarily closed, cutting off the city from the rest of the world. The cyclone also badly damaged Beira's several hotels, cutting off power, water supply infrastructure, roads, and bridges, making the city inaccessible to tourists as the area turned into a disaster zone. The iconic Beira Terrace Hotel was one such hotel affected by the sustained winds, which were about 167 km/hr, and flooding from a combination of wave run and rainfall. The tropical cyclone also degraded most beaches and caused enormous land pollution making it temporarily unfit for use by tourism and recreation. Two national parks were affected by Tropical Cyclone Idai, in namely Chimanimani National Park and the Cheringoma Plateau in Gorongosa National Park (see also Nhamo et al., 2021). Due to the impacts of the tropical cyclones, the parks were flooded, resulting in massive destruction of infrastructure that tourism stakeholders, including tourists, use.

Secondary data and field reports revealed that in the 2021 run-up to the arrival of Tropical Cyclone Eloise, the Kruger National Park, one of Africa's most populous parks, issued a warning and announced the closure of several camps and access roads as it prepared for the worst. During the cyclone on 23 and 24 January 2021, several camps were evacuated, and bridges and roads were extensively damaged or washed away altogether. Cyclones create anxiety and panic amongst tourists and other tourism stakeholders, with far-reaching implications for international tourists with tight itineraries. Earlier, Cyclone Irina affected St Lucia Wetland Park, a World Heritage Site, marking the end of the drought that had previously

96 *Handbook on tourism and conservation*

affected the park. Evidence also shows that during Cyclone Eline in 2000, the Kruger National Park was severely flooded, causing business disruption and property loss. Gonarezhou National Park lost one of its camps in the Savé Valley Conservancy game reserve.

Impact of Global Warming on Tenting Accommodation

The increase in temperature also poses significant threats to some types of accommodation. With the increased frequency of sweltering days in some national parks and several destinations across the continent, in some areas camping in ordinary tents can expose occupants to various heat-related medical challenges. In the desert and semi-desert areas, the demand for cooling increases along with the rise of global average temperatures. Since tents do not have cooling facilities, they generally become uninhabitable, particularly during summer periods when temperatures in some areas skyrocket to the upper 40°Cs. A study by Dube and Nhamo (2018) indicated that temperature adversely affected tourists even in conventional accommodations during the peak summer heat season in Victoria Falls.

Global Warming and Water-based Tourism Resorts

Increased temperature is set to result in challenges for water-based tourism resorts across the continent, such as waterfalls, dams, lakes, deltas, rivers, and estuaries. Increased water temperature amidst increasing droughts will undoubtedly affect tourism and recreation at water-dependent resorts. The increase in temperature is always associated with a demand for water amidst increasing evapotranspiration rates. As a consequence of global warming, both human and natural abstraction have a long-lasting impact on tourism, water points, and water attractions. A combination of droughts, evapotranspiration, human water demand, and abstraction has been blamed for fears of the Victoria Falls drying up in the hot summer months of October and November, when water levels drop and, in some years, to record lows, much to the displeasure of tourists and tour operators alike (Dube & Nhamo, 2018; Dube & Nhamo, 2019). Simultaneously, the decline in water levels during the early months of summer suggests that climate change has further worsened in recent years.

Other destinations like the Okavango Delta are equally vulnerable to global warming's direct and indirect impacts. Several authors have written extensively about the Okavango Delta's vulnerability to climate change (Hambira et al., 2013, 2020). According to a study by Moses and Hambira (2018), the Okavango loses more than 98 per cent of its water to evaporation. The area has also been seeing more than the usual level of evaporation, which can affect the attractiveness and activities at the destination. A study by Dube et al. (2018) revealed that tourists increasingly worry about climate change's impact on the delta's activities and attractiveness. A compounded effect of droughts and increased evaporation reportedly affects Lake Kariba's tourism along the Zambezi River downstream of the Victoria Falls. There are also fears that the increased global warming may alter water's physical and component qualities, affecting fish and fishing activities at some water resorts.

CONCLUSION AND RECOMMENDATIONS

This study aimed to examine climate change vulnerabilities of tourism destinations across Africa, focusing on the vulnerabilities of increased global warming. The study found that global warming presents national parks with a number of challenges as increased temperature challenges flora and fauna. Birds have been particularly vulnerable and populations are expected to decline due to increased heatwaves, which often result in bird die-offs. Apart from national parks, the study found that mountainous destinations risk losing some of their attractiveness due to glacier melting, particularly at Mount Kenya and Mount Kilimanjaro, with fears that in the foreseeable future, glaciers might disappear from the latter. Snow-based tourism activities are equally at risk due to increased warming, with Afriski mountain resorts facing a more significant threat.

Other risks were observed in coastal areas where coral bleaching and rising sea level are among the most significant threats. In tourism destinations in small island states and south-east Africa, intense tropical cyclones have emerged as a significant threat that disrupts business and damages infrastructure. Increased evaporation rates are also a threat to water-based tourism on the continent.

In light of the vulnerabilities, the study recommends a climate de-risked approach anchored on low carbon development across the tourism sector to reduce the pace of climate change. Regarding coastal areas, there might be a need to invest in beach nourishment programmes and relocate some of the tourism infrastructure in direct danger of constant flooding to reduce losses and the cost of such damage. Adequate insurance for tourism infrastructure is a must to protect against damage and loss caused by extreme weather events. Climate change adaptation, which is activity and destination based, has to be adopted by each sector to ensure tourism resilience. In cases where vulnerability leads to human loss and property damage, there is a need to adopt and implement early warning systems, invest in insurance, and adopt a building back better approach to ensure that the infrastructure is climate resilient.

REFERENCES

Boateng, I., Wiafe, G., & Jayson-Quashigah, P. N. (2017). Mapping vulnerability and risk of Ghana's coastline to sea level rise. *Marine Geodesy*, 40(1), 23–39.

Chang, Y. C., Ku, C. H., & Chen, C. H. (2019). Social media analytics: Extracting and visualising Hilton hotel ratings and reviews from TripAdvisor. *International Journal of Information Management*, 48, 263–279.

Chen, A. A., Wang, N. L., Guo, Z. M., Wu, Y. W., & Wu, H. B. (2018). Glacier variations and rising temperature in the Mt. Kenya since the Last Glacial Maximum. *Journal of Mountain Science*, 15(6), 1268–1282.

Cunningham, S. J., Kruger, A. C., Nxumalo, M. P., & Hockey, P. A. (2013). Identifying biologically meaningful hot-weather events using threshold temperatures that affect life-history. *PloS ONE*, 8(12), e82492.

Downing, T. A., Imo, M., & Kimanzi, J. (2017). Fire occurrence on Mount Kenya and patterns of burning. *GeoResJ*, 13, 17–26.

Dube, K. (2022). Nature-based tourism resources and climate change in Southern Africa: Implications for conservation and sustainable development. In L. Stone, M. Stone, P. Mogomotsi, & G. Mogomotsi (eds), *Protected Areas and Tourism in Southern Africa: Conservation Goals and Community Livelihoods* (pp. 160–173). New York: Routledge.

Dube, K., Mearns, K., Mini, S., & Chapungu, L. (2018). Tourists' knowledge and perceptions on the impact of climate change on tourism in Okavango Delta, Botswana. *African Journal of Hospitality, Tourism and Leisure*, 7(4), 1–18.

Dube, K., & Nhamo, G. (2018). Climate variability, change and potential impacts on tourism: Evidence from the Zambian side of the Victoria Falls. *Environmental Science & Policy*, 84, 113–123.

Dube, K., & Nhamo, G. (2019). Climate change and potential impacts on tourism: Evidence from the Zimbabwean side of the Victoria Falls. *Environment, Development and Sustainability*, 21(4), 2025–2041.

Dube, K., & Nhamo, G. (2020). Evidence and impact of climate change on South African national parks: Potential implications for tourism in the Kruger National Park. *Environmental Development*, 33, 1–11.

Dube, K., Nhamo, G., & Chikodzi, D. (2021). Rising sea level and its implications on coastal tourism development in Cape Town, South Africa. *Journal of Outdoor Recreation and Tourism*, 33, 100346.

Dube, K., Nhamo, G., & Mearns, K. (2020). &Beyond's response to the twin challenges of pollution and climate change in the context of SDGs. In G. Nhamo, G. O. A. Odularu, & V. Mjimba (eds), *Scaling Up SDGs Implementation* (pp. 87–98). Cham: Springer.

El-Raey, M. (2010). Impacts and implications of climate change for the coastal zones of Egypt. *Coastal Zones and Climate Change Report* (pp. 31–50). Stimson Center.

Erlingsson, C., & Brysiewicz, P. (2017). A hands-on guide to doing content analysis. *African Journal of Emergency Medicine*, 7(3), 93–99.

Etongo, D. (2019). Climate change adaptation in Seychelles: Actors, actions, barriers and strategies for improvement. *Seychelles Research Journal*, 1(2), 43–66.

Evaristus, I. M. (2014). Global change and sustainable mountain tourism: The case of Mount Kenya. In V. I. Grover, A. Borsdorf, J. Breuste, P. C. Tiwari, and F. W. Frangetto (eds), *Impact of Global Changes on Mountains: Responses and Adaptation* (pp. 187–207). London: CRC Press.

Fang, Y., Yin, J., & Wu, B. (2018). Climate change and tourism: A scientometric analysis using CiteSpace. *Journal of Sustainable Tourism*, 26(1), 108–126.

Fitchett, J. M. (2018). Recent emergence of CAT5 tropical cyclones in the South Indian Ocean. *South African Journal of Science*, 11–12(114), 1–6.

Fitchett, J., & Hoogendoorn, G. (2018). An analysis of factors affecting tourists' accounts of weather in South Africa. *International Journal of Biometeorology*, 62(12), 2161–2172.

Hall, C. M. (2008). Tourism and climate change: Knowledge gaps and issues. *Tourism Recreation Research*, 33(3), 339–350.

Hambira, W. L., Saarinen, J., Manwa, H., & Atlhopheng, J. R. (2013). Climate change adaptation practices in nature-based tourism in Maun in the Okavango Delta area, Botswana: How prepared are the tourism businesses? *Tourism Review International*, 17(1), 19–29.

Hambira, W. L., Saarinen, J., & Moses, O. (2020). Climate change policy in a world of uncertainty: Changing environment, knowledge, and tourism in Botswana. *African Geographical Review*, 39(3), 252–266.

Helama, S. (2015). Ernest Hemingway's description of the mountaintop in 'The Snows of Kilimanjaro' and climate change research. *The Hemingway Review*, 34(2), 118–123.

Hemp, A. (2009). Climate change and its impact on the forests of Kilimanjaro. *African Journal of Ecology*, 47, 3–10.

Hilmi, N., Safa, A., Reynaud, S., & Allemand, D. (2012). Coral reefs and tourism in Egypt's Red Sea. *Topics in Middle Eastern and African Economies*, 14, 416–434.

Hoogendoorn, G., & Fitchett, J. M. (2018). Tourism and climate change: A review of threats and adaptation strategies for Africa. *Current Issues in Tourism*, 21(7), 742–759.

Hoogendoorn, G., Stockigt, L., Saarinen, J., & Fitchett, J. M. (2021). Adapting to climate change: The case of snow-based tourism in Afriski, Lesotho. *African Geographical Review*, 40(1), 92–104.

IPCC (2018). *The Special Report on Global Warming of 1.5°C* (SR15). Intergovernmental Panel on Climate Change. https://www.ipcc.ch/sr15/, accessed 20 January 2022.

IPCC (2019). *Special Report on the Ocean and Cryosphere in a Changing Climate*. Intergovernmental Panel on Climate Change. https://www.ipcc.ch/srocc/, accessed 20 January 2022.

Karani, P., & Failler, P. (2020). Comparative coastal and marine tourism, climate change, and the blue economy in African large marine ecosystems. *Environmental Development*, 36, 100572.

Kebede, A. S., Nicholls, R. J., Hanson, S., & Mokrech, M. (2012). Impacts of climate change and sea-level rise: A preliminary case study of Mombasa, Kenya. *Journal of Coastal Research*, 28(1A), 8–19.

Kilungu, H., Leemans, R., Munishi, P. K., Nicholls, S., & Amelung, B. (2019). Forty years of climate and land-cover change and its effects on tourism resources in Kilimanjaro National Park. *Tourism Planning & Development*, 16(2), 235–253.

Knutson, T. R., McBride, J. L., Chan, J., Emanuel, K., Holland, G., Landsea, C., & Sugi, M. (2010). Tropical cyclones and climate change. *Nature Geoscience*, 3(3), 157–163.

Lin, Y., Zhao, M., & Zhang, M. (2015). Tropical cyclone rainfall area controlled by relative sea surface temperature. *Nature Communications*, 6(1), 1–7.

Mather, A. A., & Stretch, D. D. (2012). A perspective on sea level rise and coastal storm surge from Southern and Eastern Africa: A case study near Durban, South Africa. *Water*, 4(1), 237–259.

Mgadle, A., Dube, K., & Lekaota, L. (2022). Conservation and sustainability of coastal city tourism in the advent of sea level rise in Durban, South Africa. *Tourism in Marine Environments*, 17(3), 179–196.

Mieczkowski, Z. (1985). The tourism climatic index: A method of evaluating world climates for tourism. *Canadian Geographer/Le Géographe Canadien*, 29(3), 220–233.

Miller, R. G., Tangney, R., Enright, N. J., Fontaine, J. B., Merritt, D. J., Ooi, M. K., & Miller, B. P. (2019). Mechanisms of fire seasonality effects on plant populations. *Trends in Ecology & Evolution*, 34(12), 1104–1117.

Moses, O., & Hambira, W. L. (2018). Effects of climate change on evapotranspiration over the Okavango Delta water resources. *Physics and Chemistry of the Earth*, 105(Parts A/B/C), 98–103.

Murakami, H., Wang, B., Li, T., & Kitoh, A. (2013). Projected increase in tropical cyclones near Hawaii. *Nature Climate Change*, 3(8), 749–754.

Mushawemhuka, W. J., Fitchett, J. M., & Hoogendoorn, G. (2020). Towards quantifying climate suitability for Zimbabwean nature-based tourism. *South African Geographical Journal*, 103(4), 443–463.

NASA (2012). *Kilimanjaro's Shrinking Ice Fields*. https://earthobservatory.nasa.gov/images/79641/kilimanjaros-shrinking-ice-fields.

Ngazy, N., Jiddawi, N., & Cesar, H. (2005). Coral bleaching and the demand for coral reefs: A marine recreation case in Zanzibar. In M. Ahmed, C. Chong, & H. Cesar (eds), *Economic Valuation and Policy Priorities for Sustainable Management of Coral Reefs* (pp. 118–125). Penang: World Fish Center.

Nhamo, G., & Dube, K. (eds) (2021). *Cyclones in Southern Africa, Volume 2: Foundational and Fundamental Topics*. Cham: Springer.

Nhamo, G., Dube, K., & Chikodzi, D. (eds) (2020a). *Counting the Cost of COVID-19 on the Global Tourism Industry*. Cham: Springer.

Nhamo, G., Dube, K., & Chikodzi, D. (2020b). Impact of COVID-19 on the global network of airports. In G. Nhamo, K. Dube, & D. Chikodzi (eds), *Counting the Cost of COVID-19 on the Global Tourism Industry* (pp. 109–133). Cham: Springer.

Nhamo, G., Dube, K., & Saurombe, T. (2021). Impact of Tropical Cyclone Idai on tourism attractions and related infrastructure in Chimanimani, Zimbabwe. In G. Nhamo & D. Chikodzi (eds), *Cyclones in Southern Africa, Volume 3: Implications for the Sustainable Development Goals* (pp. 245–264). Cham: Springer.

NOAA (2021). *Sea Level Trends*. https://tidesandcurrents.noaa.gov/sltrends/sltrends.html.

Noome, K., & Fitchett, J. M. (2019). An assessment of the climatic suitability of Afriski Mountain Resort for outdoor tourism using the Tourism Climate Index (TCI). *Journal of Mountain Science*, 16(11), 2453–2469.

Nyongesa, K. W., & Vacik, H. (2018). Fire management in Mount Kenya: A case study of Gathiuru Forest Station. *Forests*, 9(8), 481.

Ouma, Y. O., & Tateishi, R. (2005). Optical satellite-sensor based monitoring of glacial coverage fluctuations on Mount Kenya, 1987–2000. *International Journal of Environmental Studies*, 62(6), 663–675.

Republic of Mauritius Ministry of Tourism (2021). *Adapting to the Changing Global Environment Strategic Plan 2018–2021*. Republic of Mauritius Ministry of Tourism. https://tourism.govmu.org/Documents/publication/STRATEGIC_Plan_2018-2021.pdf.

Sagoe-Addy, K., & Addo, K. A. (2013). Effect of predicted sea level rise on tourism facilities along Ghana's Accra coast. *Journal of Coastal Conservation*, 17(1), 155–166.

100 *Handbook on tourism and conservation*

Scott, D., Gössling, S., & Hall, C. M. (2012). International tourism and climate change. *Wiley Interdisciplinary Reviews: Climate Change*, 3(3), 213–232.

Scott, D., Hall, C. M., & Gössling, S. (2019). Global tourism vulnerability to climate change. *Annals of Tourism Research*, 77, 49–61.

Sibitane, Z., Dube, K., & Lekaota, L. (2022). Global warming and its implications on nature tourism at Phinda private game reserve, South Africa. *International Journal of Environmental Research and Public Health*, 19(9), 5487.

Taecharungroj, V., & Mathayomchan, B. (2019). Analysing TripAdvisor reviews of tourist attractions in Phuket, Thailand. *Tourism Management*, 75, 550–568.

Twumasi-Ampofo, K., Oppong, R. A., & Quagraine, V. K. (2020). The state of architectural heritage preservation in Ghana: A review. *Cogent Arts & Humanities*, 7(1), 1812183.

Van Wilgen, N. J., Goodall, V., Holness, S., Chown, S. L., & McGeoch, M. A. (2016). Rising temperatures and changing rainfall patterns in South Africa's national parks. *International Journal of Climatology*, 36(2), 706–721.

Wang, G., Cai, W., Gan, B., Wu, L., Santoso, A., Lin, X., & McPhaden, M. J. (2017). Continued increase of extreme El Niño frequency long after 1.5 C warming stabilisation. *Nature Climate Change*, 7(8), 568–572.

World Meteorological Organization (2021). 2020 was one of three warmest years on record. https:// public.wmo.int/en/media/press-release/2020-was-one-of-three-warmest-years-record, accessed 20 January 2022.

World Wildlife Fund for Nature (2010). Ten interesting facts about Mt. Kilimanjaro. https://www .worldwildlife.org/blogs/good-nature-travel/posts/ten-interesting-facts-about-mt-kilimanjaro, accessed 20 January 2022.

8. Tourism and climate change: consequences, adaptation and mitigation

Esraa A. El-Masry

INTRODUCTION

Managing tourism to achieve sustainable development in Africa is an issue of significant concern. While Africa has a tremendous and untapped tourism potential, it is not the world's largest tourist receiving region although tourism impacts the natural, cultural and social-economic assets in tourist destination areas (UNWTO, 2018). Tourism has grown continuously in Africa over the past few decades. In 2017, the tourism industry's contribution to Africa's economy reached 8 percent of the continent's Gross Domestic Product (GDP) valued at USD 177 billion which is equivalent to 9.7 percent of the continent's total exports (UNWTO, 2018). During the same year, tourism covered about 5.7 percent of the total investments valued at 48.2 billion and provided over 22 million jobs (6.5 percent of total employment) (Menefov, 2020).

Tourism is recognized as a highly climate-sensitive industry in which the nexus between climate change and tourism is highly complex (McKercher & Prideaux, 2020). Africa encompasses many climate regimes with its geographical area and location as key factors. The continent straddles the equator, and it is the only one that stretches from the northern to southern temperate zones (Menefov, 2020). Accordingly, climate damages (as a percentage of GDP) are higher in Africa than any other region in the world (African Development Bank, 2011). Climate change is a significant risk for the tourism industry, and that risk is greatest in countries where tourism growth is projected to be strongest. Scott and Gössling (2018) concluded that Western Europe, Central Asia and Canada fall into the low-risk category, while countries in the Mediterranean are at higher risk. While Africa, the Middle East, South Asia and small island states face the greatest threats, particularly those countries where tourism is a substantial part of the economy, all countries would be affected by reduced GDP growth worldwide.

Over the past decade, significant progress has been made in the field of climate change and its possible consequences for the natural and socioeconomic components of the tourism sector. The United Nations World Tourism Organization (UNWTO & UNEP, 2008) recognized that climate change is the greatest challenge to the sustainability of tourism in the twenty-first century (Scott, 2008). In 2018, the World Travel and Tourism Council (WTTC, 2018) joined the United Nations Framework Convention on Climate Change (UNFCCC) Climate Neutral Now initiative, committing to becoming climate neutral by 2050, and collaborating on accelerating sector-wide climate action (Scott & Gössling, 2018).

Since then, a variety of actions have been undertaken by tourism stakeholders worldwide but there is still limited public information on CO_2 emissions by tourism businesses and destinations and the integration of climate strategies in tourism policies is low. Therefore, furthering the engagement of the tourism sector in the adoption, implementation and monitoring of adaptation and mitigation measures and strategies has become essential to support

102   *Handbook on tourism and conservation*

addressing global warming and ensure the long-term sustainability of the sector (UNWTO & ITF, 2019). Accordingly, this chapter first provides a brief overview of recent advances in our understanding of global climate change and an overview of the state of climate change science, as outlined in the Intergovernmental Panel on Climate Change's Sixth Assessment Report (IPCC AR6). The chapter also situates the recent concern about the consequences of climate change for tourism and the role tourism plays in global environmental change; both as a contributor to and recipient of impacts. In addition the chapter uses the Tourism Climate Index (TCI) to assess the present and projected tourism suitability of some African destinations. Subsequently, it proposes and discusses the potential adaptation and mitigation measures to achieve sustainability for the tourism sector.

METHODS

This chapter studies the impact of climate change on the tourism sector in Africa. The theoretical part comprises two main tracks: the first presents a review of (IPCC AR6) scenarios for the tourism-related climatic variables, and the second discusses the most recent papers to address the current state of tourism, highlighting the key themes and issues that confront the industry across the continent.

The quantitative assessment of the present and the projected climatic conditions in the study area is based on the evaluation of the impact of temperature and precipitation – (T) & (P) – increases employing the TCI. The TCI was developed by Mieczkowski (1985) using the temporal analysis of climatic data as a method to quantitatively evaluate the climate suitability of a location for general tourism activities (for the detailed method, see El-Masry et al., 2022). For the TCI, the present work selects four countries – Libya, Djibouti, Cameroon and South Africa – which represent North, East, West and South Africa, respectively. Climatic data were obtained from NASA Langley Research Center (https://power.larc.nasa.gov) for the year 2020. The IPCC AR6 SSP5-8.5 climate scenario was applied to calculate projected TCI scores for the period 2030–60. Based on the adopted scenario, the projection will be assessed in relation to driver variables such as temperature and precipitation which will increase by 2.3°C and 9.5 percent, respectively.

RESULTS AND DISCUSSION

Climate Change

The IPCC AR6 defines climate change as "A change in the state of the climate that can be identified (e.g. by using statistical tests) by changes in the mean and/or the variability of its properties and that persists for an extended period, typically decades or longer."

The report titled *Climate Change 2021: The Physical Science Basis* (IPCC, 2021) gives an updated account of the scientific consensus on climate change. It states that a median global warming of 1.5°C could already be reached by 2030 – with big consequences in terms of extreme weather events. Research in the sciences has produced an almost overwhelming amount of data that has been used to build models that show how the drivers of climate change will affect global temperatures, precipitation, wind and sea-level rise (SLR). These models can be used to investigate change at a system level, with the most widely used being those that appear in the Intergovernmental Panel on Climate Change (IPPC) reports. These models will become increasingly important in coming decades as the tourism industry begins to grapple with the impact of climate change at enterprise, destination and global levels (McKercher & Prideaux, 2020).

Weather extremes like heat, heavy precipitation and drought have been increasing with scientific certainty in many regions in the world (Figure 8.1). Europe and East Asia emerge as the regions where the clearest climate change patterns have been proven by scientists. Western and Central Europe, Eastern Asia and Eastern Central Asia as well as Western Central Asia are among the places where all three weather extremes have been on the rise. Only one region experienced a decrease in extreme weather events. Tropical Northern Australia was found to be experiencing less drought due to climate change, while also seeing more hot extremes and more heavy precipitation.

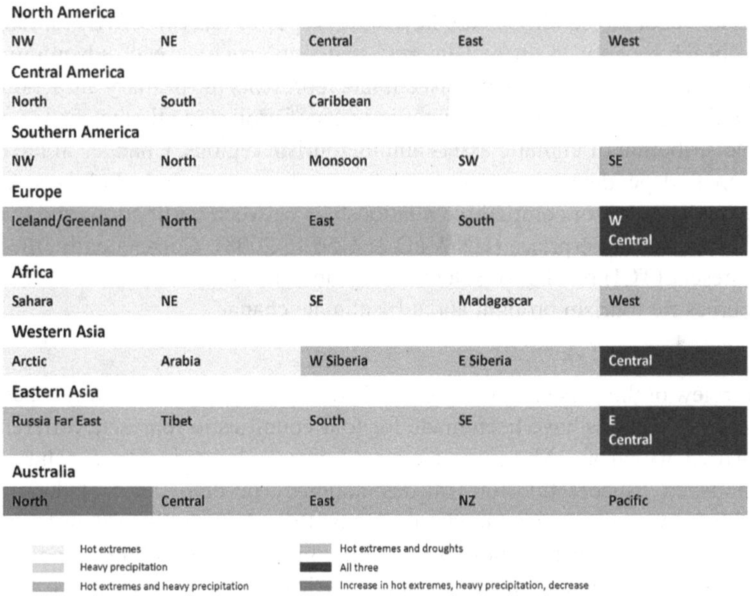

North America

NW	NE	Central	East	West

Central America

North	South	Caribbean		

Southern America

NW	North	Monsoon	SW	SE

Europe

Iceland/Greenland	North	East	South	W Central

Africa

Sahara	NE	SE	Madagascar	West

Western Asia

Arctic	Arabia	W Siberia	E Siberia	Central

Eastern Asia

Russia Far East	Tibet	South	SE	E Central

Australia

North	Central	East	NZ	Pacific

Hot extremes	Hot extremes and droughts
Heavy precipitation	All three
Hot extremes and heavy precipitation	Increase in hot extremes, heavy precipitation, decrease

Source: IPCC AR6 (2021).

Figure 8.1   Effects of climate change in world regions

Furthermore, all other regions in Europe and East Asia have been experiencing a rise in two weather extremes – which mostly means more heat and rain, but in the case of Mediterranean Europe (as well as Mediterranean Africa) translates into extreme heat and drought. Additional regions prone to heat and drought were Western North America, Northeast South America, Southern Australia as well as West and Central Africa. Both Africa and South America had limited data on the increase in heavy rains, which means that changes to precipitation patterns may simply be too little explored.

Climate Change Consequences and the Tourism Sector

Climate and climate-related variables have a profound impact on the selection of tourist destinations, and in determining the season, the time of travel and the length of stay (Hamilton and Tol, 2007; Scott and Lemieux, 2010). Therefore, this section examines the climate change variables and their direct and indirect consequences for the tourism sector while recognizing that these influences will vary substantially by market segment and geographic region. Figure 8.2 outlines climate change variables and both their direct (biophysical resources-related) and indirect (socioeconomic status-related) consequences that have a significant impact on the global tourism industry (UNWTO & UNEP, 2008; Grimm et al., 2018; IPCC, 2019; Brecha, 2020).

Direct climate change consequences for tourism activity

Climate itself is a principal resource for tourism, as it co-determines the suitability of locations for a wide range of tourist activities and is a principal driver of the seasonality of demand. In general, adequate climatic conditions are key for all types of tourism activities, ranging from conventional beach tourism to special interest segments, such as eco, adventure and sport tourism. Furthermore, at some destinations, climate represents the primary attraction on which tourism is predicated. One of the most direct impacts of projected climate change on tourism will be the redistribution of climatic assets among tourism regions. Changes in the length and quality of climate-dependent tourism seasons (i.e., sun-and-sea or ski holidays) could have considerable implications for competitive relationships between destinations and therefore the profitability of tourism enterprises (UNWTO & UNEP, 2008). Consequently, the following section will present (TCI) results for selected African countries to explore the changes in the climatic resources for tourism brought about by climate change.

Tourism Climate Index for Africa

To get an overview of the present situation, the TCI values are calculated for the baseline year of 2021. These calculations have been made for four countries to represent different climatic regions of Africa which are Morocco, Kenya, Ghana and South Africa. These countries are well-established, "successful" tourism destinations. The dominance of the countries of North Africa, e.g. Morocco, is explained not only by the sub-region's proximity to the major European generating markets but, more importantly, by its long-standing economic, political and other ties with these areas. Southern (South Africa) and eastern (Kenya) Africa are, in tourism terms, significant, because they had a vigorous expatriate community which sought to advance foreign commercial interests, including tourism (Dieke, 2020). Going forward, tourism in West (Ghana) and Central Africa is projected to grow at approximately 5 percent per annum through 2030, nearly twice the global average and among the best forecasts of any world region (UNWTO, 2014).

Results of the present study TCI results are presented in Figure 8.3. Africa is ranked, at present, according to the results and analysis of the TCI, as acceptable to excellent climatically suitable for tourism.

For the selected destinations, bimodal distribution is common which is characterized by a situation in which scores in both winter and autumn are higher than those in summer. The TCI revealed that the area ranges from 21 (very unfavorable) to 62 (very good) for the present and projected state in terms of tourism climatic suitability; however, in the future it will be less

Tourism and climate change: consequences, adaptation and mitigation 105

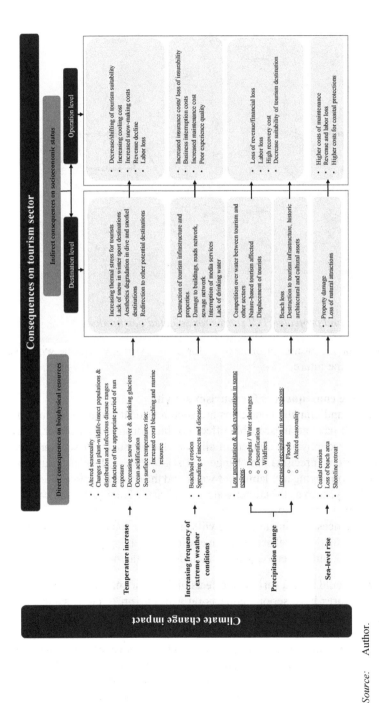

Source: Author.

Figure 8.2 Framework of climate change impacts and consequences for the tourism industry

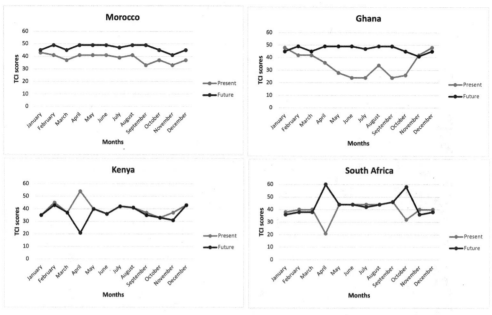

Source: Author.

Figure 8.3   Present and projected TCI scores for selected destinations

suitable for each destination for all seasons except for South Africa, where the winter season will be more suitable in the future.

Indirect climate change consequences for tourism activity

Because environmental and climatic conditions are such a critical resource for tourism, any subsequent changes will have an inescapable effect on the industry. As a sector, tourism is highly exposed to the direct physical impacts of climate change, such as rising temperatures, extreme weather conditions, precipitation change and sea-level rise. It is also threatened by indirect impacts, such as changing availability of water and the spread of some diseases. Some positive impacts are likely, such as the attractiveness to tourists of new geographical areas and opportunities for so-called "last chance" tourism.

The indirect consequences of climate change will impact the socioeconomic status of the tourism sector at the destination level and at the operational level.

At the destination level, indirect consequences include the following:

- *Rising temperature* could have a variety of effects on the tourism sector (Brecha, 2020):
  - Higher temperatures have serious impacts on eco-tourism, such as safari operators, with nature reserves increasingly isolated geographically. In sub-Saharan Africa, up to 40 percent of species in national parks are likely to become endangered by 2080.
  - Variable snowfall, retreating glaciers and milder winters have reduced visitor numbers in winter sports areas. Warming would reduce the number of resorts that are "snow reliable", especially those at low altitude, as well as shortening the skiing season.
  - Changes in the length and quality of climate-dependent tourism seasons (i.e.,

Tourism and climate change: consequences, adaptation and mitigation 107

sun-and-sea or winter sports holidays) could have considerable implications for competitive relationships between destinations (UNWTO & UNEP, 2008).

- Higher temperatures could lead to more, and more intense, forest fires in parts of the world.
- Climate change could lead to a reduction in the redistribution of wealth from rich to poor countries that tourism currently provides. The flow of tourists from cold, rich countries to warm, poorer ones could slow, as more tourists holiday nearer to home.
- Warming will increase the thermal stress for tourists and cause aesthetic degradation.
- *Increasing frequency of extreme weather conditions* will cause destruction of tourism infrastructure and properties, damage to buildings, road networks and sewage networks and interruption of media services (UNWTO & UNEP, 2008; Grimm et al., 2018). Severe storms, with associated winds, waves, rain and storm surges, can disrupt the transport, power and water supplies that the industry relies on (Brecha, 2020).
- *Precipitation change (increasing or decreasing)* will lead to beach loss and destruction of tourism infrastructure, historic architectural and cultural assets, in addition to property damage and loss of natural attractions.
- *Rising sea levels* will have profound and multiple impacts on coastal tourism (Brecha, 2020). Rising sea levels will damage tourism infrastructures and devalue coastal tourism destinations (El-Masry et al., 2022).

At the operational level, indirect consequences include the following:

- *Rising temperature* has an important influence on operating costs, such as cooling, snow-making, irrigation, food and water supply, and insurance costs. In addition a decrease in tourism suitability leads to labor loss and revenue decline (UNWTO & UNEP, 2008; El-Masry et al., 2022).
- *Increasing frequency of extreme weather conditions* is likely to become more common. The effects include disruption of travel, insurance is likely to become more expensive or even unavailable, and maintenance costs will increase which will affect businesses and result in poor customer experience in terms of quality (UNWTO & UNEP, 2008; Brecha, 2020).
- *Precipitation change (increasing or decreasing)* will affect socioeconomic status through increasing maintenance and recovery costs, business cost interruption, revenue decline, labor loss and decreasing security for tourists.
- *Rising sea levels* could have serious implications for tourism operators, especially in coastal areas due to the high cost of protection and maintenance, revenue deficit and labor loss.

It is worth noting that the tourism sector could also face impacts caused by climate change mitigation and adaptation responses in other sectors (Scott et al., 2012). For instance, tourist mobility and behavior are likely to be impacted by national or international mitigation policies in the transport sector, which aim to reduce greenhouse gas emissions through an increase in air-travel cost (Urbance et al., 2002). Long-haul destinations may be particularly affected and government officials with highly tourism dependent economies have expressed concern that such policies could negatively impact their tourism industry (Scott et al., 2012). Adaptation policies related to water rights (i.e., continued use by tourism) or insurance costs (i.e., for

108   *Handbook on tourism and conservation*

coastal resorts), could also impact upon tourism development and operating costs (UNWTO & UNEP, 2008).

The contribution (footprint) of tourism to climate change
The carbon contribution of tourism should include the carbon emitted directly during tourism activities (for example, combustion of petrol in vehicles) as well as the carbon embodied in the commodities purchased by tourists (for example, food, accommodation, transport, fuel and shopping). Carbon dioxide emissions from tourism, including transport, shopping, food, accommodation and other activities, account for about 8 percent of global CO_2 emissions, thus totaling at 4.5 Gt CO_2e the carbon footprint of global tourism (Lenzen et al., 2018).

Transport
Transport connectivity is an important prerequisite for tourism. Among all tourism sectors, according to the UNWTO & ITF (2019), transport is considered to be the most polluting, accounting for 75 percent of total emissions by tourism, with emphasis on air travel as the biggest contributor. The benefits of better transport links often spill over to local communities, making goods, services and jobs more accessible (UNWTO & ITF, 2019). Planes, cars, trains, boats and even hot air balloons allow us to explore destinations all around the world. However, all of our jet-setting and road-tripping comes with a hefty carbon footprint (Lenzen et al., 2018). In recent years, the number of people traveling internationally skyrocketed as airfares became more affordable. Similarly, between 2005 and 2016, transport-related tourism emissions increased by more than 60 percent (UNWTO & ITF, 2019).

Lodging
Accommodation accounts for a substantial part of tourism's greenhouse gas emissions: UNWTO & UNEP (2008) concluded that the sector is responsible for 21 percent of CO_2 emissions, if comparing transportation, accommodation and activities. It is estimated that each guest night generates emissions of 14 kg CO_2 on global average (direct emissions related to energy-throughput, excluding gastronomy) (Gössling & Peeters, 2015). Calculation by Lenzen et al. (2018) suggests that worldwide, accommodation currently causes annual greenhouse gas emissions of about 282 Mt CO_2 (input–output analysis, without gastronomy).

Many accommodations rely on heating and air conditioning to keep guest rooms at a pleasant temperature in hot or cold weather. These energy-intensive systems create CO_2, as do the water heaters used to warm showers, pools and spas. Electricity used to power lights, TVs, refrigerators, laundry machines and other equipment is also a big contributor, especially in areas with dated or inefficient systems. Emissions from lodgings tend to be highest in resorts and hotels that offer modern services, while smaller lodgings such as homestays and guest houses have lower emissions for the most part.

Construction
Resorts, airports and other tourism facilities can produce massive amounts of carbon even before they open their doors to tourists. Constructing a new building is an energy-intensive process – manufacturing the materials, transporting everything to the site and constructing the building all generate carbon emissions. And it's not just buildings that leave behind a footprint – the development of roads and other infrastructure for tourism also contributes to climate change.

Destruction of carbon sinks

Along with the construction process, tourism development emits carbon through the clearing of natural areas. Ecosystems, such as forests, act as carbon sinks by absorbing and storing emissions. When this carbon-rich vegetation is removed, CO_2 is released back into the atmosphere.

The mangrove forests that grow along coastlines in many tropical destinations have a tremendous capacity to store carbon. Studies show that they can store up to four times more carbon than most other tropical forests around the world. Sadly, vast areas of mangroves are often cleared to make way for tourist infrastructure such as seaside resorts, beaches, marinas and entertainment areas.

Food and drink

Food production is responsible for roughly 25 percent of the world's greenhouse gas emissions. Getting food from farm to table means growing, processing, transporting, packaging, refrigerating and cooking – all of which require energy and contribute to your meal's carbon footprint. Travel often multiplies this footprint since people tend to indulge more while on vacation.

Food waste

With hotel oversized restaurant portions, a substantial amount of the food produced for tourism ends up being thrown away. When food is wasted, all of the emissions that were generated by its production were unnecessary. Globally, less than half of hotels compost their food waste. When this food decomposes in landfills it creates methane which is 21 times more potent than carbon dioxide.

Throughout the food chain, food is lost, thrown away or wasted for technological, economic and societal reasons. Food waste (unconsumed food) emits 3.3 gigatons of carbon dioxide equivalent per year. Sub-Saharan Africa's food waste averages 210 kg of carbon dioxide per person per year, which is equivalent to 87 percent of worldwide road transport emissions (Balineau et al., 2021). Therefore, developing and improving food storage and processing practices are crucial for limiting the organic and inorganic waste produced by the increased demand for food. This requires policies to reduce overall food waste and to valorize organic food waste (Galanakis, 2018).

Shopping

Shopping is now a travel experience in itself, from street markets to high-end boutiques. Whether it is jewelry or electronics, the carbon footprint of an item must be calculated with production, manufacturing and shipping in mind. There is something special about purchasing an item that was made locally in the destination, yet oftentimes souvenirs and other products are mass-produced in factories far away. Travelers' buying habits are often different from locals', thus increasing production emissions.

ADAPTATION AND MITIGATION OF CLIMATE CHANGE IMPACTS

In 2015, the international community adopted the Paris Agreement with the objective to limit global temperature increase in this century to well below 2°C compared to preindustrial levels, and given the serious risks, to strive for 1.5°C. The Paris Agreement marked a historic

110   *Handbook on tourism and conservation*

turning point for global climate action connected to the urgent need to decouple economic growth from resource use and emissions in order to counteract the impacts of climate change. Moreover, climate action is included in the 2030 Agenda for Sustainable Development as a stand-alone Sustainable Development Goal (SDG), SDG 13, which provides a road map to reduce emissions and build climate resilience (UNWTO & ITF, 2019).

In efforts to counter climate change, mitigation and adaptation are proposed. Mitigation is an important policy response to climate change (however, it may not address the immediate risks posed by climate change), but adaptation is urgent (Njoroge, 2017). Adaptation and mitigation can be complementary, substitutable or independent of each other. If complementary, adaptation reduces the costs of climate change impacts and thus reduces the needs for mitigation (Furini & Bosello, 2021). Adaptation and mitigation are substitutable up to a point, but mitigation will always be required to avoid irreversible changes to the climate system and adaptation will still be necessary due to the irreversible climate change resulting from current and historic rises in greenhouse gases and the inertia in the climate system. In practice some tourism stakeholders are already engaging with either planned adaptation or autonomous adaptation. However, concerns have been raised on the sustainability of adaptation options and approaches. Given that sustainable adaptation has received little interest in tourism research and the fact that tourism is considered behind in many aspects of climate change, this chapter aims to fast forward this important theme in tourism climate research (Njoroge, 2017).

Adaptation Measures

Regardless of relative vulnerability to climate change, tourism and travel businesses and destinations will need to adapt to climate change in order to minimize associated risks and capitalize upon new opportunities, in an economically, socially and environmentally sustainable manner (OECD and UNEP, 2011). Table 8.1 summarizes the adaptation measures and actions related to the climate change impacts to achieve sustainability of the tourism industry.

Mitigation Policies and Measures

Climate change mitigation relates to technological, economic and sociocultural changes that can lead to reductions in greenhouse gas emissions. According to UNWTO & UNEP (2008), tourism-induced GHG emissions could be mitigated by reducing energy use, improving energy efficiency, increasing the use of renewable energy and sequestering carbon through sinks. In order to be effective, tourism climate mitigation policy must be integrated on different levels. Globally, there is a need for all countries to join the UNFCCC and to agree on binding emission reduction targets, which is a precondition for emission trading schemes to work efficiently. Nationally, more ambitious overall climate mitigation objectives need to be defined, and policies implemented to achieve them, notably including the monitoring of progress. In this context, regions and communities should be given the political freedom to go above and beyond national climate policy measures to develop role models for low carbon societies. Finally, organizations and businesses need to support these processes through developing their own climate targets. To move towards low carbon tourism, a five-fold climate policy would seem favorable (Table 8.2) (OECD and UNEP, 2011).

Tourism and climate change: consequences, adaptation and mitigation 111

Table 8.1 *Adaptation measures and actions related to climate change impacts to achieve sustainable tourism*

Adaptation measures	Adaptation actions/techniques
Temperature increase	
Promoting alternative economic sectors/tourism destinations	● Provide new employment opportunities ● Introduce new types of tourism such as event tourism
Sea-level rise	
Development of innovative pattern of coastal and infrastructure protection means based on future scenarios of SLR	● Zoning 　● Stipulate the legal framework that governs the use and development of the vulnerable areas 　● Promote zoning maps and tourism-based maps, based upon the types of uses that are permitted (e.g. residential and coastal resorts) 　● Aggregate each zone according to its design requirements (e.g. setbacks, building heights, building densities) ● Building codes and resilient design 　● Building constructions must be designated to maximize protection from the impact of SLR (e.g. elevations, building techniques and building materials) ● Setbacks/buffers ● Soft and/or hard armoring permits
Increasing frequency of extreme weather conditions	
Promoting alternative economic sectors/tourism destinations	● Promoting a disaster reduction plan ● Hard and soft protection structures ● Slope contouring ● Cyclone-proof building design and structures ● Weather forecasting and early warning systems ● Require advanced building design or material (fire resistant) standards for insurance ● Enable access to early warning equipment to tourism operators ● Redirect clients away from impacted destinations ● Hurricane interruption guarantees ● Comply with regulation (e.g. building code)
Water shortages/droughts/desertification	
	● Rainwater harvesting and water recycling systems ● Efficient use of water and energy ● Reservoirs and desalination plants ● Free structures for water consumption ● Water conservation plans
Melting glaciers/variability in snowfall	
	● Artificial snowmaking ● Slope contouring and landscaping
Coral reef bleaching	
Development of the Coral Bleaching Response plan	This plan is developed in order to: ● Improve ability to predict bleaching risk ● Provide early warnings of major coral bleaching events, measure the extent of bleaching ● Assess the ecological impacts of bleaching ● Involve the community in monitoring the health of the reef, communicate and raise awareness about bleaching ● Evaluate the implications of bleaching events for tourism management policy and strategies

Source:　Based on UNWTO & UNEP (2008); El-Masry et al. (2022).

112 *Handbook on tourism and conservation*

Table 8.2   *Mitigation measures and policies related to climate change impacts to achieve sustainable tourism*

Component	Mitigation measures/policies
Knowledge and awareness	• Development of entrepreneurial policies for efficient use of energy (e.g. lighting, air conditioning, heating, etc.)
	• Incorporate an assessment of GHG emissions and corrective measures on the applications for new licenses for tourism activities
Good practices and technological improvements	• Promotion of the use of more energy efficient technologies and their monitoring in tourism accommodation
	• Promotion of good practices by medium and small businesses and certificate of best practices
	• Introduction of the "Ecolodge" concept
Carbon pricing and emission trading	• Taxes, emission trading and other economic instruments are needed to steer energy use and emissions, conveying clear, long-term market signals
Laws and regulations	• Creation of a law to promote energy savings in tourism establishments
	• Building codes
	• Actual enforcement of existing environmental regulation needs to be ensured
Incentives	• An ecological tax reform, for instance, could shift tax burdens from labor to energy and natural resources, and thus "reward" users of low-carbon technology
	• Financial reward mechanisms or awards

Source: OECD and UNEP (2011).

CONCLUSIONS AND RECOMMENDATIONS

The tourism sector in Africa is a highly climate dependent economic sector given that most tourism activities are nature based. Therefore, it will be influenced by climate change impacts in terms of temperature and precipitation increase and the quality of the coastal environment. However, the tourism sector, which represents a significant component of most African countries' GDP, is already suffering the effects of climate change. These issues intensify the need for advancing research on climate change and tourism in Africa, significantly reflecting on the balance in projecting and planning for a changing climate. The IPCC AR6 (2021) promotes the urgency of this sectoral-focused research.

The tourism climatic suitability in the study area, at present, ranges from unfavorable to good; however, according to the TCI model, there is the prospect for a decrease in suitability of tourism activities for the three selected destinations (Morocco, Kenya and Ghana). South Africa shows improvement in tourism suitability for the projected TCI values during the winter season. Nevertheless, additional research on tourist comfort is also urgently needed. There is very little known about tourists' likely behavioral reactions to the projected climatic changes, a topic that can be addressed using a qualitative or traditional survey-based approach. New studies should consider other elements such as diversity in activities and environmental quality. For other tourism segments there is a need for a better integration of suitability assessments and tourism demand models. Tourism and climate change cannot be considered as separate issues and they are closely connected to other factors in the socioeconomic system. Therefore, a comprehensive, integrated and communicative approach is necessary to adapt to and mitigate the impacts of climate change.

REFERENCES

African Development Bank. 2011. *The Cost of Adaptation to Climate Change in Africa*. www.afdb .org/fileadmin/uploads/afdb/Documents/Project-and-Operations/Cost of Adaptation in Africa.pdf, accessed October 2011.

Balineau, G., Bauer, A., Kessler, M., & Madariaga, N. 2021. *Food Systems in Africa: Rethinking the Role of Markets*. Washington, DC: World Bank.

Brecha, R. 2020. The physical science of climate change. In V. J. Miller (ed.), *The Theological and Ecological Vision of Laudato Si': Everything is Connected*. London: Bloomsbury, pp. 29–50.

Dieke, P. U. C. 2020. Tourism in Africa: Issues and prospects. In T. Baum & A. Ndiuini (eds), *Sustainable Human Resource Management in Tourism: African Perspectives*. Cham: Springer, pp. 9–28.

El-Masry, E. A., El-Sayed, M. K., Awad, M. A., El-Sammak, A. A., & El Sabarouti, M. A. 2022. Vulnerability of tourism to climate change on the Mediterranean coastal area of El Hammam–El Alamein, Egypt. *Environment, Development and Sustainability*, 24(1), 1145–1165.

Furini, F. & Bosello, F. 2021. Accounting for adaptation and its effectiveness in international environmental agreements. *Environmental Economics and Policy Studies*, 23(2), 467–493.

Galanakis, C. M. 2018. Food waste recovery: Prospects and opportunities. In C. M. Galanakis (ed.), *Sustainable Food Systems from Agriculture to Industry*. London: Academic Press, pp. 401–419.

Gössling, S. & Peeters, P. 2015. Assessing tourism's global environmental impact 1900–2050. *Journal of Sustainable Tourism*, 23(5), 639–659.

Grimm, I. J., Alcântara, L. C. S., & Sampaio, C. A. C. 2018. O turismo no cenário das mudanças climáticas: impactos, possibilidades e desafios. *Revista Brasileira de Pesquisa em Turismo*, 12(3), 1–22.

Hamilton, J. M. & Tol, R. S. J. 2007. The impact of climate change on tourism and recreation. In M. E. Schlesinger et al. (eds), *Human-Induced Climate Change: An Interdisciplinary Assessment*. Cambridge: Cambridge University Press, pp. 147–155.

IPCC. 2019. Chapter 4: Sea level rise and implications for low lying islands, coasts and communities. *IPCC SR Ocean and Cryosphere. IPCC Special Report on the Ocean and Cryosphere in a Changing Climate* (ed. H.-O. Pörtner, D. C. Roberts, V. Masson-Delmotte, P. Zhai, M. Tignor, E. Poloczanska, K. Mintenbeck, M. Nicolai, A. Okem, J. Petzold, B. Rama, & N. Weyer). Geneva: Intergovernmental Panel on Climate Change.

IPCC. 2021. *Climate Change 2021: The Physical Science Basis* [IPCC AR6]. Contribution of Working Group I to the Sixth Assessment Report of the Intergovernmental Panel on Climate Change (ed. V. Masson-Delmotte, P. Zhai, A. Pirani, S. L. Connors, C. Péan, S. Berger, N. Caud, Y. Chen, L. Goldfarb, M. I. Gomis, M. Huang, K. Leitzell, E. Lonnoy, J. B. R. Matthews, T. K. Maycock, T. Waterfield, O. Yelekçi, R. Yu, & B. Zhou). Cambridge: Cambridge University Press.

Lenzen, M., Sun, Y. Y., Faturay, F., Ting, Y. P., Geschke, A., & Malik, A. 2018. The carbon footprint of global tourism. *Nature Climate Change*, 8(6), 522–528.

McKercher, B. & Prideaux, B. 2020. Climate change, resilience and transition to a carbon-neutral economy. In *Tourism Theories, Concepts and Models*. Oxford: Goodfellow Publishers, pp. 275–292.

Menefov, R. 2020. Perspectives of tourism development in Azerbaijan. *Scientific Bulletin*, 1(1), 108–114.

Mieczkowski, Z. 1985. The tourism climatic index: A method of evaluating world climates for tourism. *Canadian Geographer/Le Géographe Canadien*, 29(3), 220–233.

Njoroge, J. 2017. Tourism and climate change. In *Climate Change and Tourism in Sustainable Tourism Report: East Africa 7th issue*. Nairobi: Sustainable Travel & Tourism Agenda, pp. 34–39.

OECD and UNEP. 2011. *Climate Change and Tourism Policy in OECD Countries*. www.oecd.org/ industry/tourism/48681944.pdf, accessed July 2011.

Scott, D. 2008. Climate change and tourism: Responding to global challenges. CTO / CRSTDP Regional Workshop on Tourism.

Scott, D. & Gössling, S. 2018. *Tourism and Climate Change Mitigation: Embracing the Paris Agreement*. Brussels: European Travel Commission.

Scott, D. & Lemieux, C. 2010. Weather and climate information for tourism. *Procedia Environmental Sciences*, 1(1), 146–183.

Scott, D., Simpson, M. C., & Sim, R. 2012. The vulnerability of Caribbean coastal tourism to scenarios of climate change related sea level rise. *Journal of Sustainable Tourism*, 20(6), 883–898.

114 *Handbook on tourism and conservation*

UNWTO. 2014. *Yearbook of Tourism Statistics. Data 2008–2012*. Madrid, UNWTO. www.e-unwto.org/doi/pdf/10.18111/ 9789284415915, accessed February 2014.

UNWTO. 2018. *UNWTO Tourism Highlights 2018 Edition*. www.unwto.org/doi/pdf/10.18111/ 9789284419876, accessed October 2018.

UNWTO & ITF. 2019. *Transport-Related CO$_2$ Emissions of the Tourism Sector*. Madrid: UNWTO Publishing.

UN World Tourism Organization and United Nations Environment Programme [UNWTO & UNEP]. 2008. *Climate Change and Tourism: Responding to Global Challenges*. Madrid: UNWTO Publishing. https://doi.org/10.18111/9789284412341.

WTTC (2018). *Travel & Tourism Economic Impact, South East Asia*. www.wttc.org/-/media/files/reports/economic-impact-research/regions-2018/southeastasia2018.pdf, accessed September 2019.

Urbance, R. J., Field, F., Kirchain, R., Roth, R., and Clark, J. 2002. Market model simulation: The impact of increased automotive interest in magnesium. *Journal of Operations Management*, 54(8), 25–33.

9. Evaluating climate change communication for sustainable environmental conservation in the tourism sector

Sharon Tshipa and Olekae T. Thakadu

INTRODUCTION

Globally perceived in the form of droughts, cyclones, sea level rise, melting glaciers and heat-waves, among other severe threats to biodiversity and human rights, climate change is considered the fastest-growing peril to the natural and cultural heritage (Day et al., 2020; Eckstein et al., 2019; Hambira, 2018; Hambira & Mbaiwa, 2020; IPCC, 2019; Samuels & Platts, 2020). Effects that have since compelled the tourism sector to take action against climate change include the decline in reef health in Australia's Great Barrier Reef (see Goldberg et al., 2018), ocean acidification in Thailand's dive industry (see Tapsuwan & Rongrongmuang, 2015) and rainfall season shift at the Victoria Falls (Dube & Nhamo, 2019). These, among other threats registered worldwide, have attracted empirical studies, and media coverage.

Locally, this past decade has seen Botswana striving to wean the economy off its over reliance on diamonds (Jefferis et al., 2019). Tourism, especially nature based, thus seemed a sustainable alternative as Botswana is endowed with natural resources in the form of near pristine wetlands of international importance such as the Okavango Delta Ramsar and World Heritage Site, Lake Ngami, the Chobe Linyanti River systems, as well as desert landscapes and wildlife boasting the big five (Hambira, 2018; Moswete & Dube, 2013; Saarinen et al., 2020). The government's resolve to concentrate on the tourism industry is justified in that, worldwide, the sector plausibly contributes to GDP. According to the World Travel & Tourism Council (WTTC), in 2019 it contributed 10.3 per cent to global GDP and experienced a 3.5 per cent growth (WTTC, 2019). During that same year, the World Economic Forum (WEF) highlights that the share of the travel and tourism sector to Botswana's GDP was 4.7 per cent, while international tourist arrivals were 1,574,000, and the sector also generated 30,700 jobs (WEF, 2019). Apart from the 2019 catastrophic drought year and 2020, given the Covid-19 pandemic, Botswana has experienced a steady rise of tourist arrivals since 2015 (Statistics Botswana, 2016, 2018, 2019, 2020c; Gössling et al., 2020). The industry's potential exponential growth and its consequent ability to significantly contribute to Botswana's socio-economic transformation is nevertheless riddled by challenges ranging from population growth, poor waste management systems and desertification, to climate change related risks (Ahmadalipour et al., 2019; Burns & Bibbings, 2009; Chanza et al., 2020; Cil & Jones-Walters, 2011; Harrington & Otto, 2020; Anup, 2018; Lequechane et al., 2020; Moswete & Dube, 2013).

Besides increasing temperatures as stated by Hoogendoorn and Fitchett (2018, 2019), drought is the greatest hazard to the sector's long-standing salient role in poverty eradication and development (Dube et al., 2020; SADC, 2010; Scott et al., 2012). This is evidenced by the fact that between 2009 and 2019, Botswana only had one non-drought year, being the

116 *Handbook on tourism and conservation*

2013–14 period, and it had one partial drought year, which was the 2016–17 period (Statistics Botswana, 2020b). According to the Botswana Press Agency (BOPA), though the succeeding 2019–20 season was declared a non-drought year, geographical areas such as Tswapong, North-East District, Bobirwa, Boteti, Tutume and Tonota that did not receive adequate rainfall were declared drought stricken (BOPA, 2020). So roughly, this unrelenting climate change induced impact has left Botswana battling decadal long economic, social and environmental downsides (Statistics Botswana, 2020b).

To ensure that a purely nature-based tourism industry becomes attainable, industry stakeholders have intensified their environmental conservation efforts, as they are also conscious of the sector's dichotomous role – as a growing contributor to anthropogenic climate change (Anup, 2018; Moswete & Dube, 2013; Samuels & Platts, 2020). To achieve holistic results, dialogue and engagement in environment and climate change issues became paramount. Therefore, this chapter evaluates climate change communication for sustainable environmental conservation in Botswana's tourism sector. Specifically, the study: (i) identifies actors/sources of climate change news in nature-based tourism, and (ii) explores their role in influencing and/or undertaking sustainable environmental conservation activities in the sector's social-ecological system/subsystems.

This is because rapid drought impacts on Botswana's national food security have for too long made climate change an agricultural problem. Climate change communication therefore largely raises awareness and delivers adaptation and mitigation measures to crop and livestock farmers.

Scholarship on the tourism–climate change nexus may have begun in the 1960s, but research on climate change communication, especially in the context of tourism, has been limited (Kaján & Saarinen, 2013). This is because, among others, most studies concerned themselves with reviewing climate change threats and adaptation strategies (see Hambira, 2017; Hoogendoorn et al., 2021; Kaján & Saarinen, 2013), and with assessing the climate vulnerability of the world's natural and cultural heritage (see Day et al., 2020). However, this climate change and tourism link focused mostly on mitigation and adaptation issues in the Global North despite the fact that the Global South also faces dire climatic threats (Coldrey & Turpie, 2020; Hambira et al., 2020a; Ma & Kirilenko, 2020; Saarinen et al., 2012).

In the Southern African region, studies on tourism and climate change are said to have commenced in 2005 with the publication of a book chapter by Preston-Whyte and Watson entitled 'Nature Tourism and Climatic Change in Southern Africa' (Hoogendoorn & Fitchett, 2019). Since then, a number of publications have been recorded (see Dube et al., 2018; Hambira et al., 2013, 2020a; Hoogendoorn & Rogerson, 2015; Pandy & Rogerson, 2021), yet Hoogendoorn and Fitchett (2019) claim that the tourism and climate change topic remains under-reported. Hence, by evaluating climate change communication for sustainable environmental conservation in Botswana's tourism sector, this study adds to the body of knowledge.

Ma and Kirilenko (2020) suggest that the climate change dialogue in mass media has been thoroughly researched; however, this chapter particularly fills a research gap in climate change communication in the context of tourism. This gap was highlighted by empirical studies which view communication as a potential tool that can remove misconceptions that hinder people's adaptive capacity, a tool that can avail well packaged information on the subject in order to raise awareness among the affected stakeholders, and a tool that can bridge the gap between science and publics, to cite a few foreseen advantages (Bolsen & Shapiro, 2018; Cristobal-Fransi et al., 2018; Hambira, 2018; Hambira & Mbaiwa, 2020; Ma & Kirilenko,

2020; Rhomberg & Rhomberg, 2016; Tapsuwan & Rongrongmuang, 2015). This chapter, therefore, employs the framing theory bearing in mind that experts like scientists employ frames to simplify technical details and make them persuasive, and journalists use frames to craft thought-provoking and interesting articles, while end-users rely on those frames to make sense of and discuss an issue, and policymakers also apply frames to define policy options and reach decisions, as highlighted by Nisbet (2009).

Guided by the framing theory, and the Social-Ecological System Framework (SESF), the chapter identifies actors of climate change news in nature-based tourism and explores their role in influencing and/or undertaking sustainable environmental conservation activities in the sector's social-ecological system/subsystems. The Governance subsystem of the SESF constitutes actors that are often sources of news, while the Users subsystem points to individuals/ tourists who have a role to play in sustainable conservation. As for the framing theory introduced by Erving Goffman in 1974, researchers have utilised it to understand how print and other media present information (Linström & Marais, 2012). Leading media scholars apply the concept of framing to explain how the media structure their delivery of news, promoting certain interpretations of events by selecting certain facts (Linström & Marais, 2012). In this chapter, the framing theory provided for fluent document analysis of local and international news articles published during Botswana's 2019 catastrophic drought.

METHODOLOGY

Study Focus

·The study focus is Botswana, a developing country situated in southern Africa, with a landmass of about $581,730$ km^2. Thirty-seven per cent of Botswana's landmass is gazetted as protected areas for wildlife conservation, of which 17 per cent are national parks and game reserves, with the remaining 20 per cent being wildlife management areas (Maude & Reading, 2010). Less than 1 per cent (0.8 per cent) is gazetted for forest reserves in the Chobe district (Central Statistics Office, 2004). Thus, Botswana is a hub of tourism activities, mainly the northern part, primarily focusing on wildlife, wilderness and water. Botswana actively pursues innovative conservation and management approaches such as community based natural resources management (CBNRM) programmes, transfrontier conservation areas (TFCA) or transboundary natural resources management (TBNRM) and recently the Community Management of Protected Areas for Conservation (COMPACT). Tourism is the second contributor to the national economy, followed by agriculture. Botswana has a dry, semi-arid climate with dry winters from May to early August (Statistics Botswana, 2020a). The rainy season is normally between November and April, though early rains from September are not unusual (Statistics Botswana, 2020a). Rainfall in Botswana is characteristically unreliable as evidenced by the recurrence of droughts.

Methods

A document analysis of local and international news articles published during Botswana's 2019 catastrophic drought was undertaken. Publications identified were classified by genre, and their contents' human story, sources, tone, craft, jargon, relatability, credibility and uti-

118 *Handbook on tourism and conservation*

lisation of the power of visual media were assessed, to cite a few elements. These elements formed the basis of the codes created, and influenced subsequent themes realised. Following Corbin and Strauss's (2008) approach to analysing and reporting qualitative research, articles were organised, coded and interpreted using Atlas.ti, a computer program commonly used in data analysis.

Data collection and sampling
The study focused on coverage by local media outlets with a heavy online presence (Okon, 2015, 2017). Twelve media houses were purposively sampled. The criteria for sampling were based on media outlets with Facebook followership of more than 100,000. Of the 12 sampled, tourism and climate change news articles published between January and December 2019 were searched in the *Mmegi Newspaper, Weekend Post, Botswana Guardian, Sunday Standard, The Botswana Gazette, The Patriot on Sunday, The Voice Newspaper, Mmegi Monitor, The Midweek Sun, The Parrot Online, The Argus Online* and *Daily News*. The search was conducted using keywords such as 'climate change and tourism', 'global warming', 'conservation' and 'tourism and drought'. The search for news articles was conducted using the Google search engine, mostly by accessing website archives of sampled news publications. Articles based on the keywords were subsequently found in *Mmegi Online, Sunday Standard, The Botswana Gazette, The Patriot on Sunday* and *Daily News*.

The search yielded 37 articles. Articles included were those relevant to the current research topic. These addressed tourism in the context of climate change. Many articles that addressed tourism but made no mention of climate change were omitted. If they focused solely on wildlife conservation, they were also excluded. After the selection of highly relevant articles and the removal of duplicate and near duplicate articles the sample size fell to 21. The content of these largely midmarket publications was classified as 'general', except for the *Sunday Standard* which was classified as 'general with a special political focus'.

Data analysis
Qualitative data analysis was conducted using Atlas.ti. Data was analysed using methods such as content coding and thematic analysis. The data's human story, sources, tone, craft, jargon, relatability, credibility and utilisation of the power of visual media were coded, ensuring a thorough assessment of the effectiveness of climate change communication in nature-based tourism. Subsequently, burning issues such as climate sceptics and acceptors, consensus messaging, media shortcomings, scientist–media relations, polarisation of climate change and apolitical ways of communicating climate change, as highlighted by several empirical studies carried out on the African continent and the world at large, were explored.

RESULTS AND DISCUSSION

Results in Table 9.1 show that private news publications are leading the climate change discourse. The reason for this could largely be that most media outlets in the country are privately owned as opposed to countries like China where the vast majority of media is state-run and critical articles are subject to scrutiny and censorship (Tolan & Nan, 2007). On the other hand, a paper by Poberezhskaya (2015) that explored which actors and factors influence media coverage of climate change in Russia found that regardless of the ownership structure of the

Evaluating climate change communication in the tourism sector 119

Table 9.1   Local media articles

News publications	Number of articles	Percentage (%)
Sunday Standard	3	14.2
Mmegi Online	12	57.1
Botswana Daily News	3	14.2
The Botswana Gazette	2	9.5
The Patriot on Sunday	1	4.7
Total	21	100

newspapers or their dependence on advertising, there is little difference in quantity and quality of overall coverage on climate change. The omission of climate change issues from discussion in national newspapers Poberezhskaya (2015) further highlighted as becoming a greater problem than biased coverage, as the lack of commentary decidedly prevents these issues from entering the public debate. To ensure that state-owned publications significantly cover the climate change story, in their study which analysed the extent to which Tanzanian newspapers paid attention to climate change information between January 2006 and December 2015, Siyao and Sife (2018) suggested the revisitation of the scope of the government-owned newspapers which can be done by having a section specifically dedicated to the climate change issue. If considered in Botswana, the proposed solution is likely to increase *Daily News*'s coverage of tourism and climate change.

Main News Coverage Themes

In evaluating climate change communication for sustainable environmental conservation in the tourism sector, results of this study reveal that even during Botswana's 2019 cataclysmic drought year, media coverage covering the tourism sector was meagre. This chapter therefore shares Siyao and Sife's (2018) view that the volume of coverage devoted to climate change by newspapers is very low and disproportionate to the level of threat. In Botswana, the little that was done regarding tourism and climate change mostly focused on the government's controversial decision to lift the hunting ban. These results are consistent with findings by Ma and Kirilenko (2020) who found that mass media attention to climate change and tourism is trailing the overall discussion of climate change in mass media.

Regardless, the seemingly insubstantial tourism and climate change coverage analysed dealt with a wide number of themes such as biodiversity loss, wildlife migration, human migration, conservation, cultural heritage loss, economic implications, ecosystem degradation, greenhouse gases, coal, heatwaves, human and wildlife conflicts, limited rainfall, renewable energy, rising temperatures and wildfires, to cite a few. Of these, biodiversity loss was the most prevalent issue extracted from the retrieved local articles, as well as international news publications which were collected and studied separately for argument's sake. Biodiversity loss was followed by economic implications, limited rainfall, wildlife migration, conservation, human and wildlife conflicts, greenhouse gases and rising temperatures. These pressing concerns were valid given that the entire nation and its two neighbouring countries – Angola and Namibia that have for years gifted Botswana with massive volumes of water that flows along the Okavango River basin into the magnificent wetland – were experiencing a hydrological drought due to the lowest rainfall on record ever recorded (Wolski et al., 2006).

120 *Handbook on tourism and conservation*

To communicate climate change fully, journalists and other contributing actors had to underscore the causes of climate change by alluding to greenhouse gases and stressing the impacts of climate change such as rising temperatures, as well as pointing to the dire need for conservation as a solution to the planet's recurrent and heightening problem. An issue that local media outlets overlooked but which international news publications stressed has the potential to inspire sustainable conservation among nature-based tourism industry stakeholders is climate and health in the context of wildlife, as highlighted by the following article excerpts:

> AFPA10: "Due to the severe drought, elephants end up ingesting soil while grazing and get exposed to the anthrax bacteria spore," the ministry said in a statement.

> XA5: "Raphaka said wildlife go long distances in search for water and vegetation since the whole of the northern Botswana is affected by the climate change induced drought hence getting in contact with diseases and ultimately end up dying."

All zones considered, they almost represent the length and breadth of Botswana. This shows how widespread the effects of climate on the tourism sector were felt and seen during the drought year under scrutiny. International news articles sampled mainly focused on the Okavango and Chobe areas.

Media Framing of Tourism and Climate Change Issues

In covering climate change news pertaining the tourism sector, *Mmegi Online* emerged as the leading national news outlet. Findings also reveal that journalists in all five media houses studied are not particularly proactive; rather, conferences, press releases and reports inspire most coverage. Similar results were noted by Tolan and Nan (2007) who indicated that most coverage relied on recycled reports from Western science. Of the total articles analysed in this current study, results indicate that 47.6 per cent (n=10) were reactive while 42.8 per cent (n=9) were proactive, and 9.5 per cent (n=2) were articles contributed to the publications by experts and bloggers. Journalists also failed to directly highlight the connection between climate change and tourism, considering the anthropogenic activities of the sector. But their efforts in raising climate change awareness, and in relaying climate change impacts such as drought, heatwaves, rising temperatures, cyclones and wildfires were plausible. More so that 52.3 per cent (n=11) of them could be classified as Solution Journalism pieces, as they not only exhibited the threats presented by the acute drought, but also proffered solutions, given the nation's resilience and its resolution to adapt at the time. The extracts below illustrate a few adaptation activities employed by communities in the period studied:

> MA3: "Sandenbergh provides a daily supply of bales of hay to the hippos. We use our money to buy this hay for the hippos to help their plight," he says.

> SA1: Boreholes which have been drilled some years ago are being turned on to water the wildlife. Most of the 60 boreholes are in the Chobe area. These are strategically placed in animals' corridors. He said the aim was to ensure that there are no mass mortalities.

The realisation of the adaptation frame by this study is contrary to research results by Molek-Kozakowska (2018) whose study traced popularity-driven coverage of climate change

Evaluating climate change communication in the tourism sector 121

in *New Scientist* with the special aim of identifying which aspects of the issue have been backgrounded. Ultimately, the paper discovered that mitigation and adaptation frames were relatively rare (Molek-Kozakowska, 2018). The mitigation frame is, however, also rare in this current study.

Results from further evaluating the effectiveness of climate change communication for sustainable environmental conservation in the tourism sector show that articles collected made use of various framing tools that resonated and engaged with the readers. Given the fact that the 2019 drought year resulted in biodiversity loss, it was a consequent commercial and newsworthy agenda that both local and international news outlets concentrated on. Most writers took advantage of the parched rivers, the harrowing deaths of wildlife species and the feasting scavengers in order to pen compelling articles that had the potential to appeal to extensive demographics. Flanking the articles were striking images that backed and abridged issues highlighted by the texts, evidencing the realities of life. Some images, however, did not depict the shocking effects of climate change but rather, they coerced readers to reminisce on what once was while the text painted the drought picture.

Taking research findings by O'Neill and Nicholson-Cole (2009) into consideration, this mix may be commendable. In their study which explored the assertion that using fearful representations of climate change may be counterproductive, the authors posit that dramatic, sensational, fearful, shocking and other climate change representations of a similar ilk can successfully capture people's attention and promote a general sense of the importance of the issue (O'Neill & Nicholson-Cole, 2009). However, they also noted that this can overwhelm and render individuals helpless, consequently distancing them from the climate change fight. These types of representations, the authors conclude, have a common presence in the mass media and wider public domain, an assertion true to this current study as most of the images reviewed have the fear-inducing factor/frame, and the few that did not fit into this category ranked low in efficacy as they lacked the capacity to make readers feel that they can fight climate change (O'Neill & Nicholson-Cole, 2009).

International news publications take the lead in the utilisation of the power of visual media. The picture composition, quality and relevance of the photographs to the subject matter is testimony to big newsroom budgets which can afford highly skilled photographers, high end equipment and transport costs for travel to affected geographical areas. Local articles were supplemented by 0–1 image each, while those published by international media houses had 0–13; some even went as far as linking videos to the articles. Of the 21 local news articles assessed, only 42.8 per cent had images, while only 14.2 per cent of these were closely linked to the tourism and climate change topic.

All in all, the text and image frames attempted to effectively define the drought problem, outline its causes, its risks, its threat and the uncertainties surrounding the phenomenon, but gave minute hope and diffidently proffered conservation as a solution in nature-based tourism (Ardèvol-Abreu, 2015).

This study did not only focus on the resultant main topic themes when employing the framing theory, but it also scrutinised specific ways the main issues in discussion were presented. It did so by exploring the tone, craft, jargon, relatability, credibility, sources and human-interest aspects of the articles.

Subsequent findings revealed that the majority of the works published were hard news pieces (90.4 per cent, n=19). Given the main issues already presented in this chapter, this suggests that media houses considered the timeliness, climate emergency and seriousness of

122 *Handbook on tourism and conservation*

the disastrous effects of the climate event on biodiversity. While no human-interest articles were discovered from the worldwide data, only two constituted the local batch. This is because less attention was given to the story behind the story, and the effects of drought on tourism stakeholders such as operators, though some had to retrench their employees, and/or relocate their boats, among other coping mechanisms. The few articles that acknowledged the impacts of climate change on industry players did so by referring to the economic implications of the drought. This chapter therefore concludes that this unbalanced focus made tourism and climate change a wildlife story – as also affirmed by the word frequency count of the noun *wildlife*, which is 111. This will be potentially problematic when trying to encourage citizenry to play an active role in the fight against climate change through sustainable environmental conservation as they may feel detached from the subject, and not so much directly under threat. Nevertheless, the majority of the corpus of articles reviewed were relatable given the circumstances at the time and the fact that Batswana have lived alongside wildlife for centuries, and that droughts, though acute and now happening at shorter intervals, are synonymous with the semi-arid Botswana. Yet, it can be argued that articles could have been more relatable to the masses had journalists taken advantage of the human-interest angle.

The fact that the majority of the articles were almost devoid of scientific, climate change jargon helped, as coverage meaningfully communicated the climate phenomena to lay publics (Ardèvol-Abreu, 2015). Moreover, frames used communicated effectively through craft and tone. Of the local articles analysed 52.3 per cent (n=11) were crafted in a way that pulled readers into the story. The imagery, the word choice, the figurative language and sentence structures were captivating, though there is need for improvement. A good example is this extract by MA3:

> During its heydays, Thamalakane River bustled with water sport activities. There were constant roaring engines of boats with ecstatic screams of passengers enjoying the rides. There would be silent mekoro gliding past the river's shallow sides, riding slow and smooth. There would be many at Matlapana beach doing the shaora – swimming. But everything is dead now.

Given the main issue themes highlighted by this chapter, such as biodiversity loss, economic implications, and human and wildlife conflicts, the texts examined were also loaded with tones that evoked sympathy and urged readers to ponder and/or comprehend a certain issue. This excerpt by GA1 which depicts strong cultural linkages between the journalist, the primary source of information and the news recipients is a fitting illustration:

> These challenges have also resulted in Batswana losing their identity as it was traditionally known that the month of October (Phalane) for instance ushered in the birth of the impala antelope (phala). "Are we still seeing the same pattern? Our indigenous knowledge is now becoming irrelevant. Our ATM [Automated Teller Machine] used to be our nature but now the stock of natural goods is on a decline and we need to do something," he said.

Despite all this, in boosting the credibility of the articles, some jargon had to be utilised. Only 9.5 per cent of the local stories were riddled with jargon as depicted by the sample extract below:

> MA1: Wolski maintains that, "This is all part of what we call a multi-decadal variability in regional climate, likely linked to processes such as Pacific Decadal Oscillation or similar, which are essentially large scale anomalies in sea surface temperatures".

In bolstering the credibility of their news articles, journalists also opted for reliable, professional sources that can easily be verified. A diverse selection of actors such as researchers, government officials, industry experts and academics were attributed. Of those quoted by local publications, 88 per cent (n=22), and 90 per cent (n=9) of those referenced by international stories, spoke to the effects of climate change induced drought on the tourism sector, while the remaining percentages expressed uncertainty and doubted the link between the 2019 drought and climate change. Though the majority of the actors – most of whom were cited more than once by different articles – attributed the severity of the 2019 drought year to climate change, these sources did not attempt to inspire sustainable environmental conversation. Yet, 16 per cent (n=4) of those quoted by the local media, and 10 per cent (n=1) of those quoted by the international media made the effort, but generally failed to stipulate specific sustainable environmental conservation measures that could readily be utilised by all stakeholders, inclusive of ordinary citizens.

Despite this, the idea of conservation resounded in many articles as framed by journalists, thus the Word Cloud frequency count for the noun *conservation* is 56, *sustainable* is 22 and *management* is 51, showing that sustainable conservation is part of the tourism and climate discussion in the country. Nonetheless, it is apparent that media framing can do more in inspiring sustainable environmental conservation.

Burning Issues in Climate Change Communication

Empirical studies on climate change communication carried out across the globe have highlighted several issues including climate sceptics and acceptors, consensus messaging, media shortcomings, scientist–media relations, polarisation of climate change, and apolitical ways of communicating climate change. This study explored these elements in the corpus of articles retrieved with the aim of determining the extent to which these issues are prevalent in Botswana's tourism and climate change communication, and assessing their role in influencing dialogue and engagement for the sake of sustainable environmental conservation. Results indicate that these issues are not prevalent in either local or international news. Yet, a few sceptics who argued that climate change was not the cause of the 2019 drought and framed the science of climate change as being of minimal relevance were singled out. In trying to balance the MA1 article, the journalist introduced the debate as follows:

> MA1: Researchers disagree that the current drying of the Okavango Delta is due to the climate change phenomenon.

While the climate change acceptor argued that it is indeed climate change that caused the delta to dry up, and stated that she prefers to call it 'climate variation', the climate change sceptics' arguments were as follows:

> Sceptic 1: "Is what we see at the moment a result of climate change? It is not very likely. I think we mostly see the effect of decadal climate variability. I don't think the recent decline in rainfall and this year's rainfall being the lowest since 1981 is the effect of climate change. Climate change might have played a role in the recent trend, but it was minimal at best."

> Sceptic 2: Professor Paul Skelton from the South African Institute of Aquatic Biodiversity also argues that the current drought is unlikely due to climate change.

124 *Handbook on tourism and conservation*

As much as the journalist tried to balance the article, the debate itself was tilted as sources against and/or stressing uncertainty outweighed the sources for climate change. This article therefore has the potential to sway readers, and lead them into believing that climate change is overrated, and that droughts are a historical normalcy. Though this negative did not outweigh the good coverage rendered, a study by Lee et al. (2013) on trends in reports on climate change in the South Korean press generally found that articles vague about climate change (lack of precise data, negative or sceptical tone, and improper use of terminology) were much more common than the articles presenting accurate knowledge. Contrary to both these study results are findings that the Indian press entirely endorses climate change as a scientific reality (Billett, 2010).

Since some policymakers are said to remain sceptical about climate change and its impacts on tourism despite growing evidence from regional scientific research (Hambira et al., 2020b), this chapter is of the view that such media framing could worsen the status quo. In their study, Hambira et al. (2020b) suggested that constraints that hamper progress in policy response measures include inadequate knowledge of, and the extent to which, climate science can be trusted. Consider the tone and normalcy frames in the excerpt below:

> MA11: Historical records interestingly show that the Lake dried exactly a decade ago in 1819. In 1820 (long before David Livingstone reached there) Kwebe, the first capital of Batawana, located on the foothill of Kwebe Hills and peripheries of Lake Ngami, dried up forcing the then Kgosi Moremi to move to Namanyana, which is the current day Toteng. The most recent dry spell of the lake was from the 1990s and it lasted for over a decade. In 2010, Lake Ngami was swollen to its glory days and sustained life around the area.

While this article maintained that the 2019 drought was unlikely to be a result of climate change but a normal natural cycle, one other article argued that carbon dioxide "(CO_2) is beneficial for our environment and is not a pollutant, as it benefits plant life by increasing biomass and thus improves the basis for all human life on earth. So, producing and burning coal using state of the art technology can still be a sustainable development solution." Further justification by this participant is as follows:

> MA6: The present warm period has lasted over 8,000 years longer than any of the three prior ones, giving the oceans a much longer time to warm up and release more CO_2 into the atmosphere, which would also contribute to the current level of 400 ppm. This means that coal does not carry all the blame as is stated by socio-environmentalist groups and politicians.

The fact that this participant works in Botswana's fossil fuel industry – as per the disclaimer at the bottom of the article – may suggest that his/her arguments are subject to bias. That said, this chapter concludes that these two articles are framed in a way that fosters uncertainty and polarisation among publics, as to the causes, risks and climate change influences on drought, ultimately disadvantaging sustainable environmental conservation in the sector already said to have poor adaptive capacity and inadequate information for informed decision making (Hambira & Mbaiwa, 2020). If those in the tourism industry and community members (whose benefits and participation are often minimal – see Hambira et al., 2020b; Saarinen et al., 2020) do not realise how their lifestyle choices and daily activities contribute to global warming, they will see no need to reduce their carbon footprint. Actions that the sector had adopted in the fight against climate change may also be abandoned if media frames continue to contradict themselves. Some actions registered included energy efficient and water conservation efforts.

In their paper, Hambira et al. (2020b) also suggested that low awareness levels may lead to inaction, thus it is not surprising that Hoogendoorn et al. (2021) noted that "considering the current global environmental change projections for southern Africa, greater adaptive action is necessary".

CONCLUSION

This chapter's findings have significant implications for policy, environmental management, and climate adaptation and mitigation. Overall, the study suggests that media coverage of climate change and tourism is insufficient, and it trails the general discussion of climate change in mass media. Hence there is need for development stakeholders to work in unity in raising climate change awareness in the context of tourism, proffering applicable and practical solutions to citizens. The fact that journalists also failed to sufficiently highlight the connection between climate and tourism, considering the anthropogenic activities of the tourism sector (as the industry's activities exert increased pressure on fossil fuels, worsening forms of pollution such as air emissions), is proof that media houses alone cannot carry the burden. Government officials, researchers and scientists among other relevant actors need to relay the unattended frames to journalists, who will then creatively disseminate them to the masses. This is bearing in mind that the efficacy of science communication is constrained by framing conventions and argumentative positionings that reflect editorial lines and market forces, and in consequence this may result in a questionable or unintended climate change discourse (Molek-Kozakowska, 2018).

As sources of news content, these actors should not only attribute the severity of drought to climate change but should attempt to inspire sustainable environmental conversation activities by stipulating specific sustainable environmental conservation measures that can be readily adopted by all. As much as balanced debates and dialogue of the climate change phenomena are imperative, this chapter suggests that precaution should be taken to ensure that platforms availed and/or employed by actors do not influence uncertainty and polarisation in any manner. Doing so has the potential to dissuade people from considering conservation. All things said, it is quite apparent that media framing can do more in inspiring sustainable environmental conservation in the tourism sector, as journalists ultimately decide what will become news, and subsequently make certain aspects of that news salient. Nonetheless, this chapter views it imperative to reiterate what was emphasised by Bolsen and Shapiro (2018) which all development stakeholders should be aware of, which is the fact that framing is not an elixir. Framing efforts are purported to have often failed to produce support for policy action on climate change or personal engagement on the issue, or even backfire in some instances (Bolsen & Shapiro, 2018). Therefore, diverse strategies of communication must be employed.

Recommendations

Most local newsrooms do not have a specific climate change reporting desk, hence development stakeholders should consider assisting media houses with the necessary resources (financial and/or training) that can ensure that each outlet has an environment and climate change reporter. This will increase news coverage. But stakeholders should not consequently treat

126 *Handbook on tourism and conservation*

reporters as advocates of climate change issues; when dealing with journalists, they should afford them neutrality.

The 2021 approval of the long-awaited climate change policy by the Parliament of Botswana is a welcome development that will go a long way in influencing coverage. More so given that the draft acknowledged climate change as a factor that has the tendency to shift the nature-based tourism dynamics to follow the tourist attraction elements (mainly wild animals), consequently affecting the revenue generation from tourism activities including loss of jobs (Koboto et al., 2019). The policy's intention to implement sustainable climate change response measures particularly on education and awareness, information dissemination, climate change activism and climate good governance is a positive approach (Koboto et al., 2019). However, to communicate climate change effectively, this chapter sees the need for a sector specific climate change strategy, especially because most climate change messages in the country target farmers.

REFERENCES

Ahmadalipour, A., Moradkhani, H., Castelletti, A., & Magliocca, N. (2019). Future drought risk in Africa: Integrating vulnerability, climate change, and population growth. *Science of the Total Environment, 662*, 672–686.

Anup, K. C. (2018). Tourism and its role in environmental conservation. *Journal of Tourism and Hospitality Education, 8*, 30–47.

Ardèvol-Abreu, A. (2015). Framing theory in communication research: Origins, development and current situation in Spain. *Revista Latina de Comunicación Social, 70*(1053), 423–450.

Billett, S. (2010). Dividing climate change: Global warming in the Indian mass media. *Climatic Change, 99*, 1–16.

Bolsen, T., & Shapiro, M. A. (2018). The US news media, polarization on climate change, and pathways to effective communication. *Environmental Communication, 12*(2), 149–163.

BOPA (2020). Govt declares 2019/2020 non-drought year. Botswana Press Agency. www.dailynews .gov.bw/news-details.php?nid=58236 (accessed 2 February 2021).

Burns, P., & Bibbings, L. (2009). The end of tourism? Climate change and societal challenges. *Twenty-First Century Society, 4*(1), 31–51.

Central Statistics Office (2004). *Forestry Statistics*. Environment Statistics Unit, Department of Printing and Publishing Services, Gaborone, Botswana.

Chanza, N., Siyongwana, P. Q., Williams-Bruinders, L., Gundu-Jakarasi, V., Mudavanhu, C., Sithole, V. B., & Manyani, A. (2020). Closing the gaps in disaster management and response: Drawing on local experiences with Cyclone Idai in Chimanimani, Zimbabwe. *International Journal of Disaster Risk Science, 11*(5), 655–666.

Cil, A., & Jones-Walters, L. (2011). Biodiversity action plans as a way towards local sustainable development. *Innovation, 24*(4), 467–479.

Coldrey, K. M., & Turpie, J. K. (2020). Potential impacts of changing climate on nature-based tourism: A case study of South Africa's national parks. *Koedoe, 62*(1), 1–12.

Corbin, J., & Strauss, A. (2008). *Basics of Qualitative Research: Techniques and Procedures for Developing Grounded Theory*. Thousand Oaks, CA: Sage.

Cristobal-Fransi, E., Daries, N., Serra-Cantallops, A., Ramón-Cardona, J., & Zorzano, M. (2018). Ski tourism and web marketing strategies: The case of ski resorts in France and Spain. *Sustainability (Switzerland), 10*(8), 1–24.

Day, J. C., Heron, S. F., & Markham, A. (2020). Assessing the climate vulnerability of the world's natural and cultural heritage. *Parks Stewardship Forum, 36*(1), 142–153. https://doi.org/10.5070/p536146384.

Dube, K., Mearns, K., Mini, S. E., & Chapungu, L. (2018). Tourists' knowledge and perceptions on the impact of climate change on tourism in Okavango Delta, Botswana. *African Journal of Hospitality, Tourism and Leisure*, *7*(4), 1–18.

Dube, K., & Nhamo, G. (2019). Climate change and potential impacts on tourism: Evidence from the Zimbabwean side of the Victoria Falls. *Environment, Development and Sustainability*, *21*(4), 2025–2041.

Dube, K., Nhamo, G., & Chikodzi, D. (2020). Climate change-induced droughts and tourism: Impacts and responses of Western Cape province, South Africa. *Journal of Outdoor Recreation and Tourism*, *39*. https://doi.org/10.1016/j.jort.2020.100319.

Eckstein, D., Künzel, V., Schäfer, L., & Winges, M. (2019). *Global Climate Risk Index 2020: Who Suffers Most from Extreme Weather Events?* Germanwatch. www.germanwatch.org/en.

Goldberg, J., Birtles, A., Marshall, N., Curnock, M., Case, P., & Beeden, R. (2018). The role of Great Barrier Reef tourism operators in addressing climate change through strategic communication and direct action. *Journal of Sustainable Tourism*, *26*(2), 238–256.

Gössling, S., Scott, D., & Hall, C. M. (2020). Pandemics, tourism and global change: A rapid assessment of COVID-19. *Journal of Sustainable Tourism*, *29*(1), 1–20.

Hambira, W. L. (2017). *Botswana Tourism Operators and Policy Makers' Perceptions and Responses to the Tourism-Climate Change Nexus: Vulnerabilities and Adaptations to Climate Change in Maun and Tshabong area*. Oulu: Nodia Geographical Publications.

Hambira, W. L. (2018). Botswana tourism operators' and policy makers' perceptions and responses to climate change. *Matkailututkimus*, *14*(1), 55–59.

Hambira, W. L., & Mbaiwa, J. E. (2020). Tourism and climate change in Africa. In M. Novelli, E. A. Adu-Ampong, & M. A. Ribeiro (eds), *Routledge Handbook of Tourism in Africa* (pp. 98–116). New York: Routledge.

Hambira, W. L., Saarinen, J., Atlhopheng, J. R., & Manwa, H. (2020a). Climate change, tourism, and community development: Perceptions of Maun residents, Botswana. *Tourism Review International*, *25*(2), 105–117.

Hambira, W. L., Saarinen, J., Manwa, H., & Atlhopheng, J. R. (2013). Climate change adaptation practices in nature-based tourism in Maun in the Okavango Delta area, Botswana: How prepared are the tourism businesses? *Tourism Review International*, *17*(1), 19–29.

Hambira, W. L., Saarinen, J., & Moses, O. (2020b). Climate change policy in a world of uncertainty: Changing environment, knowledge, and tourism in Botswana. *African Geographical Review*, *39*(3), 252–266.

Harrington, L. J., & Otto, F. E. L. (2020). Reconciling theory with the reality of African heatwaves. *Nature Climate Change*, *10*(9), 796–798.

Hoogendoorn, G., & Fitchett, J. M. (2018). Tourism and climate change: A review of threats and adaptation strategies for Africa. *Current Issues in Tourism*, *21*(7), 742–759.

Hoogendoorn, G., & Fitchett, J. M. (2019). Fourteen years of tourism and climate change research in southern Africa: Lessons on sustainability under conditions of global change. In M. T. Stone, M. Lenao, & N. Moswete (eds), *Natural Resources, Tourism and Community Livelihoods in Southern Africa: Challenges of Sustainable Development*. New York: Routledge.

Hoogendoorn, G., & Rogerson, C. M. (2015). Tourism geography in the global South: New South African perspectives. *South African Geographical Journal*, *97*(2), 101–110.

Hoogendoorn, G., Stockigt, L., Saarinen, J., & Fitchett, J. M. (2021). Adapting to climate change: The case of snow-based tourism in Afriski, Lesotho. *African Geographical Review*, *40*(1), 92–104.

IPCC (2019). *Global Warming of 1.5°C*. An IPCC Special Report on the impacts of global warming of 1.5°C above pre-industrial levels and related global greenhouse gas emission pathways, in the context of strengthening the global response to the threat of climate change, sustainable development, and efforts to eradicate poverty (V. Masson-Delmotte et al. eds). Geneva: Intergovernmental Panel on Climate Change.

Jefferis, K., Sojoe, S., & Mokhurutshe, K. (2019). *Economic Review*. http://econsult.co.bw/tempex/file/Econsult%20Review%202019_1st%20Quarter%20Final.pdf (accessed 2 March 2021).

Kaján, E., & Saarinen, J. (2013). Tourism, climate change and adaptation: A review. *Current Issues in Tourism*, *16*(2), 167–195.

Koboto, O., Lesolle, D., Sisay, L., Chambwera, M., Gopolang, B., Goakgethelwe, N., & Molefhe, N. (2019). *Draft Climate Change Response Policy Version 2.*

Lee, J., Hong, Y. P., Kim, H., Hong, Y., & Lee, W. (2013). Trends in reports on climate change in 2009–2011 in the Korean press based on daily newspapers' ownership structure. *Journal of Preventive Medicine and Public Health, 46*(2), 105–110.

Lequechane, J. D., Mahumane, A., Chale, F., Nhabomba, C., Salomão, C., Lameira, C., Chicumbe, S., & Semá Baltazar, C. (2020). Mozambique's response to cyclone Idai: How collaboration and surveillance with water, sanitation and hygiene (WASH) interventions were used to control a cholera epidemic. *Infectious Diseases of Poverty, 9*(1), 1–4.

Linström, M., & Marais, W. (2012). Qualitative news frame analysis: A methodology. *Communitas, 17,* 21–38.

Ma, S. D., & Kirilenko, A. P. (2020). Climate change and tourism in English-language newspaper publications. *Journal of Travel Research, 59*(2), 352–366.

Maude, G., & Reading, R. (2010). The role of ecotourism in biodiversity and grassland conservation in Botswana. *Great Plains Research, 20,* 109–119

Molek-Kozakowska, K. (2018). Popularity-driven science journalism and climate change: A critical discourse analysis of the unsaid. *Discourse, Context & Media, 21,* 73–81.

Moswete, N. N., & Dube, N. N. (2013). Wildlife-based tourism and climate: Potential opportunities and challenges for Botswana. In L. D'Amore & P. Kalifungwa (eds), *Meeting the Challenges of Climate Change to Tourism: Case Studies of Best Practice.* Newcastle upon Tyne: Cambridge Scholars Publishing, pp. 395–416.

Nisbet, M. (2009). Communicating climate change: Why frames matter for public engagement. *Environment, 51*(2), 12–23.

Okon, G. B. (2015). Promoting tourism in West Africa through newspaper constructionism, framing and salience: A content analysis of select newspapers in Nigeria. *International Journal of Humanities and Social Science, 5*(8), 1–15.

Okon, G. B. (2017). Role of newspaper in promoting tourism in Nigeria: Content analysis. *Mass Communicator: International Journal of Communication Studies, 11*(2), 4–12. https://doi.org/10.5958/0973-967x.2017.00009.6.

O'Neill, S., & Nicholson-Cole, S. (2009). "Fear won't do it": Promoting positive engagement with climate change through visual and iconic representations. *Science Communication, 30*(3), 355–379.

Pandy, W. R., & Rogerson, C. M. (2021). Climate change risks and tourism in South Africa: Projections and policy. *GeoJournal of Tourism and Geosites, 35*(2), 445–455.

Poberezhskaya, M. (2015). Media coverage of climate change in Russia: Governmental bias and climate silence. *Public Understanding of Science, 24*(1), 96–111.

Rhomberg, M., & Rhomberg, M. (2016). Climate change communication in Austria. *Oxford Research Encyclopedia of Climate Science.* https://doi.org/10.1093/acrefore/9780190228620.013.449.

Saarinen, J., Hambira, W. L., Atlhopheng, J., & Manwa, H. (2012). Tourism industry reaction to climate change in Kgalagadi South District, Botswana. *Development Southern Africa, 29*(2), 273–285.

Saarinen, J., Moswete, N., Atlhopheng, J. R., & Hambira, W. L. (2020). Changing socio-ecologies of Kalahari: Local perceptions towards environmental change and tourism in Kgalagadi, Botswana. *Development Southern Africa, 37*(5), 855–870.

SADC (2010). *Southern Africa Sub-Regional Framework of Climate Change Programmes.* Gaborone: Southern African Development Community.

Samuels, K. L., & Platts, E. J. (2020). An ecolabel for the world heritage brand? Developing a climate communication recognition scheme for heritage sites. *Climate, 8*(3). https://doi.org/10.3390/cli8030038.

Scott, D., Gössling, S., & Hall, C. M. (2012). International tourism and climate change. *Wiley Interdisciplinary Reviews: Climate Change, 3*(3), 213–232.

Siyao, P. O., & Sife, A. S. (2018). Coverage of climate change information in Tanzanian newspapers. *Global Knowledge, Memory and Communication, 67*(6–7), 425–437.

Statistics Botswana (2016). *Tourism Statistics Annual Report 2015.* www.statsbots.org.bw.

Statistics Botswana (2018). *Tourism Statistics Report 2016 Revision 2 1. Tourism Statistics Report 2016 Revision 2.* www.statsbots.org.bw.

Statistics Botswana (2019). *Annual Report 2017 Tourism Statistics.* www.statsbots.org.bw.

Statistics Botswana (2020a). *Botswana Environment Statistics Climate Digest September*. www.statsbots.org.bw.

Statistics Botswana (2020b). *Botswana Environment Statistics: Natural and Technological Disasters Digest 2019*. http://www.statsbots.org.bw.

Statistics Botswana (2020c). *Tourism Statistics Annual Report 2018*. www.statsbots.org.bw/sites/default/files/publications/Tourism%20Statistics%20Report%202018%20REVISED.pdf (accessed 25 January 2021).

Tapsuwan, S., & Rongrongmuang, W. (2015). Climate change perception of the dive tourism industry in Koh Tao island, Thailand. *Journal of Outdoor Recreation and Tourism, 11*, 58–63. https://doi.org/10.1016/j.jort.2015.06.005.

Tolan, S., & Nan, W. (2007). *Coverage of Climate Change in Chinese Media*. http://environment.newscientist.com/article/dn11707-chinas- (accessed 30 June 2021).

WEF (2019). *Travel and Tourism Competitiveness Report 2019 – Reports – World Economic Forum*. http://reports.weforum.org/travel-and-tourism-competitiveness-report-2019/country-profiles/#economy=BWA (accessed 2 March 2021).

Wolski, P., Savenije, H. H. G., Murray-Hudson, M., & Gumbricht, T. (2006). Modelling of the flooding in the Okavango Delta, Botswana, using a hybrid reservoir-GIS model. *Journal of Hydrology, 331*(1–2), 58–72.

WTTC (2019). *Travel & Tourism Economic Impact, World Travel & Tourism Council*. https://wttc.org/Research/Economic-Impact.

10. Perspectives on the effects of environmental change in northern Botswana and its implications for CBNRM

Maduo Mpolokang and Jeremy Perkins

INTRODUCTION

The collapse of the world's largest herbivore (Ripple et al., 2015) and carnivore (Ripple et al., 2014; Everatt et al., 2019) populations is widely predicted (Moritz and Agudo, 2013). However, the severity of the looming biodiversity and extinction crisis, especially in Africa, appears to be largely unrecognised (Andermann et al., 2020) outside of the scientific community in which it is discussed. The key threats of land use and land cover change, persecution, diseases, invasive species and climate change are all ongoing, such that there appears to be little reason for optimism. The current Covid-19 pandemic appears likely to accentuate the conservation crisis further, not least by drastically reducing, if not temporarily halting altogether, income streams from visiting tourists (Lindsey et al., 2020). The pandemic is widely believed to have originated from wildlife trade and our exploitative relationship with nature. It was predicted that as people reduce biodiversity they are increasing the risk of disease pandemics (Jones et al., 2008; Tollefson, 2020), emphasising the need to change our behaviour to one that works with nature rather than against it.

Climate variability over southern Africa is predicted to become more extreme this century (Engelbrecht et al., 2015). The region is predicted to become hotter and drier with increased 'heatwaves', 'mega droughts' and flooding events (Engelbrecht et al., 2015). More frequent dry days or more persistent dry sequences (Pohl et al., 2017) will have profound effects on both the ecology of the system and the tourists that visit it (Gössling & Hall, 2006; Hoogendoorn & Fitchett, 2018; Tervo-Kankare et al., 2018). Growing attention has been given to this topic in countries such as Mexico (Scott & Lemieux, 2010), Vietnam (Koubi et al., 2016), South Africa (Brook et al., 2010; Hoogendoorn & Fitchett, 2018) and Finland (Saarinen & Tervo, 2006).

The influence of environmental change on wildlife has been shown to affect their distribution and composition in several tourism destination areas and/or Protected Areas such as the Otago Peninsula (Kutzner, 2019), Abijata-Shalla Lakes National Park (ASLNP) (Tessema et al., 2010), the Okavango Delta (Tsheboeng et al., 2014) and Chobe National Park (Mosugelo et al., 2002; Wolf, 2009). In almost all of these, climate change has been identified as a new phenomenon exacerbating environmental change, which has the potential to disrupt the benefits accrued to stakeholders from landscape and other natural resources which form the basis of wildlife tourism, for example, a loss of aesthetic value, key habitats, available water and rare species.

Extreme environmental variability may also undermine attempts to increase the involvement of local communities in wildlife conservation. This is because environmental variability concentrates wildlife populations into a few areas where the habitat requirements of key wildlife species are met. The provision of artificial water points (AWPs) is a major intervention in the ecological functioning of semi-arid savannah ecosystems (Owen-Smith, 1996). AWPs have been widely adopted in Botswana, particularly in Protected Areas (Perkins, 2020). It is also often assumed that AWP provision will reduce human–wildlife conflict (HWC) in the neighbouring local communities. This is because AWPs serve to constrain the distribution of problem animals, such as elephants and predators, by keeping them far from human settlements. It is, however, important to emphasise that such a role of AWPs is not proven and may even serve to accentuate HWC by inadvertently boosting populations of problem animals and creating new foci from which to disperse towards the very people and communities they were trying to protect (Emily Bennitt, a research scholar in Okavango Research Institute (ORI) in Maun, Botswana., pers. comm.). Such concentrations could also limit opportunities to increase the geographic spread of key tourist activities and so the potential to deliver benefits related to wildlife conservation to local communities (Adams & Hutton, 2007; Massé & Lunstrum, 2016).

To analyse a spectrum of conservation approaches and environmental change implications for wildlife-based tourism, this chapter is guided by the Sustainable Livelihood Framework (SLF). This aptly captures how people use natural resources to sustain their well-being and how this can be affected by extreme events. Community sustainability is valued through coping and recovery from stress and shocks (Chambers & Conway, 1992). Therefore, this theoretical underpinning is used to understand the relationship between nature and people. The literature involving environmental change is sparse with the few available studies (e.g. Herrero et al., 2016; Mogende & Moswete, 2018; Moswete et al., 2017) not adequately exploring the influence of environmental change on the tourism industry. In the context of wildlife-based dependency on the environment, this chapter aims to assess the effects of environmental change on wildlife-based tourism and wildlife conservation in the Chobe National Park (CNP).

This chapter addresses environmental variability and wildlife conservation by assessing the perceptions of vehicle and boat guides along the Chobe Riverfront of north-eastern Botswana, as well as key tourism stakeholders, regarding their perception of the impacts environmental change will have on wildlife-based tourism. The eastern Chobe Riverfront of Botswana is one of the premier sites in Africa for viewing a great diversity of wild ungulates and their predators in a natural setting. Along with the Okavango Delta the CNP accounts for most game viewing tourists to Botswana and the revenue that accrues from it. This chapter contributes to the debate on how to improve policies and management plans guiding authorities focusing on the environment, tourism and wildlife areas.

ENVIRONMENTAL CHANGE

The term environmental change is used to refer to changes that alter and transform the ecosystem, especially animal and vegetation distribution, through activities that affect climate and land surfaces (Vitousek et al., 1997). Environmental drivers such as drought, fire, temperature changes and disease can all be exacerbated by climate change exerting pressure on savannah landscapes and can result in unexpected outcomes (Sekonya, 2016). These stressors, together with the Earth's terrestrial surface, may strongly influence the ecosystem and climate systems (Scheiter et al., 2018) and livelihoods (Sekonya, 2016).

132   *Handbook on tourism and conservation*

Cumulative environmental changes have been documented indicating an increase in ocean temperature, and insufficient food supply resulting in devastating outcomes that threaten the sustainability of tourism businesses (Lambert et al., 2010) which ultimately impact upon community development and livelihoods. A study conducted in Waterton Lakes National Park (Canada) by Scott et al. (2007) revealed that environmental change transforms wildlife and vegetation compositions, thus affecting livelihoods. Gössling et al. (2006) revealed increased precipitation and humidity, and frequent storms resulting from environmental change are exacerbated by climate change in Tanzania. Similarly, in South Africa, the threats include degradation of environmental resources such as wild animals and heritage sites, water scarcity, biodiversity loss and reduced landscape aesthetic beauty (Rogerson, 2016).

Climate variability usually alters the flora and fauna, especially in parks and reserves thereby affecting park-based tourism resources, especially during prolonged dry climatic conditions (Hoogendoorn and Fitchett, 2016). The altered flora and fauna reduce the aesthetic value suitable for tourist attractions. Climate change and variability have been found to affect tourism mobility and flows to South Africa (Rogerson, 2016). Nevertheless, the potential contribution of environmental change to wildlife-based tourism and community development has rarely been examined though it poses a greater threat to ecosystems especially in Southern Africa due to their vulnerability.

WILDLIFE-BASED TOURISM

Studies reveal that tourism and climate change and environmental variability are closely intertwined (Saarinen, 2013). Key environmental change drivers such as heatwaves, droughts and floods result in unfavourable environmental conditions that detrimentally affect tourist flows. Such variability will create environmental settings that support the emergence of persistent diseases such as malaria and vector-borne diseases which can have substantial implications for destination competitiveness (Simpson, 2008; Scott et al., 2007; Scott, 2011). Hambira et al. (2013) argue that alteration in the wildlife tourism resources may create complex and dynamic responses from both wildlife tourism managers and visitors.

Wildlife-based tourism has over recent decades been regarded as an integral component of the economy in Botswana, just as it has for much longer in eastern African countries, such as Kenya and Tanzania. Wildlife-based tourism systems are dynamic as they are influenced by natural resources such as wildlife species composition and other physical features at a particular tourist destination area (Yurco et al., 2017). When linked to Community-Based Natural Resource Management (CBNRM) in surrounding communities, wildlife conservation in Protected Areas can simultaneously ensure ecological sustainability of natural resource use and the socio-economic empowerment of local people (Paksi & Pyhälä, 2018). The contribution of wildlife-based tourism to community development and Botswana's economy has been regarded as a key policy issue (Mbaiwa, 2017) for decades. It is integral to the achievement of the United Nations' Sustainable Development Goals (UN SDGs) by 2030. Environmental variability seems likely to accentuate current threats to wildlife conservation and decrease opportunities for the CBNRM programme to achieve human–wildlife conflict mitigation. This is the case because as human–wildlife conflict increases due to the competition for scarce natural resources such as surface water and available forage, the likelihood of achieving the SDGs decreases.

TOURISM AND COMMUNITY DEVELOPMENT

Existing literature on tourism and community development emphasises concerns about the relationship between the environment, communities and sustainability (Roseland, 2000; Richards & Hall, 2002). Tourism philosophies such as ecotourism, pro-poor tourism, community tourism and sustainable tourism (Goodwin, 1996; Hall & Brown, 2006; Simpson, 2008) have been used to emphasise the interconnectedness of tourism and benefits to local communities. These include employment creation, income generation and increases in the capacity of local communities to manage the tourism resources within their area (Forstner, 2004; Mbaiwa & Stronza, 2010; Mamba et al., 2020). However, CBNRM is not without its challenges, such as the exclusion of local communities in the decision-making process and the domination of the tourism industry by elites (Stone & Stone, 2020).

Saarinen (2006, p. 1132) states that, 'sustainability should primarily be connected with the needs of people—not a certain industry—and the use of natural and cultural resources in a way that will also safeguard human needs in the future'. This concept allows for proper balancing between tourism and community development. It also encourages the development of management plans that mitigate environmental change and maintain the sustainability of community livelihoods and that of wildlife resources.

CONCEPTUAL APPROACH

This study is informed by the Sustainable Livelihood Framework (SLF) (Pandey et al., 2017). Community sustainability is valued as a way of coping with and recovering from stress and shocks, maintaining its abilities and assets without compromising opportunities for the future (Chambers & Conway, 1992). Communities are dependent on natural resources and can be affected by extreme events, but through their prosperity, or better coping strategies, they are less likely to be critically damaged by extreme events (Pandey et al., 2017). Thus, the main components of sustainable development and livelihoods are located within the nexus between nature and people. These components are interdependent and, in the event of environmental change, are likely to affect the well-being of the community. In the context of this study, SLF is used to understand the impacts of environmental change on wildlife-based tourism and their implications for community livelihood and/or sustainability.

MATERIALS AND METHODS

Study Site

The study was conducted in the northern part of Botswana in the Chobe National Park in Chobe District. The park was established in 1961 and is the second largest in the country popularly known for its rich wildlife resources. Chobe National Park covers approximately 10,590km^2 (DWNP, 2008; GOB, 2001), which consists of floodplains, swamps and woodland (BTO, 2016). The park supports a unique diversity and concentration of wildlife, with one of the largest concentrations of fauna and flora in Africa (BTO, 2016). The socio-economic setting of Chobe District is mostly dominated by wildlife and tourism activities especially at CNP,

134 *Handbook on tourism and conservation*

Chobe enclave and the Chobe Riverfront (CRF). Tourism is considered the most important sector in Chobe as it addresses both conservation and development issues. According to CSO (2011), Chobe District has a human population of 23,347, with 12,023 being males and 11,324 being females. Some 21 per cent of this population is made up from the Chobe enclave villages of Mabele, Parakarungu, Kavimba, Kachikau and Satau and 46 per cent from Kasane. The main economic activities in the area include crop production, livestock production and wage employment. These economic activities are complemented by small-scale businesses such as selling baskets, game meat and thatching grass (Jones, 1999). Crop production is favoured by the climatic and soil conditions which are important factors in food production (Gupta, 2013), especially in the Pandamatenga area (Alemaw & Simalenga, 2015). Chobe enclave residents carry out rain-fed and floodplain recession (*molapo*) farming activities. Stone (2013) contends that tourism in CNP is a developing trend worldwide, and this has the potential to constrain community and park relationships.

Data Collection

A cross-sectional design was employed where observations of the sampled population were made at a single point in time (Babbie, 2014). This study draws largely from a combination of survey and in-depth interviews. The two combined techniques were complemented by a review of secondary data sources (published and unpublished documents). A standardised, semi-structured questionnaire was used to solicit data from the respondents. The study participants were drawn from a 'pool' of mobile and fixed wildlife tour guides from safari lodge operators in Kasane, Kazungula and Chobe National Park. The mobile tour guides are those that spend days travelling with tourists through the park while fixed tour guides are those that usually go out on game drives with tourists and return to the accommodation the same day. The licence for wildlife tour guides allows for both mobile and fixed guiding and it was not possible to separate mobile tour guides from the fixed tour guides, hence they were treated the same during data collection.

Questionnaires were administered to a sample of 63 wildlife/safari tour guides. A purposive sampling approach was used to select the participants for both semi-structured interviews and in-depth interviews (key informant interview). This is because there is a limited number of mobile and fixed wildlife tour guides. The DWNP issued a list of safari lodge operators that use the Chobe National Park Sedudu entrance gate; wildlife tour guides were sourced from the offered list. Tour guides were asked at the end of the interview if there were others who could be contacted for interviews. Since data was collected during a peak season, it was difficult to interview some tour guides as they were busy with tourists. Also, some tour guides could not be accessed because of the restrictions and accessibility of some establishments, thus the study could not involve some of the respondents.

The questionnaires were administered to the respondents through face-to-face interviews, conducted by one of the researchers at the interviewees' offices, homes and at convenient times. The questionnaire tapped on the following information: demographic and socio-economic information of the respondents; duration of involvement in wildlife guiding; and potential impacts of environmental change on wildlife-based tourism.

The perceived rating of environmental change was assessed through statements bearing Likert-type responses, where participants were asked to rate the level of their agreement or disagreement on a five-option point scale. Perceptions were classified into five-point ratings:

Strongly Agree (coded 1), Agree (coded 2), Neutral (coded 3), Disagree (coded 4) and Strongly Disagree (coded 5). The classification was explained to the participants in terms of frequency of events and/or drivers of environmental change, thus the description of the ratings was explained to the respondents to facilitate their rating of the overall level of environmental change in their area.

On the other hand, in-depth interviews were conducted with nine purposively sampled key informants. Interviewees were asked how they perceive the effect of environmental change on wildlife and wildlife-based tourism. Responses were recorded by the interviewer based on each participant's answer. This helped to clarify and discuss questions where necessary. The method allows for the interview to be flexible so that important information can still arise (Dawson, 2009). The informants were considered carefully based on the authors' prior knowledge of their responsibilities, knowledge and participation in community and conservation programmes. This guaranteed that applicable data came from the designated target. The key informants comprised representatives from Botswana Tourism Organisation (BTO), Department of Wildlife and National Parks (DWNP), Elephants Without Borders (EWB), Centre for Conservation of African Resources, Animals, Communities and Land-use (CARACAL), Kavango-Zambezi (KAZA), Department of Forestry and Range Resources (DFRR) and Land Board. These informants were responsible for the management and conservation of natural resources at both community level and district and/or country level. The informants provided information on communities' reliance on wildlife-based tourism. The interviews lasted between 30 and 45 minutes and all ethical protocols were observed.

Data Analysis

Data was compiled and managed using Statistical Package for Social Sciences (SPSS) version 25. Descriptive and inferential statistics were used to analyse quantitative data. Descriptive statistics included frequencies such as percentages. Inferential statistics included the use of Principal Component Analysis. Before the data was entered into the computer for analysis, the semi-structured questionnaires were checked, cleaned and coded according to the planned codes. Thematic and narrative analysis was used for analysing qualitative data. Purposive sampling selected the participants that could best provide the needed information on environmental change, wildlife-based tourism, and conservation in Protected Areas.

Thematic analysis is a method used for identifying, analysing and reporting patterns (themes) within interview notes relating to the research questions (Creswell, 2013). It minimally organises and describes the dataset detail (Braun & Clarke, 2006). In this study, a thematic analysis approach was followed for exploring salient viewpoints from the interviews. For this study data coding was done after transcribing. Therefore, the themes were collated to ease interpretation and analysis. The emerging themes that relate to the research question within the interview notes were identified (Creswell, 2013). Hence, the responses noted from the interviews were organised into themes for interpretation and analysis. All the transcripts were read and anything that was mentioned relating to the identified codes under each objective was identified. In the context of this study, themes and codes were established while interpreting the data by identifying any reoccurring themes throughout and highlighting any similarities and differences in the data. Subsequently, these were verified to compare the dataset by rechecking the transcripts and codes. Also, for key informant interviews, narratives were applied.

136   *Handbook on tourism and conservation*

RESULTS AND DISCUSSION

Perceptions on Environmental Change

Of the tour guides interviewed (n = 32) 50.8 per cent demonstrated a high level of familiarity with environmental change. They associated environmental change with transformation of vegetation and wildlife populations because of anthropogenic activities such as deforestation and also resulting from natural factors such as increased temperatures, climate change, flooding, fires as well as increase in elephant population. Key informants (CARACAL, KAZA, DWNP) indicated similar familiarity with the concept of environmental change to that of the tour guides. In an interview, a KAZA liaison officer (24 August 2018) noted that:

> Environmental change is actually transformation of both the bio-physical and the ecological environment. The bio-physical in this case refers to the landscape, vegetation, soils, and the ecology with the inclusion of living creatures.

Considering the issue of environmental change, key informants expressed their concern on the issue of climate change. For example, the Programmes Manager for EWB commented that:

> Environmental change is always ongoing, it is always changing. Generally, what is happening around the world with climate change is making a huge impact. (27 August 2018)

Overall, the majority (n = 6) of the key informants indicated the increase in elephant population, high incidence/increase in diseases such as anthrax and the dryness in CNP to be a result of climate change. Given the changes in vegetation and surface water availability in the park, crowding of elephants (*Loxondanta africana*) and buffalo (*Syncerus caffer*) is usually experienced along the Sedudu Island. In the view of the tour guides and key informants these situations suggest that the impact of environmental change is high in relation to wildlife since it influences migration due to competition for food/forage. However, participants pointed out that some animals such as elephants are likely to survive while others are more likely to die such as impala (*Aepyceros melampus*). Such environmental change may have implications for community development especially those people living within the vicinity of CNP.

The tour guides' level of perceptions towards environmental change was measured with items created using a 5-point Likert scale. In order to better understand the influence of climate variability, tour guide participants were asked to describe the region's climate. Over two-thirds (84.1 per cent) of the tour guides referred to the climate as having recently changed in their region (Chobe District) and in the CNP. Tour guides' views about climate variability indicate that it has affected wildlife-based tourism since some species such as waterbuck (*Kobus ellipsiprymnus*) are decreasing, and there is also evidence of stunted growth of vegetation as a result of frequently recurring fires. It should be noted that a sizeable percentage of tour guides (84.1 per cent) agreed that climate variability has affected wildlife-based tourism, while only 3.2 per cent disagreed. While almost all tour guides (85.7 per cent) agreed that climate variability will affect wildlife-based tourism in future, a few (3.2 per cent) disagreed with the statement.

To better understand the relative importance of climate, participants were then asked to indicate the role of climate in environmental change. A total of 79.3 per cent (n = 50) of the tour guides indicated that they strongly agree that climate change scenarios negatively influence

Table 10.1   *Principal Component Analysis matrix*

	Component	
	1	2
Loss of biodiversity	0.795	
Displacement of wildlife due to climate change	0.689	
Loss of land aesthetics (attractiveness)	0.665	
Lack of surface water availability	0.637	-0.470
Climate change scenario (more rains, storms)	0.573	
High volume flooding	0.517	0.667

travel decisions for tourism. About 87 per cent agreed that climate change could lead to loss of land attractiveness. In addition, a total of 99 per cent of the key informants from DWNP succinctly supported the tour guides by stating that environmental change, in particular climate change, will result in loss of aesthetic value of the area, especially some of the prime areas such as the Chobe Riverfront and Sedudu Island. Similarly, 96.8 per cent agreed that climate change could result in a loss of biodiversity especially in the wetlands environment. Also, the majority (98.5 per cent) of the tour guides agreed that climate change may negatively influence displacement and also have an effect on wildlife-based tourism activities. About 74.6 per cent felt that high volume flooding due to climate change would not be favourable for wildlife-based tourism. However, 4.8 per cent disagreed with the statement that high flooding would not favour wildlife-based tourism. Furthermore, 98.4 per cent of the tour guides agreed that climate change leads to lack of surface water availability due to drought which ends up affecting wildlife, while 1.6 per cent (n = 1) was neutral (neither agree nor disagree) with the statement.

ENVIRONMENTAL CHANGE EFFECTS ON WILDLIFE-BASED TOURISM

Principal Component Analysis

This study's Principal Component Analysis (PCA) showed that landscape (scenery) and water availability are highly influenced by environmental and climate change. At best, the landscape component had a higher correlation coefficient, with loss of biodiversity, displacement of wildlife and loss of land aesthetics with values of 0.795, 0.689 and 0.665 respectively (see Table 10.1). The grouping of climate change factors indicate that they have an effect on wildlife-based tourism. The study results showed that landscape/scenery is mostly defined by the availability of biodiversity, wildlife and land aesthetics. This indicates that landscape/scenery effect is important although the changing climate jeopardises the experience of watching animals and spectacular bird life especially in their undisturbed natural environment. It is worth noting that environmental change, in particular climate change, can have serious effects on biodiversity, wild animals, land aesthetics and consequently the future growth of wildlife-based tourism.

Similarly, another component was set which highly related to water availability. The high volume flooding was seen to be positively related (0.667) to the theme, with a lack of surface water availability indicated as negatively related (−0.470). High volume flooding and lack of

138　*Handbook on tourism and conservation*

surface water availability are linked to increased rainfall and severe droughts and increased temperatures which continue to challenge CNP. This suggests a scarcity of water will reduce the appeal of the area leading to less tourist attraction. In addition, the climate change scenario had little or no effect on the travelling decisions of wildlife-based tourists as it was not related to any of the extracted components.

Environmental Change Implications for Community Development

A sizeable percentage of participants agreed that climate variability has affected wildlife-based tourism while 23 per cent disagreed. Almost all participants (94 per cent) agreed that climate variability will affect wildlife-based tourism in future. Furthermore, 44.4 per cent agreed that climate change could lead to loss of land attractiveness. Similarly, 52.4 per cent agreed that climate change could result in loss of biodiversity especially in the wetlands environment. The majority (55.6 per cent) of participants agreed that displacement of wildlife due to climate change may lead to increased competition for forage, hence affecting wildlife-based tourism activities. About 44.4 per cent felt that high volume flooding due to climate change would not be favourable for wildlife-based tourism. More than half (69 per cent) of the respondents strongly agreed that climate change leads to lack of surface water availability due to drought which ends up affecting wildlife.

Given an interplay of myriad factors as a result of environmental change, diversification of wildlife-based tourism as a product is essential for community involvement and livelihood improvement. For example, one key informant from EWB suggested that:

> There is need to diversify tourism products especially at the Chobe Riverfront through the utilisation of Forest Reserves, horse-riding safaris, cultural products in the enclave villages, opening of the Nogatshaa area and uplifting of the whole Kasane and Kazungula area. (27 August 2018)

Several participants were confident that environmental change brings mostly negative impacts particularly on wildlife-based tourism. However, participants noted that there are positive impacts as well resulting from climate change. As illustrated in the following quote, a CARACAL director shared some views on environmental change and its positive impacts in general as key factors in the ecosystem:

> African ecosystems are dynamic, they change constantly, therefore, without those dynamics you cannot have biodiversity, so to some extent we need to accept environmental change and very often that's important in the functioning of these ecosystems and maintaining them. (29 August 2018)

As a result, environmental change may positively impact on wildlife:

> We are seeing certain areas where there is loss of woodlands due to the increasing elephant population. However, a study we conducted five years ago while I was working for the Department of Wildlife and National Parks, observed visual changes taking place. Elephants were also creating more diverse habitat and species diversity. Therefore, not all these changes are negative, some of them are positive. (29 August 2018)

DISCUSSION

The examination of perceptions of environmental change within the wildlife-based tourism nexus has shed light on concerns surrounding the sustainability of the industry in the face of increasing extreme environmental variability in rainfall and temperature, due to climate change. The latter seems likely to accentuate many key challenges that exist already within Chobe National Park and include:

1. The over concentration of wildlife (elephants in particular) and visitors (boats and game drive vehicles) along the eastern Chobe Riverfront, to the detriment of the overall quality of the game viewing experience.
2. The under-utilisation of extensive areas of Chobe National Park, the dry interior, by water-dependent wild ungulates (elephant, buffalo and zebra) and their associated predators.
3. A tendency to fall into increasingly 'reactive' forms of wildlife management such as the provision of AWPs which are relatively easily implemented, as compared to more definitive solutions such as the provision of migratory corridors to increase the connectivity between conservation landscapes at the regional (KAZA-TFCA) scale, but require a cross-sectoral approach (Perkins, 2020).
4. The increased sedentarisation of mobile ungulates and predators due to increasing constraints to movement through wildlife corridors and into outlying dispersal areas, despite the fact migratory systems are both more productive and a major attraction to tourists (Larsen et al., 2020).
5. More extensive and severe fires (Bowman et al., 2020; Cassidy et al., 2022) within both the surrounding Forest Reserves and the Chobe National Park and consequent reduction in tree cover and habitat diversity.
6. Increasing isolation of local communities from meaningful participation within the tourism industry and the benefits of wildlife conservation, as game viewing becomes increasingly concentrated in a few select areas already dominated by well-established corporate players within the tourism industry.
7. Increased centralisation of key wildlife management activities within the Department of Wildlife and National Parks and other key government departments (Department of Forestry and Range Resources) regarding the monitoring and management of key factors such as fire, invasive and alien species, poaching, human–wildlife conflict, rare and endangered species and habitats, disease, problem animal control and consumptive use of natural resources around the Protected Area (e.g. trophy hunting).
8. Increased concentration of wildlife populations within the confines of the Protected Area landscape and the closure of opportunities for wildlife dispersal into 'shared landscapes' with people and agriculture in the broader outlying areas (such as north-western Botswana/ Namibia) due to a dominant 'command and control' style of management.
9. Increased 'boom and bust' population dynamics within the ecosystem due to the increased frequency and severity of droughts, floods, disease, fires, invasive species (Liu et al., 2020; e.g. *Salvinia molesta*; Kurugundla et al., 2016), pest species (e.g. locusts), poaching and illegal trade and trafficking of wildlife and wildlife products (Lunstrum & Givá, 2020).

As Southern Africa becomes drier and hotter this century the above factors will become even more important, with the opening of the new road and rail bridge across the Zambezi

140   *Handbook on tourism and conservation*

at Kazungula delivering unprecedented numbers of visitors to Kasane and the already 'over-crowded' eastern Chobe Riverfront. Early warning signs of what is to come have already been experienced with the mysterious die-off of hundreds of elephants in the Okavango Delta due to high cyanobacteria concentrations in waterholes (BBC News, 2020), an escalation in the poaching of ivory (Schlossberg et al., 2019) within and around the Protected Area, increased encroachment of livestock and people into the margins around the Protected Area – most markedly into the floodplain areas around the Okavango Delta with a consequent increase in bush meat harvesting (Rogan et al., 2017) and human–wildlife conflict (Weise et al., 2018, 2019) – and increasing pressure on key wildlife migratory corridors between the Chobe National Park and Namibia (Naidoo et al., 2014) and Zimbabwe (Purdon et al., 2018).

Within the current structural policy constraints surrounding CBNRM in Botswana, increased environmental variability will further isolate local communities from the benefits of wildlife-based tourism and wildlife conservation. Extreme environmental variability is a relatively new influence on community development (Lapeyre, 2011; Yurco et al., 2017) and one that will require cross-sectoral solutions and transformative change to the way in which wildlife ecosystems and wilderness areas are valued and managed. Most critically there is a need to expand the spatial and temporal scale of wildlife management beyond the confines of KAZA (e.g. to Eastern Africa via KALARIVA-TFCL, the Kalahari-Rift Valley TransFrontier Landscape; Perkins, 2019, 2020) and to fully integrate sustainable development into wildlife conservation via the coexistence of wildlife and agriculture across extensive areas (Ellis, 2019).

The negative effects experienced at CNP are consistent with those observed elsewhere, particularly where environmental issues are accompanied by inappropriate policies and monitoring issues (Gössling et al., 2006; Hambira et al., 2013). Similarly, our analysis also indicated that environmental change, in particular climate change, will continue to cause significant stress on the wildlife ecosystem and the future development of tourism, without transformative change that places local communities at the very heart of the benefits accruing from wildlife conservation and the associated tourist industry (Thondhlan et al., 2020).

Similarly, the observation is consistent with Moswete and Mavondo (2003), who observed that diversification may be used as a form of investment especially for tourism products with increased employment opportunities which may certainly improve the lives of the local communities. Moreover, our tour guide interviews demonstrated that although communities have to benefit from wildlife and wildlife-based tourism, social and political issues act as a hindrance to allow them to benefit from the abundant wildlife resources. This is in line with Stone's (2013) findings which emphasise that such changes may influence new economic patterns, and in turn shift community value systems further away from wildlife. While some of the arguments found here are similar to those expressed in other studies elsewhere in the country (see Garekae et al., 2020; Mbaiwa & Hambira, 2019; Sebele, 2010), Chobe National Park is a unique wildlife and tourism resource and a key refuge area for wildlife within the broader KAZA-TFCA. Botswana is a signatory to the United Nations SDGs, the achievement of which is embedded in Vision 2036 and the National Development Plan (NDP11). It remains to be seen whether broader scale policy commitments can be translated to transformative change on the ground, particularly as it involves CBNRM and adaptive management to climate change.

CONCLUSION AND IMPLICATIONS FOR COMMUNITY DEVELOPMENT

Chobe National Park, like Protected Areas throughout Africa, is under unprecedented pressure from a number of key threats primarily driven by a growing population and expanding economy. Chobe National Park is, along with the Serengeti-Mara ecosystem of eastern Africa, the premier destination for viewing African wildlife in a wilderness setting and has developed a thriving tourism market. However, as is the case throughout the world the benefits of an expanding economy – and the costs – are unequally distributed and have been for decades. The result has been the convergence of a number of related threats that have collectively caused a decline in the integrity and resilience of the ecosystem upon which Chobe National Park depends.

There is general evidence that the environment plays an integral role for humankind through its functions that highly support development. While the ecosystem services are seen as involving many factors, the majority of respondents perceived environmental change negatively. This perception highlighted that community development is threatened by many factors such as environmental degradation and climate change. Without transformational change occurring that enables local communities to benefit from living with wildlife in shared landscapes, and actively opting to do so, it is difficult to see how the current deteriorating situation for conservation and sustainable development in and around CNP can be averted. Interestingly, study findings suggested possible options for diversification at both small scale and large scale. Such options are believed to have a role to play in saving the vulnerable wildlife-based tourism system. Essentially, the findings conform to the propositions of the conceptual approach that community sustainability is valued for its important aspects of coping and recovery of resources without compromising opportunities for future use. This study also concludes that the environment can manage itself through its own dynamics to achieve sustainability. Interestingly, there is a need to assess the behaviour of people towards the ecosystem for purposes of sustainable livelihoods in a systematic way. It would be entirely fitting if Botswana is once again able to take the global community on a new trajectory that could lead to the realisation of the UN SDGs and the saving of humanity. There is certainly a need for future studies to juxtapose people's perceptions about environmental change with trend analysis data to provide a vivid insight on environmental and/or climate change implications for community development.

REFERENCES

Adams, W. M., & Hutton, J. (2007). People, parks and poverty: Political ecology and biodiversity conservation. *Conservation and Society, 5*(2), 147–183.

Alemaw, B. F., & Simalenga, T. (2015). Climate change impacts and adaptation in rainfed farming systems: A modeling framework for scaling-out climate smart agriculture in sub-Saharan Africa. *American Journal of Climate Change, 4*(4), 313–329.

Andermann, T., Faurby, S., Turvey, T., Antonelli, A., & Silvestrol, D. (2020). The past and future human impact on mammalian diversity. *Science Advances, 6*(36). https://doi.org/10.1126/sciadv.abb2313.

Babbie, E. R. (2014). *The Basics of Social Research* (6th edn). Belmont, CA: Wadsworth/Cengage Learning.

BBC News (2020). Botswana: Mystery elephant deaths caused by cyanobacteria. www.bbc.com/news/world-africa-54234396, accessed 11 October 2021.

142 *Handbook on tourism and conservation*

Bowman, D. M. J. S., Kolden, C. A., Abatzoglou, J. T., Johnston, F. H., van der Werf, G. R., & Flannigan, M. (2020). Vegetation fires in the Anthropocene. *Nature Reviews Earth & Environment*, *1*(10), 500–515.

Braun, V., & Clarke, V. (2006). Using thematic analysis in psychology. *Qualitative Research in Psychology*, *3*(2), 77–101.

Brook, G. A., Scott, L., Railsback, L. B., & Goddard, E. A. (2010). A 35 ka pollen and isotope record of environmental change along the southern margin of the Kalahari from a stalagmite and animal dung deposits in Wonderwerk Cave, South Africa. *Journal of Arid Environments*, *74*(7), 870–884.

BTO [Botswana Tourism Organisation] (2016). *The Chobe National Park*. www.botswanatourism.co.bw/chobeNationalpark.php, accessed 3 March 2021.

Cassidy, L., Perkins, J. S., & Bradley J. (2022). Too much, too late: Fires and reactive wildfire management in northern Botswana's forests and woodland savannas. *African Journal of Range and Forage Science*, *39*(1), 160–174.

Chambers, R., & Conway, G. (1992). *Sustainable Rural Livelihoods: Practical Concepts for the 21st Century* (IDS Discussion Paper Number 296). Brighton: Institute of Development Studies.

Creswell, J. W. (2013). *Research Design: Qualitative, Quantitative, and Mixed Methods Approaches*. London: Sage Publications.

CSO (2011). *Botswana Population and Housing Census*. Central Statistics Office, Gaborone, Botswana.

Dawson, C. (2009). *Introduction to Research Methods: A Practical Guide for Anyone Undertaking a Research Project* (4th edn). Oxford: How to Books.

DWNP [Department of Wildlife and National Parks] (2008). *Chobe Management Plan, 2008*. Ministry of Environment, Wildlife and Tourism, Department of Wildlife and National Parks, Gaborone, Botswana.

Ellis, E. C. (2019). Sharing the land between nature and people. *Science*, *364*(6447), 1226–1228.

Engelbrecht, F., Adegoke, J., Bopape, M. J., Naidoo, M., Garland, R., Thatcher, M., McGregor, J., Katzfey, J., Werner, M., Ichoku, C., & Gatebe, C. (2015). Projections of rapidly rising surface temperatures over Africa under low mitigation. *Environmental Research Letters*, *10*(8), 085004.

Everatt, K., Kokes, R., & Pereira, C.L. (2019). Evidence of a further emerging threat to lion conservation: Targeted poaching for body parts. *Biodiversity and Conservation*, *28*(14), 4099–4114.

Forstner, K. (2004). Community ventures and access to markets: The role of intermediaries in marketing rural tourism products. *Development Policy Review*, *22*(5), 497–514.

Garekae, H., Lepetu, J., & Thakadu, O. T. (2020). Forest resource utilisation and rural livelihoods: Insights from Chobe enclave, Botswana. *South African Geographical Journal*, *102*(1), 22–40.

GOB [Government of Botswana] (2001). *Botswana National Atlas*. Gaborone: Department of Surveys and Mapping, Government Printer.

Goodwin, H. (1996). In pursuit of ecotourism. *Biodiversity & Conservation*, *5*(3), 277–291.

Gössling, S., Bredberg, M., Randow, A., Sandström, E., & Svensson, P. (2006). Tourist perceptions of climate change: A study of international tourists in Zanzibar. *Current Issues in Tourism*, *9*(4–5), 419–435.

Gössling, S., & Hall, C. M. (2006). An introduction to tourism and global environmental change. In S. Gössling & C. M. Hall (eds), *Tourism and Global Environmental Change*. New York: Routledge, pp. 1–34.

Gupta, A. C. (2013). Elephants, safety nets and agrarian culture: Understanding human–wildlife conflict and rural livelihoods around Chobe National Park, Botswana. *Journal of Political Ecology*, *20*(1), 238–254.

Hall, D. R., & Brown, F. (2006). *Tourism and Welfare: Ethics, Responsibility and Sustained Well-Being*. Wallingford: Cabi.

Hambira, W. L., Saarinen, J., Manwa, H., & Atlhopheng, J. R. (2013). Climate change adaptation practices in nature-based tourism in Maun in the Okavango Delta area, Botswana: How prepared are the tourism businesses? *Tourism Review International*, *17*(1), 19–29.

Herrero, H. V., Southworth, J., & Bunting, E. (2016). Utilizing multiple lines of evidence to determine landscape degradation within protected area landscapes: A case study of Chobe National Park, Botswana from 1982 to 2011. *Remote Sensing*, *8*(8), 623–640.

Hoogendoorn, G., & Fitchett, J. M. (2016). Tourism and climate change: A review of threats and adaptation strategies for Africa. *Current Issues in Tourism*, *21*(7), 742–759.

Hoogendoorn, G., & Fitchett, J. M. (2018). Perspectives on second homes, climate change and tourism in South Africa. *African Journal of Hospitality, Tourism and Leisure*, *7*(2), 1–18.

Jones, K. E., Patel, N. G., Levy, M. A., Storeygard, A., Balk, D., Gittleman, J. L., & Daszak, P. (2008). Global trends in emerging infectious diseases. *Nature*, *451*(7181), 990–993.

Jones, T. B. (1999). Community-based natural resource management in Botswana and Namibia: An inventory and preliminary analysis and progress. Report submitted to International Institute for Environment and Development.

Koubi, V., Spilker, G., Schaffer, L., & Böhmelt, T. (2016). The role of environmental perceptions in migration decision-making: Evidence from both migrants and non-migrants in five developing countries. *Population and Environment*, *38*(2), 134–163.

Kurugundla, C. N., Mathangwane, B., Sakuringwa, S., & Katorah, G. (2016). Alien invasive aquatic plant species in Botswana: Historical perspective and management. *The Open Plant Science Journal*, *9*(1), 1–40.

Kutzner, D. (2019). Environmental change, resilience, and adaptation in nature-based tourism: Conceptualizing the social-ecological resilience of birdwatching tour operations. *Journal of Sustainable Tourism*, *27*(8), 1142–1166.

Lambert, E., Hunter, C., Pierce, G. J., & MacLeod, C. D. (2010). Sustainable whale-watching tourism and climate change: Towards a framework of resilience. *Journal of Sustainable Tourism*, *18*(3), 409–427.

Lapeyre, R. (2011). Governance structures and the distribution of tourism income in Namibian communal lands: A new institutional framework. *Tijdschrift voor Economische en Sociale Geografie*, *102*(3), 302–315.

Larsen, F., Grant, J., Hopcraft, C., Hanley, N., Hongoa, J. R., Hynes, S., Loibooki, M., Mafuru, G., Needham, K., Shirima, F., & Morrison, T. A. (2020). Wildebeest migration drives tourism demand in the Serengeti. *Biological Conservation*, *248*. https://doi.org/10.1016/j.biocon.2020.108688.

Lindsey, P., Allan, J., & Brehony, P. (2020). Conserving Africa's wildlife and wildlands through the COVID-19 crisis and beyond. *Nature Ecology and Evolution*, *4*(10), 1300–1310.

Liu, X., Blackburn, T. M., Song, T., Wang, X., Huang, C., & Li, Y. (2020). Animal invaders threaten protected areas worldwide. *Nature Communications*, *11*(1), 1–9. https://doi.org/10.1038/s41467-020 -16719-2.

Lunstrum, E., & Givá, N. (2020). What drives commercial poaching? From poverty to economic inequality. *Biological Conservation*, *245*, 108505. https://doi.org/10.1016/j.biocon.2020.108505.

Mamba, H. S., Randhir, T. O., & Fuller, T. K. (2020). Community attitudes and perceptions concerning rhinoceros poaching and conservation: A case study in Eswatini. *African Journal of Wildlife Research*, *50*(1), 1–7.

Massé, F., & Lunstrum, E. (2016). Accumulation by securitization: Commercial poaching, neoliberal conservation, and the creation of new wildlife frontiers. *Geoforum*, *69*, 227–237.

Mbaiwa, J. E. (2017). Poverty or riches: Who benefits from the booming tourism industry in Botswana? *Journal of Contemporary African Studies*, *35*(1), 93–112.

Mbaiwa, J. E., & Hambira, W. L. (2019). Enclaves and shadow state tourism in the Okavango Delta, Botswana. *South African Geographical Journal*, *102*(1), 1–21.

Mbaiwa, J. E., & Stronza, A. L. (2010). The effects of tourism development on rural livelihoods in the Okavango Delta, Botswana. *Journal of Sustainable Tourism*, *18*(5), 635–656.

Mogende, E., & Moswete, N. (2018). Stakeholder perceptions on the environmental impacts of wildlife-based tourism at the Chobe National Park River Front, Botswana. *PULA: Botswana Journal of African Studies*, *32*(1), 48–67.

Moritz, C., & Agudo, R. (2013). The future of species under climate change: Resilience or decline? *Science*, *341*(6145), 504–508.

Mosugelo, D. K., Moe, S. R., Ringrose, S., & Nellemann, C. (2002). Vegetation changes during a 36-year period in northern Chobe National Park, Botswana. *African Journal of Ecology*, *40*(3), 232–240.

Moswete, N., & Mavondo, F. (2003). Problems facing the tourism industry of Botswana. *Botswana Notes and Records*, *35*(1), 69–77.

Moswete, N., Nkape, K., & Tseme, M. (2017). Wildlife tourism safaris, vehicle decongestion routes and impact mitigation at Chobe National Park, Botswana. In I. Borges de Lima & R. J. Green (eds), *Wildlife Tourism, Environmental Learning and Ethical Encounters*. Cham: Springer, pp. 71–88.

Naidoo, R., Du Preez, P., Stuart-Hill, G., Beytell, P., & Taylor, R. (2014). Long-range migrations and dispersals of African buffalo (Syncerus caffer) in the Kavango–Zambezi Transfrontier Conservation Area. *African Journal of Ecology, 52*(4), 581–584.

Owen-Smith, N. (1996). Ecological guidelines for waterpoints in extensive protected areas. *South African Journal of Wildlife Research, 26*(4), 107–112.

Paksi, A., & Pyhälä, A. (2018). Socio-economic impacts of a national park on local indigenous livelihoods: The case of the Bwabwata National Park in Namibia. *Senri Ethnological Studies, 99*, 197–214.

Pandey, R., Jha, S. K., Alatalo, J. M., Archie, K. M., & Gupta, A. K. (2017). Sustainable livelihood framework-based indicators for assessing climate change vulnerability and adaptation for Himalayan communities. *Ecological Indicators, 79*, 338–346.

Perkins, J. S. (2019). 'Only connect': Restoring resilience in the Kalahari ecosystem. *Journal of Environmental Management, 249*, 109420.

Perkins, J. S. (2020). Take me to the river along the African drought corridor: Adapting to climate change. *Botswana Journal of Agriculture and Applied Sciences, 14*(1), 60–71.

Pohl, B., Macron, C., & Monerie, P. A. (2017). Fewer rainy days and more extreme rainfall by the end of the century in Southern Africa. *Scientific Reports, 7*(1), 1–7.

Purdon, A., Mole, M. A., Chase, M. J. et al. (2018). Partial migration in savanna elephant populations distributed across southern Africa. *Scientific Reports, 8*, 11331. https://doi.org/10.1038/s41598-018 -29724-9.

Richards, G., & Hall, D. (2002). The community: A sustainable concept in tourism development? In D. Hall & G. Richards (eds), *Tourism and Sustainable Community Development*. London: Routledge, pp. 19–32.

Ripple, W. J., Estes, J. A., Bescht, R. L., Wilmers, C. C., Ritchie, E. G., Hebblewhite, M., Berger, J., Elmhagen, B., Letnic, M., Nelson, M. P., Schmitz, O. J., Smith, D. W., Wallach, A. D., & Wirsing, A. J. (2014). Status and ecological effects of the world's largest carnivores. *Science, 343*(6167), 1241484.

Ripple, W. J., Newsome, T. M., Wolf, C. C., Dirzo, R., Everatt, K., Galetti, M., Kerley, G. I. H., Levi, T., Lindsey, P. A., Macdonald, D., Malhi, Y., Painter, L. E., Sandom, C., Terborgh, J., and Valkenburgh, B. V. (2015). Collapse of the world's largest herbivores. *Science Advances, 1*(4), e1400103.

Rogan, M. S., Lindsey, P. A., Tambling, C. J., Golabek, K. A., & McNutt, J. W. (2017). Illegal bushmeat hunters compete with predators and threaten wild herbivore populations in a global tourism hotspot. *Biological Conservation, 210*(A), 233–242.

Rogerson, C. M. (2016). Climate change, tourism and local economic development in South Africa. *Local Economy, 31*(1–2), 322–331.

Roseland, M. (2000). Sustainable community development: Integrating environmental, economic, and social objectives. *Progress in Planning, 54*(2), 73–132.

Saarinen, J. (2006). Traditions of sustainability in tourism studies. *Annals of Tourism Research, 33*(4), 1121–1140.

Saarinen, J. (2013). Critical sustainability: Setting the limits to growth and responsibility in tourism. *Sustainability, 6*(1), 1–17.

Saarinen, J., & Tervo, K. (2006). Perceptions and adaptation strategies of the tourism industry to climate change: The case of Finnish nature-based tourism entrepreneurs. *International Journal of Innovation and Sustainable Development, 1*(3), 214–228.

Scheiter, S., Gaillard, C., Martens, C., Erasmus, B. F. N., & Pfeiffer, M. (2018). How vulnerable are ecosystems in the Limpopo province to climate change? *South African Journal of Botany, 116*, 86–95.

Schlossberg, S., Chase, M. J., & Sutcliffe, R. (2019). Evidence of a growing elephant poaching problem in Botswana. *Current Biology, 29*(13), 2222–2228.

Scott, D. (2011). Why sustainable tourism must address climate change. *Journal of Sustainable Tourism, 19*(1), 17–34.

Scott, D., Jones, B., & Konopek, J. (2007). Implications of climate and environmental change for nature-based tourism in the Canadian Rocky Mountains: A case study of Waterton Lakes National Park. *Tourism Management, 28*(2), 570–579.

Scott, D., & Lemieux C. (2010). Weather and climate information for tourism. *Procedia Environmental Sciences, 1*, 146–183.

Sebele, L. S. 2010. Community-based tourism ventures, benefits and challenges: Khama rhino sanctuary trust, central district, Botswana. *Tourism Management, 31*(1), 136–146.

Sekonya, G. J. (2016). Mopane worm use, livelihoods and environmental change in Limpopo Province, South Africa. Master's dissertation, University of Cape Town, South Africa.

Simpson, M. C. (2008). Community benefit tourism initiatives: A conceptual oxymoron? *Tourism Management, 29*(1), 1–18.

Stone, M. T. (2013). Protected areas, tourism and rural community livelihoods in Botswana. PhD dissertation, Arizona State University, USA.

Stone, M. T., & Stone, L. S. (2020). Challenges of community-based tourism in Botswana: A review of literature. *Transactions of the Royal Society of South Africa, 75*(4), 1–13.

Tervo-Kankare, K., Kaján, E., & Saarinen, J. (2018). Costs and benefits of environmental change: Tourism industry's responses in Arctic Finland. *Tourism Geographies, 20*(2), 202–223.

Tessema, M. E., Lilieholm, R. J., Ashenafi, Z. T., & Leader-Williams, N. (2010). Community attitudes toward wildlife and protected areas in Ethiopia. *Society and Natural Resources, 23*(6), 489–506.

Thondhlan, G., Redpath, S. M., Vedeld, P. O., van Eeden, L., Pascual, U., Sherren, K., & Murata, C. (2020). Non-material costs of wildlife conservation to local people and their implications for conservation interventions. *Biological Conservation, 246*, 108578. https://doi.org/10.1016/j.biocon.2020.108578.

Tollefson, J. (2020). Why deforestation and extinctions make pandemics more likely. *Nature, 584*(7820), 175–176.

Tsheboeng, G., Bonyongo, M., & Murray-Hudson, M. (2014). Flood variation and soil nutrient content in floodplain vegetation communities in the Okavango Delta. *South African Journal of Science, 110*(3-4), 1–5.

Vitousek, P. M., Mooney, H. A., Lubchenco, J., & Melillo, J. M. (1997). Human domination of Earth's ecosystems. *Science, 277*(5325), 494–499.

Weise, F. J., Hauptmeier, H., Stratford, K. J., Hayward, M. W., Aal, K., Heuer, M., Tomeletso, M., Wulf, V., Somers, M. J., & Stein, A. B. (2019). Lions at the gates: Transdisciplinary design of an early warning system to improve human–lion coexistence. *Frontiers in Ecology and Evolution, 6*, 242. https://doi.org/10.3389/fevo.2018.00242.

Weise, F. J., Hayward, M. W., Casillas Aguirre, R., Tomeletso, M., Gadimang, P., Somers, M. J., & Stein, A. B. (2018). Size, shape, and maintenance matter: A critical appraisal of a global carnivore mitigation strategy: Livestock protection kraals in northern Botswana. *Biological Conservation, 225*, 88–97.

Wolf, A. (2009). Preliminary assessment of the effect of high elephant density on ecosystem components (grass, trees, and large mammals) on the Chobe riverfront in northern Botswana. University of Florida.

Yurco, K., King, B., Young, K. R., & Crews, K. A. (2017). Human–wildlife interactions and environmental dynamics in the Okavango Delta, Botswana. *Society & Natural Resources, 30*(9), 1112–1126.

PART III

SUSTAINABLE AGRITOURISM

11. Potentials and challenges of sustainable agritourism in Fortín, Veracruz, Mexico

Karina Nicole Pérez-Olmos, Noé Aguilar-Rivera and Carlos Enrique Villanueva-González

INTRODUCTION

Agritourism represents an opportunity for agricultural producers to commit to work under a sustainable approach by offering products and services designed for agritourists interested in conserving the environment and agriculture (Flanigan et al., 2014). Gil et al. (2013) indicate that agritourism is the set of activities related to a working farm or other agricultural establishments, which have the purpose of educating or entertaining visitors. Likewise, the diversification in rural spaces as a complement to the traditional practices of farming creates a space with a new vision of rurality to improve the quality of life of local people (Barbieri, 2010; Gao et al., 2013; Reyes-Aguilar et al., 2019). Viewing tourism as a diversification strategy in Latin America's rural areas, this activity is among those that most enhance poverty alleviation and it also serves as a protector of natural resources (Blanco and Riveros, 2010; Kieffe, 2018; Lyon, 2013).

Agricultural landscapes mainly of crops such as sugarcane, coffee, flowers, foliage and the banana leaf (belillo) characterize Fortín (García-Albarado et al., 2018). Agritourism can be a multifunctional strategy of the rural space to strengthen the socioeconomics of the producers and create an image of rural tourism based on agricultural resources (Barbieri, 2010; Gao et al., 2013; LaPan and Barbieri, 2014). Despite this, the agritourism offer in Fortín has not had any major advances due to different challenges that must be addressed to achieve the integral growth of the subsector. Likewise, there is scant information about agritourism in Fortín. Therefore, this chapter aims to analyze the potentialities and challenges related to agritourism in Fortín from the perspective of local actors who are decision-makers in the tourism sector and/or who seek to improve their livelihoods through this livelihood option.

AGRITOURISM

Agritourism represents a strategy of multi-functionality landscape and, at the same time, can help preserve farms. Additionally, agritourism is a means of economic diversification in agricultural territories undergoing restructuring and an alternative option for facing an increasingly growing demand for rural recreational uses (Che et al., 2005). In this sense, the literature indicates that one of the main reasons for the adoption of agritourism by producers is to increase farm income (Tew and Barbieri, 2012); allowing them to maintain a lifestyle that helps the preservation of the agricultural use of their land and the conservation of nearby natural resources (Veeck et al., 2016). Likewise, through agritourism activities, producers add services to the agricultural production of the farm and cultural ecosystem services (e.g.

148   *Handbook on tourism and conservation*

enjoyment of the landscape, fresh air and open spaces to practice sports). In addition, other activities can be associated with agritourism such as recreational fishing, hunting grounds, rural accommodation, rural restaurants, local crafts, home industries and other leisure activities that revalue the rural lifestyle (Templado Isasa and Lucarelli, 2020).

Some countries that are globally recognized as success stories in the practice of agritourism are France, the United States and the United Kingdom (Rauniyar et al., 2021). In Latin America, some countries such as Colombia, Costa Rica and Mexico have developed agritourism offers based on visits to coffee, cocoa, pineapple, banana and grape plantations, among others (Blanco and Riveros, 2010).

Agritourism in Mexico

In Mexico, the tourism sector is one of the most dynamic areas of the national economy, representing 8.5 percent of the gross domestic product, providing about 2.3 million jobs and it is one of the trendy areas in recent years for the creation of public policies (OECD, 2017). Despite this, tourism policies have been focused on creating basic infrastructure for new destinations since the mid-1960s (Sandoval Quintero et al., 2017), working on mass tourism destinations of sun and beach (Rioja Peregrina et al., 2019).

Since the 1970s, the Mexican Government has been integrating rural tourism into its planning and management. One of the important components of this integration was the creation and implementation of regulatory frameworks and support programs for rural populations (Sandoval Quintero et al., 2017). Despite the various initiatives implemented in the national territory that in theory seek to promote rural tourism, public policies are mainly oriented to reflect foreign models that do not correspond to the needs, priorities and specific problems that this tourist modality seeks to resolve. Consequently, it is difficult to carry out successful participatory work in all phases of tourism programs and projects in the Mexican rural space (Thomé Ortiz et al., 2014).

Mexican rural tourism is generally based on the use of old haciendas, crop areas and typical Mexican products (e.g. tequila, mezcal, cocoa and vanilla); and foreign products that have been incorporated into Mexican culture (e.g. cheese, vines and coffee). A type of rural tourism is agritourism, which in Mexican territory is characterized by an offer of agri-food routes whose main attractions are regional crops and agricultural products where visitors participate in agricultural tasks, learn about the production process, consume the local gastronomy and buy handicrafts from the place (Alcalá Escamilla and López López, 2017). Agritourism in Mexico is still in an incipient development and the public tourism policies do not give the required attention to the subsector to enhance the welfare of the communities (Pérez-Olmos and Aguilar-Rivera, 2021a).

Agriculture and Agritourism in Fortín

Fortín is one of the regional urban centers that offer employment in factories and the construction industry to temporary residents who emigrate from nearby municipalities (Rodríguez, 2016). Agricultural activity is still latent in the rural areas of Fortín. Currently, the main crops present in the municipality according to the area cultivated in hectares are coffee (1,790), sugarcane (1,099), banana leaf (420) and chayote (53). Sugarcane ranks first in production with 103,905 tons and coffee in fourth place with 1,443 tons (SIAP, 2019). The low production of coffee is due to various problems in the coffee sector in Mexico including a drop in its prices and the attack of pests such as coffee berry borer and rust (Escamilla and Landeros, 2016). The production of piloncillo (a product of sugarcane juice) represents one of the traditional agro-industrial activities (Cortés et al., 2013). However, the piloncillo sector is currently affected by lack of technical assistance and financing (Cabrera Martínez and Aguilar-Rivera, 2019).

During the last years, there has been a significant increase in crops dedicated to tropical ornamental horticulture (García-Alonso et al., 2014). Flower cultivation is a traditional economic activity of the municipality and the gardenia flower (*Gardenia jasminoides* Ellis) is the representative image of Fortín. Despite this, today agricultural production continues in Fortín but to a lesser extent, giving way to an economy based mainly on the services sector (Ayuntamiento de Fortín, 2018). The agroindustry of Fortín faces various challenges that impact the competitiveness of the sector, such as difficulty in accessing credit, high costs of various inputs, lack of technical assistance, phytosanitary and deforestation problems, erosion and environmental pollution, and land use change from agricultural to residential areas (Cabrera Martínez and Aguilar-Rivera, 2019; García-Alonso et al., 2014; Rodríguez Deméneghi et al., 2020). Likewise, the sector is strongly affected by the use of local intermediaries for sale (Martínez, 2012).

The greatest efforts made on the development of agritourism in Fortín during the last two decades have been through a few local actors from private companies (rural accommodation and tour operators) who offer activities where visitors can learn about the agricultural aspects of Fortín. However, research about agritourism in Fortín is very scant, and few studies address the topic. Some of these studies have looked at tropical floriculture as a complement to rural tourism in the central area of the state of Veracruz, Mexico (Baltazar-Bernal et al., 2018); identity elements of the region of the high mountains of Veracruz (García-Albarado et al., 2018); proposal for the development of sugarcane agritourism in Fortín (Pérez-Olmos and Aguilar-Rivera, 2019); and agritourism opportunities in the *cañada del Metlac* (Metlac canyon) (Pérez-Olmos and Aguilar-Rivera, 2021b).

Therefore, there are no specific studies on agritourism characteristics in Fortín, its challenges and its potential to metamorphose into an important socioeconomic sector, as well as an option for the conservation of the existing ecological and agricultural resources. In this sense, the chapter seeks to analyze the situation of agritourism in Fortín to provide updated information that can be used by local actors interested in the development of this subsector and by researchers who wish to carry out other studies on the subject in the municipality and region.

METHODOLOGY

Study Zone

The municipality of Fortín is located in the state of Veracruz, Mexico (Figure 11.1), which has a land area of 62 km² and a population of 66,768 inhabitants and stretches to the high mountains region (CONAPO, 2015; Ayuntamiento de Fortín, 2018). The agricultural production of Fortín is characterized by crops such as coffee, flowers, sugarcane and banana veil, among others; agriculture has a rich tradition and the region boasts beautiful agricultural landscapes suitable for the practice of rural tourism.

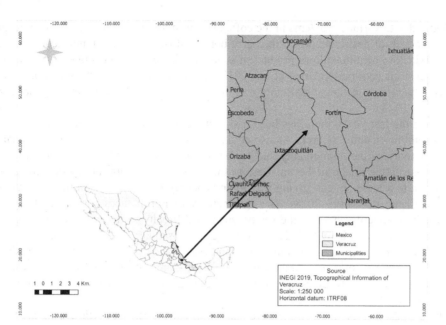

Source: Own elaboration.

Figure 11.1   Map of Fortín, Veracruz

Materials and Methods

The results presented in this chapter are part of research on "Agritourism in the multifunctionality of the landscape" that was carried out in 2017–2021. The study was developed following the exploratory study guidelines proposed by Hernández-Sampieri et al. (2014), who define it as those studies that investigate little-studied problems, and help to identify promising concepts, which serve as the basis for future studies. In this sense, the information was generated through field trips, surveys carried out among producers, a tourist inventory of the municipality and focus groups. Also, observation, field notes and informal talks were considered.

A structured survey was carried out among producers in the municipality with a non-probability sampling utilizing the snowball technique (n = 25) being analyzed using frequencies in the Infostat statistical software version 2018 to summarize the main characteristics and socioeconomic aspects. Likewise, three focus group discussions were held with 20 local actors comprising tourism entrepreneurs, local government officials and producers. Within the pre-established script, the participants were asked about the current situation of agritourism in Fortín and the challenges facing the sector. This methodology seeks to make visible, analyze and capture the perspectives and opinions of local actors about the situation of agritourism in Fortín. As indicated by Abarca Alpízar (2016), participatory methodology integrates systematic actions of analysis and study of the reality of different initiatives, leading those involved in this process to reflect on the issue that is under analysis.

In addition, an inventory of tourism resources was conducted following the methodology of the Ministry of Foreign Trade and Tourism of Peru (2018) which groups tourist resources into the following categories: (1) Natural sites; (2) Museums and cultural events; (3) Technical

Potentials and challenges of sustainable agritourism in Fortín 151

and scientific achievements; (4) Scheduled events; and (5) Tourist services. Furthermore, the possibilities of developing agritourism in the study area are evaluated according to the characteristics of agritourism presented by Phillip et al. (2010) and Flanigan et al. (2014) in their studies on a typology to define this tourist modality.

RESULTS AND DISCUSSION

Potentialities of Fortín in Strengthening Sustainable Agritourism

Agricultural resources

The offer of leisure activities on farms promotes the preservation and perpetuation of the inheritance of family farms (Barbieri et al., 2016), motivating the visitors through a touristic offer based on relaxation, natural landscapes, the authenticity of the farms and agricultural activities (Gao et al., 2013). From this perspective, the agricultural resources of Fortín distinguish it nationally, especially in the production of flowers, and can be used to promote agritourism in the area. This municipality is known for being a pioneer in the production of flowers in Mexico, becoming the largest exporter nationwide some years ago (México Desconocido, 2014). In this sense, it is worth mentioning that the state of Veracruz accounts for 1.4 percent of the national production of flowers (SAGARPA, 2018). In this context, surveys were conducted among 25 producers about the general aspects of their agricultural activity and their socioeconomics to learn more about the agricultural reality of Fortín (see Table 11.1).

Most respondents (producers) were male (56 percent) being over 46 years old (72 percent). Most of the respondents (56 percent) had only attended elementary school; these findings support several studies developed in the state of Veracruz (e.g. Pérez-Olmos and Aguilar-Rivera, 2019; Alejandre-Castellanos et al., 2020) where producers have a similar age range and have a low level of schooling. The respondents produced traditional crops, most of which are coffee and foliage (36 percent apiece). Only 16 percent of the respondents had another economic activity as the main source of income. Most producers (96 percent) had small plots of land not exceeding 5 ha. According to Tew and Barbieri (2012), agritourism can be a strategy that makes a difference in improving the economy of small producers.

Only 24 percent of the producers have an irrigation system. It is noteworthy that Fortín agriculture is characterized by low technological levels (Cabrera Martínez and Aguilar-Rivera, 2019). Most of the respondents applied organic fertilization on their crops (40 percent). Most of the respondents (96 percent) indicated that they had never received tourists on the farm. Only 8 and 4 percent of the respondents had an administration and a promotion system, respectively, for their agricultural activities. Most producers (56 percent) have their vehicle to use for farm work.

Other attractions

As Varisco (2016) indicates, the attractions or resources are the elements of the tourism system that motivate the visitor to travel to the tourist destination and these can be of a natural or

152 *Handbook on tourism and conservation*

Table 11.1   Frequency analysis of surveys of producers of Fortín (n = 25)

Variable	Categories	Absolute frequency	Relative frequency
Gender	Female	11	44
	Male	14	56
Age	36–46 years	7	28
	+46 years	18	72
Education level	None	2	8
	Elementary school	13	52
	High school	6	24
	Technical school	1	4
	University	3	12
Principal crop	Coffee	9	36
	Flowers	3	12
	Foliage	9	36
	Sugarcane	4	16
Irrigation system	Yes	6	24
	No	19	76
Land surface	Smaller than 1 ha	11	44
	1–5 ha	13	52
	Bigger than 5 ha	1	4
Fertilization	Organic	10	40
	Chemical	6	24
	Mixed (organic and chemical)	9	36
Agriculture as main economic activity	Yes	21	84
	No	4	16
Tourist reception	Yes	1	4
	No	24	96
Administrative system	Yes	2	8
	No	23	92
Promotional system	Yes	1	4
	No	24	96
Vehicle(s)	Yes	14	56
	No	11	44

Source: Own elaboration.

cultural nature. Below, we briefly describe the main tourist attractions of Fortín which could strengthen agritourism in the area (see also Figure 11.2).

1. The Tatsugoro Bonsai[1] Museum where the visitors, guided by experts, can appreciate various specimens belonging to this ancient practice.
2. San Juan Hill (The Antennas Hill) where people can practice paragliding; an important representation of adventure tourism in the municipality.
3. The Tule lagoon is a body of water located in the community of Monte Blanco and surrounded by agricultural landscapes. In its surroundings, cycling and marathon activities take place.
4. Nursery areas: there are two parks for flower growers and two areas that have different nurseries (communities of Villa Unión and Monte Blanco). Some proprietors of these establishments carry out their production in greenhouses.

Potentials and challenges of sustainable agritourism in Fortín 153

Source: First author (2018).

Figure 11.2 Main tourist attractions of Fortín: (a) Tatsugoro Museum (sample of bonsai trees), (b) San Juan Hill, (c) Chula Vista camping, (d) Tule lagoon, (e) Nurseries of ornamental plants, (f) Corazón del Metlac

5. The Metlac canyon: this natural area has dominant vegetation of medium sub-evergreen forest where the arboreal layer is between 4 and 35 meters. Also, it houses areas with crops such as sugarcane (*Saccharum officinarum*), chayote (*Sechium edule*), coffee (*Coffea arabica*), banana (*Musa paradisiaca*), sweet orange (*Citrus sinensis* Osbeck) and cattle pasture (Landero-Torres et al., 2014). Inside the canyon is the so-called *Corazón del Metlác*, which is a site of tourist visitation and has an important ecological value due to its richness in species of fauna and flora (Almaraz-Vidal, 2015).
6. Architectural elements from the *El Mexicano* train: there are historical architectural elements belonging to the coffee industry and industrial heritage in the landscape of the canyon and surrounding areas such as the old track of the Sumidero-Fortín section of the *El Mexicano* train (Rivera-Hernández et al., 2019). This architectural work has had scenic beauty value and the landscape has been painted by artists such as José María Velasco (Figure 11.3).

Source: Velasco (1881).

Figure 11.3 "Cañada de Metlac"

Tourist services

Within the region of the High Mountains, the municipality of Fortín is among the municipalities (along with Córdoba, Orizaba and Huatusco) with the greatest commercial activity and has a tourist infrastructure made up of hotels, restaurants, travel agencies and local tour operators (Aguilera and Calvario, 2014). According to the tourist inventory applied in 2021 for this study (Figure 11.4), Fortín currently has nine accommodations, 87 restaurants, coffee shops and related businesses, two local tour operators and two adventure tourism service providers (paragliding).

Concerning transport and communications, Fortín has a 27.4 km road network and the railroad crosses Fortín through the center of the municipal territory, joining Córdoba with Ixtaczoquitlán, passing through the Fortín de las Flores headland. Regarding public transport, the vehicles available for this service represent only 3.84 percent of the total vehicle fleet, where the most representative is the public car (61.40 percent) and in second place the passenger truck or bus (32.99 percent) (Ayuntamiento de Fortín, 2018).

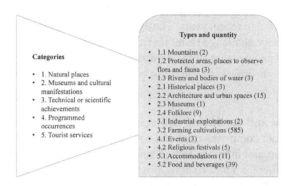

Source: Own elaboration.

Figure 11.4   Summary tourist inventory of Fortín

Challenges of Fortín in Strengthening Sustainable Agritourism

A series of limitations that hamper the sustainable growth of agritourism in Fortín were identified by the participants of the focus groups (Table 11.2). Regarding the tourist visits to the municipality, the focus group participants commented that due to the Covid-19 pandemic, tourist arrivals in Fortín have decreased alarmingly. Also, the municipality has not realized any tourism promotion strategy outside the region. As a result of this weak tourist diffusion, the few alternative tourism companies operating in the territory are at risk of disappearing. In addition, there are a few regulatory frameworks on tourism in the area and there are also no clear regulations on alternative tourism in Fortín.

Table 11.2   Responses obtained from focus groups

1. Little tourist flow	8. Problems between producers with team working
2. The tourism infrastructure needs to be strengthened	9. Low profitability in the prices of agricultural products
3. Lack of opportunities for tourism sector	10. Migration of producers' sons
4. Weak structure of tourist promotion	11. Poor management of public funds
5. Lack of safety and security	12. Little interest from the municipal government
6. Almost nonexistent framework of regulations	13. Few training opportunities
7. Producers do not have the necessary technical assistance	

Source: Own elaboration.

Agriculturally, the area suffers from some setbacks thus making the producers opine that the sector has been overlooked by the municipal governments; the reason could be that in the

156   *Handbook on tourism and conservation*

last years the municipality has concentrated on the growth of other sectors such as services. Some participants indicated that the government had not invested in rural development matters implying that there is mismanagement of funds. This is reflected in almost nonexistent technical assistance, and lack of opportunities for training leading to the migration of farm youths in the search of employment.

In another vein, many social problems confront residents of the municipality including lack of security. The participants indicated that violence linked to organized crime in both rural and urban areas had increased in recent years. The existence of only a few agricultural associations in the area is also a major challenge resulting from producers' preference for individualism at the expense of the advantages that social networks might confer on them. Based on the information obtained through different techniques, we present below the main challenges facing agritourism in Fortín.

Product development and marketing strategies
The agritourism product comprises rural and natural heritage, agricultural and industrial heritage, and tourism plant and infrastructure (Blanco and Riveros, 2010). All these elements are present in Fortín, but there is a lack of agritourism products with the appropriate standards and marketing strategies following the new market trends. This situation may be due to the limited opportunities which producers can access from the government and private sector. In Latin America, the aid of international funds for agricultural development has decreased in recent years (Lyon, 2013).

A strategy that has worked in several Latin American countries (e.g. Mexico, Costa Rica and Argentina) and that allows the promotion of products of territorial identity is the realization of agri-food routes. Through this tourism product, various modalities such as agritourism, ecotourism and rural tourism can be integrated such that the visitors could learn about agricultural processes, visit sites of natural and historical interest, and acquire local products, complemented with services such as food and lodging (Blanco and Riveros, 2010). Also, it is important to establish a brand name to perform better market segmentation and seek to satisfy the consumers' needs through the various services that agritourism offers. The establishment of the brand facilitates effective marketing and collaboration between stakeholders (Barbieri et al., 2016).

Technical assistance, education and training
According to Blanco and Riveros (2010), a big obstacle to the development of agritourism is the lack of trained personnel to interpret in an attractive way, and with sufficient scientific knowledge, the practices that most attract tourists. Although there is an interest among the producers in terms of undertaking agritourism, there is a lack of education in the area and training in the agricultural production sector (52 percent of the producers interviewed had only completed elementary school), thus engendering the low competitiveness in their production. Moreover, the majority (approximately 72 percent) of producers in the area were elderly, many of whom did not have a generational change that continues agricultural practices.

Management and entrepreneurship
Rural tourism initiatives in the center zone of Veracruz are privately owned and do not actively involve rural communities. There are few opportunities for training and a weak entrepreneurial vocation among the people (Rivera-Hernández et al., 2018; Vargas Matías, 2020). Also, the

Table 11.3   Criminal incidence in Fortín and surrounding municipalities 2011–2017

Municipality	Crime incidence	Crime incidence rate
Córdoba	2,653	127.07
Ixtaczoquitlán	852	121.23
Fortín	840	128.58
Amatlán de los Reyes	210	45.98
Atzacan	67	30.83
Chocamán	47	21.13
Naranjal	11	23.2
Veracruz de Ignacio de la Llave	49,205	60.27

Source: Adaptation of the Ayuntamiento de Fortín (2018) with data from the Executive Secretariat of the National Public Security System (2011–2017).

lack of accounting and administrative organization systems within the agricultural sector is a major challenge. Only one of the producers stated that they used an administrative system. Having one or more business incubators in the area would be an option to contribute substantially to the growth of entrepreneurs through support that allows producers to establish successful small and micro agritourism businesses. This can also be achieved hand in hand with the relevant departments of the municipal government through talks, training and workshops that help producers to strengthen their management and entrepreneurship skills while at the same time involving the youth to enable them to bring about a generational change.

Public policies
The agritourism sector presents a gap in legal frameworks and policies that support its development, which is also reflected in many world regions (Carpio et al., 2008). In contrast, agritourism in the European Union is legally defined; policies are arranged in a way that allows for the provision of a variety of resources for producers to enter agritourism (Gil et al., 2013; Sznajder et al., 2009).

Past governments in Veracruz state enacted tourism policy planning, which focused on certain catchphrases such as "diversify and reconceptualize its products" and "promote sustainable development and tourism activities harmonious with the preservation of the environment", among other similar ones. Nonetheless, these proposals aimed at promoting already existing initiatives, neglecting the creation of new tourism products (Vargas Matías, 2020). In this context, tourism in Fortín is not a priority sector for the local government as against the vision of the Municipal Development Plan (2018–2021); tourism is timidly mentioned in components such as sustainable urban development.

Safety and security
Security problems continue to increase in Mexico, representing a serious concern for the tourism sector (Rioja Peregrina et al., 2019) most especially in Fortín, which is also not insulated from the problem. This constitutes a risk factor for the development of tourism (Aguilera and Calvario, 2014). According to the Municipal Crime Incidence report of the Executive Secretariat of the National Public Security System, for the period 2011–17, Fortín presents a crime incidence rate of 128.58 crimes per 10,000 inhabitants (Table 11.3). The local tour operators in Fortín, together with the municipal authorities, have implemented some strategies that have worked temporarily, such as the reinforcement of safety on tourist excursions to natural areas using the public security forces.

Possibilities of Developing Agritourism in Fortín

According to the methodology presented by Phillip et al. (2010) and its revision (Flanigan et al., 2014), the concept of agritourism can have different typologies depending on the visitor's contact with agriculture, thus presenting different facets according to the interests and profiles of visitors on diversifying the activity (Figure 11.5). Therefore, an analysis of the agritourism typologies seen in Fortín and the possibilities for agritourism development in this area was carried out.

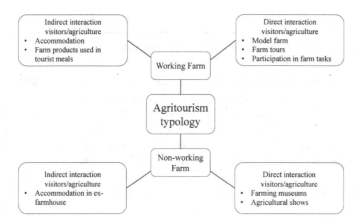

Source: Adaptation of Phillip et al. (2010) and Flanigan et al. (2014).

Figure 11.5 *Agritourism typology*

Following this typology, we can describe and analyze the agritourism typology in Fortín. The analysis is based on the three parameters presented by Phillip et al. (2010) below:

- Are the agritourism products based on a working farm (a farm where agriculture activities are currently carried out) or on a non-working farm (staged farmland settings that provide a tourist experience linked to agriculture)?
- Contact with agriculture (passive, direct or indirect).
- The authenticity of the agriculture (agricultural activities involved in tourist services are staged or authentic.

In this sense, the agritourism typologies in Fortín are:

1. Working farm, indirect interaction: there are working farms that provide agritourism services based on crops such as coffee, sugarcane and banana leaf where the suppliers offer lodging, food and beverage services, among other services.
2. Working farm, direct interaction: some of these farms belong to both the first and the second typology, because in addition to offering the services mentioned above, they offer guided tours, participation in the harvest and in the preparation of value-added products (such as jams). Moreover, the municipality has areas for ornamental plants in greenhouses where the visitors can learn about how floricultural activities are performed, buy plants or

consume typical local dishes. There are also two small factories (*trapiches*), which produce piloncillo.

3. Non-working farm, indirect interaction: there exists the Ex Hacienda Monte Blanco in Santa Lucía, which belongs to the period of the War of Independence (1810–21) and various cultural activities are performed in front of this construction (Vargas Matías, 2020).

4. Non-working farm, direct interaction: the Tatsugoro Museum, which is in the municipal capital, has a sample of various bonsai specimens and is the first venue in Mexico dedicated exclusively to bonsai. Besides, we cannot fail to mention the most outstanding tourist event linked to agriculture in the area: *la Feria de la Flor* (the Flower Fair). This fair is a cultural event where visitors can find products and activities such as ornamental plants, musical presentations and handicrafts, resulting in the promotion of the local traditions of Fortín (Pérez-Olmos and Aguilar-Rivera, 2019).

Agritourism development in Fortín based on working farms is favored by several factors. About 90 percent of land use in Fortín territory is designated for agriculture (INEGI, 2020); the municipality has regional recognition as a place to visit and acquire ornamental plants (Figueroa-Rodríguez et al., 2020); the coffee tours are all about its cultivation and artisan processing and the labor of making piloncillo in the *trapiche*. Despite this, Fortín has many but untapped possibilities that can link agritourism with the traditions and customs to create an attractive agritourism destination. For example, the visitors could learn about the relationship between agriculture and the religious customs of the place; they could participate in the elaboration of floral confections, handicrafts related to flowers; preparation of typical artisanal dishes, drinks and sweets; they could stay in the farmers' houses to share their daily life and learn about the Fortín agriculture history through an exclusive museum.

It is important to mention that the concept of authenticity for visitors could be potentially different for producers. As such, agritourism entrepreneurs could create staged experiences (that do not correspond to the daily life of the place) to receive visitors (Phillip et al., 2010). Given its resources and agricultural tradition, Fortín can transmit to the visitor a unique experience in agritourism that not only includes agricultural practices but go further by providing a differentiated product that involves the local people, their experiences, stories of the place and its people, and their way of life and customs.

CONCLUSION

Over many years, agriculture has been an important sector in the socioeconomy of Fortín, which, however, has declined in recent decades. Agritourism can be an alternative for economic diversification to enable producers to continue their agricultural activities and transmit their knowledge to new generations. Local authorities have focused their efforts on the development of some urban tourism activities, neglecting rural tourism and the promotion of Fortín's tourist offer. Despite the interest shown by the producers in undertaking agritourism activities, there is no community tourism in the municipality and the producers have few opportunities in terms of training, education and technical assistance. In this sense, it is important to mention that there is a need for training and technical assistance in the territory; the strengthening of these areas would help the continuity of the agricultural activity, which

160   *Handbook on tourism and conservation*

in turn could enhance agritourism development and attractions. In this context, information on the potential to develop agritourism in Fortín and the possible associated challenges is scant. Therefore, this chapter provides information that could assist in the development of policies on agritourism in Fortín. The development of a sustainable approach for a successful agritourism offer in Fortín that contributes to the common well-being of its inhabitants needs to be devised by all relevant stakeholders (producers, tourism entrepreneurs and local authorities).

ACKNOWLEDGMENTS

We thank the respondents who participated in the development of this research in Fortín, the Faculty of Biological and Agricultural Sciences of the Universidad Veracruzana, where the doctoral studies (from which this chapter emerged) were conducted. In addition, thanks are extended to the participating authorities of the Fortín City Council and the National Council of Science and Technology (CONACYT) for the scholarship granted.

NOTE

1.   Bonsai is an ancient art form originating from the Asian continent; it is literally a "tree in a pot", which seeks to imitate, in miniature, the semblance of an old tree in nature. The Japanese have developed this technique from the thirteenth century, collecting and potting wild trees that were naturally dwarfed; thus giving rise to the first specimens of bonsai (Relf and Close, 2020).

REFERENCES

Abarca Alpízar, F. (2016). La metodología participativa para la intervención social: Reflexiones desde la práctica. *Revista Ensayos Pedagógicos, 11*(1), 87–109.

Aguilera, A., & Calvario, Á. (2014). Medición de la competitividad turística en la Subregión Cafetalera ubicada en las Altas Montañas en el estado de Veracruz. *Ciencias Administrativas y Sociales, Handbook, 4,* 54–62.

Alcalá Escamilla, B., & López López, Á. (2017). Zonas con potencial agroturístico en la región citrícola de Nuevo León, México: Un análisis a partir del álgebra de mapas. *Cuadernos de Turismo, 39,* 17–40.

Alejandre-Castellanos, L. R., Devezé-Murillo, P., Mora-Brito, A. H., & Villagómez-Cortés, J. A. (2020). Potencial del agroturismo como actividad emergente en el municipio de Cuitláhuac, Veracruz, México. *Estudios Sociales. Revista de Alimentación Contemporánea y Desarrollo Regional, 30*(55).

Almaraz-Vidal, D. (2015). Evaluación ecológica rápida de la herpetofuna del Corazón del Metlác, Veracruz, México. *Bioma, 35,* 23–33.

Ayuntamiento de Fortín. Plan Municipal de Desarrollo 2018–2021 (2018). Municipio de Fortín, Veracruz.

Baltazar-Bernal, O., Zavala-Ruiz, J., Gaytán-Acuña, E. A., & Caamal-Velázquez, J. H. (2018). La floricultura tropical: Un complemento del turismo rural en la zona centro del estado de Veracruz, México. *Agroproductividad, 11*(8), 27–32.

Barbieri, C. (2010). An importance-performance analysis of the motivations behind agritourism and other farm enterprise developments in Canada. *Journal of Rural and Community Development, 5*(1), 1–20.

Barbieri, C., Xu, S., Gil-Arroyo, C., & Rich, S. R. (2016). Agritourism, farm visit, or ...? A branding assessment for recreation on farms. *Journal of Travel Research, 55*(8), 1094–1108.

Blanco, M., & Riveros, H. (2010). El agroturismo como diversificación de la actividad agropecuaria y Agroindustrial. In *Desarrollo de los agronegocios y la agroindustria rural en América Latina y el Caribe. Conceptos, instrumentos y casos de cooperación técnica.* San José, Costa Rica: IICA (pp. 117–128).

Cabrera Martínez, H., & Aguilar-Rivera, N. (2019). Competitividad de la agroindustria del piloncillo en la zona central de Veracruz. *Textual, 73,* 297–330.

Carpio, C. E., Wohlgenant, M. K., & Boonsaeng, T. (2008). The demand for agritourism in the United States. *Journal of Agricultural and Resource Economics, 33*(2), 254–269.

Che, D., Veeck, A., & Veeck, G. (2005). Sustaining production and strengthening the agritourism product: Linkages among Michigan agritourism destinations. *Agriculture and Human Values, 22*(2), 225–234.

Consejo Nacional de Población (CONAPO) (2015). Proyecciones de Población de los Municipios de México 2015–2030. www.gob.mx/conapo/documentos/proyecciones-de-la-poblacion-de-los -municipios-de-mexico-2015-2030, accessed 12 September 2020.

Cortés, D., Díaz, S., Cabal, A., & Del Ángel, O. (2013). Análisis del sector agroindustrial piloncillero en la región Huatusco-Fortín, Veracruz. In M. Ramos & V. Aguilera (eds), *Ciencias Agropecuarias Handbook* (pp. 13–20). Valle de Santiago: ECORFAN.

Escamilla, P. E., & Landeros, S. C. (2016). Cafés diferenciados y de especialidad. México. Centro Nacional de Investigación, Innovación y Desarrollo Tecnológico del Café (Cenacafé).

Figueroa-Rodríguez, K., Castillo-González, L. A., Fernández-Fernández, O., Mayett-Moreno, Y., & Sangerman-Jarquín, D. M. (2020). Biodiversity, management, and commercialization of ornamental plants at nurseries in Fortin de las Flores, Veracruz. *Agroproductividad, 13*(11), 87–94.

Flanigan, S., Blackstock, K., & Hunter, C. (2014). Agritourism from the perspective of providers and visitors: A typology-based study. *Tourism Management, 40,* 394–405.

Gao, J., Barbieri, C., & Valdivia, C. (2013). Agricultural landscape preferences: Implications for agri-tourism development. *Journal of Travel Research, 53*(3), 366–379.

García-Albarado, J. C., Gómez-Merino, F. C., Bruno-Rivera, A., Rosas-López, F., Servín-Juárez, R., & Muñoz-Márquez-Trujillo, R. A. (2018). Identificación de elementos identitarios en la región de las altas montañas de Veracruz. *Agroproductividad, 11*(8), 95–100.

García-Alonso, O., Gómez-Gómez, A. A., Chalita-Tovar, L. E., Brambila-Paz, J. D. J., & García-Alonso, R. (2014). Factibilidad financiera por opciones reales para la producción de Anturio (*Anthurium andreanum Lindem Ex Andre*) en Cuitláhuac, Veracruz. *Revista Mexicana de Ciencias Agrícolas, 8,* 1467–1475.

Gil, C., Barbieri, C., & Rozier, S. (2013). Defining agritourism: A comparative study of stakeholders' perceptions in Missouri and North Carolina. *Tourism Management, 37,* 39–47.

Hernández-Sampieri, R., Fernández-Collado, C., & Baptista-Lucio, P. (2014). Definición del alcance de la investigación que se realizará: exploratorio, descriptivo, correlacional o explicativo. In R. Hernández-Sampieri, C. Fernández-Collado, & P. Baptista-Lucio, *Metodología de la Investigación* (6th edn, pp. 88–101). Mexico City: McGraw-Hill.

Instituto Nacional de Estadística y Geografía (INEGI) (2020). *Compendio de información geográfica municipal 2010.* Fortín: Veracruz de Ignacio de la Llave. Instituto Nacional de Estadística y Geografía, 1–10.

Kieffe, M. (2018). Conceptos claves para el estudio del Turismo Rural Comunitario. *El Periplo Sustentable, 34,* 8–43.

Landero-Torres, I., García-Martínez, M. A., Galindo-Tovar, M. E., Leyva-Ovalle, O. R., Lee-Espinosa, H. E., Murguía-González, J., & Negrín-Ruiz, J. (2014). Diversidad Alfa de la Mirmecofauna del Área Natural Protegida Metlac de Fortín, Veracruz, México. *Southwestern Entomologist, 53*(3), 541–553.

LaPan, C., & Barbieri, C. (2014). The role of agritourism in heritage preservation. *Current Issues in Tourism, 17*(8), 666–673.

Lyon, S. (2013). Coffee tourism and community development in Guatemala. *Human Organization, 2*(3), 188–198.

Martínez, M. Á. (2012). Cambios estructurales en la economía cafetalera. In C. Del Valle (Coord.), *Crisis estructural y alternativas de desarrollo en México.* Ciudad de México: IIEc-UNAM.

México Desconocido (2014). *Special Guidebook. Revista México Desconocido-Discover Veracruz.* Impresiones Aéreas, S. A. de C. V., México.

162 *Handbook on tourism and conservation*

Ministerio de Comercio Exterior y Turismo (MINCETUR) [Ministry of Foreign Trade and Tourism of Peru]. (2018). *Manual para la elaboración y actualización para el Inventario de Recursos Turísticos*. Lima: Impresiones & Publicidad Viserza.

Organización para la Cooperación y el Desarrollo Económico (OECD) [Organisation for Economic Cooperation and Development] (2017). *Estudios de la OECD sobre Turismo. Estudio de la Política Turística de México*. Paris: Organización para la cooperación y el desarrollo económico.

Pérez-Olmos, K. N., & Aguilar-Rivera, N. (2019). Agroturismo Cañero en Fortín de las Flores, Veracruz, México. *Trayectorias, 21*(49), 28–48.

Pérez-Olmos, K. N., & Aguilar-Rivera, N. (2021a). Agritourism and sustainable local development in Mexico: A systematic review. *Environment, Development and Sustainability, 23*(12), 17180–17200.

Pérez-Olmos, K. N., & Aguilar-Rivera, N. (2021b). Oportunidades para el desarrollo del agroturismo en la barranca del Metlac. In Torres Castillo, J. A. (coord. ed.) *8vo Congreso Nacional de Ciencia y Tecnología Agropecuaria. Universidad Autónoma de Tamaulipas* (p. 9). Ciudad Victoria: SOMECTA.

Phillip, S., Hunter, C., & Blackstock, K. (2010). A typology for defining agritourism. *Tourism Management, 31*(6), 754–758.

Rauniyar, S., Awasthi, M. K., Kapoor, S., & Mishra, A. K. (2021). Agritourism: Structured literature review and bibliometric analysis. *Tourism Recreation Research, 46*(1), 52–70.

Relf, D., & Close, D. (2020). *The Art of Bonsai*. Virginia Cooperative Extension. Virginia Tech, Virginia State University.

Reyes-Aguilar, A. K., Pérez-Ramírez, C. A., Serrano-Barquín, R. D. C., & Moreno-Barajas, R. (2019). Rural tourism and environmental conservation: The participation of peasant women in the Tuxtlas Biosphere Reserve, Veracruz, Mexico. *Revista Rosa Dos Ventos – Turismo e Hospitalidade, 11*(1), 157–177.

Rioja Peregrina, L. H., Benítez López, J., & Hernández Espinosa, R. (2019). Representación social y políticas públicas en materia de turismo: Los casos de los Centros Integralmente Planeados de Cancún, Litibú, e Ixtapa-Zihuatanejo, México. *El Periplo Sustentable, 37*, 92–121.

Rivera-Hernández, J. E., Muñoz-Márquez Trujillo, R. A., Vargas-Rueda, A. F., Alcántara-Salinas, G., Real-Luna, N., & Sánchez-Páez, R. (2019). Flora, vegetación y paisaje de la región de las altas montañas de Veracruz, México, elementos importantes para el turismo de naturaleza. *Agroproductividad, 12*(12), 19–29.

Rivera-Hernández, J. E., Pérez-Sato, J. A., Alcántara-Salinas, G., Servín-Juárez, R., & García-García, C. (2018). Ecoturismo y el turismo rural en la región de las Altas Montañas de Veracruz, México: Potencial, retos y realidades. *Agroproductividad, 11*(8), 129–135.

Rodríguez, M. T. (2016). Migración en tránsito y prácticas de ayuda solidaria en el centro de Veracruz, México. *Encuentro, 103*, 47–58.

Rodríguez Deméneghi, M. V., Aguilar-Rivera, N., & Murguía González, J. (2020). Valores culturales, socioeconómicos, simbólicos e históricos y prospectiva tecnológica de la gardenia en Fortín de las Flores, Veracruz, México. *CIENCIA ergo-sum, 27*(1).

Sandoval Quintero, M. A., Pimentel Aguilar, S., Pérez Vázquez, A., Escalona Maurice, M. J., & Sancho Comíns, J. (2017). El turismo rural en México: Una aproximación conceptual al debate suscitado sobre las políticas públicas desarrolladas, la irrupción de agentes externos y las nuevas metodologías de acción endógena y participativa. *Estudios Geográficos, 78*(282), 373–382.

Secretaría de agricultura, ganadería, desarrollo rural, pesca y alimentación (SAGARPA) (2018). *Atlas agroalimentario 2012–2018*. Servicio de Información Agroalimentaria y Pesquera. https://nube.siap .gob.mx/gobmx_publicaciones_siap/pag/2018/Atlas-Agroalimentario-2018, accessed 20 September 2020.

Secretariado Ejecutivo del Sistema Nacional de Seguridad Pública [Executive Secretariat of the National Public Security System]. Datos de Incidencia Colectiva del fuero común. Municipal 2011–2017. www .gob.mx/sesnsp/acciones-y-programas/incidencia-delictiva-del-fuero--comun?idiom=es, accessed 2 October 2020.

Servicio de Información Agroalimentaria y Pesquera (SIAP) (2019). Producción agrícola. www.gob.mx/ siap/acciones-y-programas/produccion-agricola-33119, accessed 30 September 2020.

Sznajder, M., Przezbórska, L., & Scrimgeour, F. (2009). *Agritourism*. Wallingford: CABI.

Templado Isasa, M., & Lucarelli, R. (2020). El agroturismo en la educación. In O. F. von Feigenblatt & B. Peña-Acuña (eds), *Perspectivas españolas en la educación: mejores prácticas para el siglo XXI* (pp. 23–48). Miami: Catholic University.

Tew, C., & Barbieri, C. (2012). The perceived benefits of agri-tourism: The provider's perspective. *Tourism Management, 33*(1), 215–224.

Thomé Ortiz, H., Renard Hubert, M. C., Nava Bernal, G., & Souza Vlentini, A. (2014). La Ruta del Nopal (Opuntia Spp.). Turismo y reestructuración productiva en el Suelo Rural de La Ciudad de México. *Rosa Dos Ventos, 6*(3), 390–408.

Vargas Matías, S. A. (2020). Caminos de Memoria, Sendas de Progreso: Propuesta para la Creación de un Itinerario Cultural en la Zona Centro del Estado de Veracruz, México. *Trayectorias, 50*, 80–107.

Varisco, C. A. (2016). Turismo rural: Propuesta metodológica para un enfoque sistémico. *PASOS Revista de Turismo y Patrimonio Cultural, 14*(1), 153–167.

Veeck, G., Iv, L. H., Che, D., & Veeck, A. N. N. (2016). The economic contributions of agricultural tourism in Michigan. *Geographical Review, 106*(3), 421–440.

Velasco, J. M. (1881). "La cañada de Metlac" [Óleo sobre tela]. Museo Nacional de Arte (MUNAL) de la Ciudad de México. https://revistalacolmena.com/pintura/canada-de-metlac-o-el-citlaltepetl-jose-maria-velasco-pintura-historica/, accessed 1 November 2020.

12. 'Negotiating with the juggernaut': on agritourism and the paradoxes of market-driven conservation

Mikael Andéhn and Patrick J. N. L'Espoir Decosta

INTRODUCTION

> A sacred site in Western Australia that showed 46,000 years of continual occupation and provided a 4,000-year-old genetic link to present-day traditional owners has been destroyed in the expansion of an iron ore mine.[1]

> Venice's population has declined rapidly from roughly 175,000 after World War II to about 50,000 today. Remaining residents complain that their city is being overrun by tourists while they have to pick up the bill for cleaning and security.[2]

The Covid-19 pandemic that has plagued the world since 2020 continues to unravel other challenges that are exacerbating the precarious nature of the global socio-economic, political, and ecological environments. Subsequently, the whole world must significantly re-evaluate and reconfigure their economic activities by shifting towards a systemic repurposing across various ecosystems (McGehee & Kim, 2004) heralded in the social-ecological system (Berkes & Folke, 1998). One popular means of such repurposing is to convert agricultural operations, small-scale industrial operations, and even the headquarters of near-derelict business infrastructures into (hybrid) tourist attractions, where tourists come into implicit or explicit contact with some form of agriculture. Indeed, agritourism or cultural tourism is typically considered as a means to regenerate or develop an economic ecosystem in a sustainable way that safeguards values that not only underpin the tourism product but that also need to be conserved for posterity (Barbieri, 2013; Flanigan et al., 2014; Phillip et al., 2010). However, the persistent problems related to the cumulative negative impacts of tourism activities in general on natural and socio-cultural environments including the ubiquitous climate change require that we question the premise of (post-)sustainability and conservation that underpins regenerative tourism especially in a post-vaccine and post-Covid-19 return to travel (Glusac, 2020). We do so in this chapter by critically appraising and discussing the role of markets, 'marketization', and its outcomes, generally viewed as an essential element in the success of regeneration endeavours such as restoration and sequestering carbon for instance, but that we consider here as constitutive of the juggernaut of inexorable environmental and social problems.

Based on the premise that the development and trajectory of agri- and cultural tourism closely reflect their position in the lifecycle of the destination in which they are located (Petrovic et al., 2016), we argue that this track from initial 'raw' commercial potential to a quasi-terminal point of 'hyper exploitation' is equivalent to the process whereby repurposing with commercial intent jeopardizes the ability to convey experiences meant to be, at a minimum, authentic and sustainable (Andéhn & L'Espoir Decosta, 2021a; Barbieri, 2013), that is, to safeguard the place is often contingent on being able to safeguard its ability to convey experiential authentic-

'Negotiating with the juggernaut': agritourism and market-driven conservation 165

ity (Andéhn & L'Espoir Decosta, 2021a). Regenerative tourism no longer implies a particular position on, for instance, the destination lifecycle, but points rather to a regenerated relationship of tourism with life through a cycle of rebirth. The purpose of this chapter is to critically explore the dynamics at play in this regeneration process while highlighting the hurdles on its way. To achieve this purpose, we meet two objectives formulated to:

- articulate the regenerative and conservational potential of agritourism when expressed through space, place and culture; and
- highlight the challenges agritourism faces as it is still subject to market logic.

AGRITOURISM AND REGENERATION

One area that is increasingly being repackaged as tourism products is agriculture. A possible explanation lies in the fact that issues that revolve around the intersection of social and economic sustainability are more relevant in regions that once relied on agriculture to sustain their economic growth but that have since been marginalized by increasing competition (Barberi, 2013; McGehee & Kim, 2004). The significant socio-economic implications of such repurposing in terms of whether to address the long-term lack of economic development has since motivated transnational organizations such as the World Bank, as well as world governments to deploy several strategies to reinstall economic growth (see Ndaguba & Hanyane, 2019). While these strategies vary in scope and approach, they all tend to reflect an understanding of sustainability that is profoundly neoliberal (Cammack, 2004) with more emphasis on economic regeneration than sustainability itself through engaging the local populace of a disaffected region in some entrepreneurial activities aimed at creating and promoting new commercial opportunities from 'undiscovered' or 'dormant' potentials. In other words, some aspects that were not previously integrated in the economic ecology are now realigned (see Prahalad, 2005; also, Ferguson, 1995) through, for example, the recycling of derelict land and buildings. In the critical business literature, including in Marxist-inspired thought, this process embraces the ideological connotations of such practices that are sometimes referred to as 'marketization' (see Fairclough, 1993) through the restructuring into and participation in a market.

Marketization denotes the application of strategies, procedures, and indeed ideology that owes its lineage to the world of commerce and business, to new areas with appeal and promotion to the market (Fairclough, 1993; Harvey, 2001). An instance of marketization is the phenomenon of 'new public management' whereby public institutions are managed and operated as market-oriented firms (Dunleavy & Hood, 1994; Hood, 1995). Some scholars think of this process of marketization as a causal epiphenomenon of capitalism having reached an advanced, or 'late', stage (Harvey, 2002; Fisher, 2009; Jameson, 1991; Mandel, 1975). This view also emphasizes how the market logic does not only spread to new geographical areas, that is, 'horizontally', but also goes 'deeper' or expands 'vertically', making it a more palpable and encompassing influence in areas where it once might have started as a more marginal phenomenon (Mandel, 1975). In its totality, this shift in market logic thus becomes a cultural tendency with a profound impact on how we make sense of the world around us, leading some to characterize the current era as one in which any social phenomenon is understood from a position of an already established and inexorable 'business ontology' (Fisher, 2009). This chapter discusses some aspects related to this approach of promoting sustainable

166 *Handbook on tourism and conservation*

tourism growth. We do so through the lens of two opposite poles on the spectrum of market 'valuation', which include the following two cases:

- the Juukan Gorge cave shelter (now known as the Rio Tinto Brockman 4 mine expansion) in Western Australia, and
- the day-tour hotspot in central Venice, Italy.

While the former is an example of a quasi-null value, and the latter a (sign of) value over-saturation, they both constitute states that serve as a destabilizing and destructive influence on the ecology of the places in which they initially emerged as regenerative initiatives.

MARKETIZATION AS CHALLENGE TO THE EXPERIENTIAL VALUE POTENTIAL OF TOURISM

Marketization as implied in the context of regenerative agritourism relies principally on promoting tourism as an economic activity that provides value through experience provision. Everything and anything related to tourism activities is understood as translatable into components that hold some form of 'value potentiality' through their ability to generate and provide experiences through encounters with 'tourism products' (Di Domenico & Miller 2012; Smith, 1994; Sugathan & Ranjan, 2019). The transformational capacity of this simple mechanism of marketization should not be underestimated because the process by which a 'space' or 'place' becomes a 'destination' does not only entail spatial reconfigurations pertaining to access, accommodation, and various functions to support various agri- and cultural tourism products but also causes an almost metaphysical transformation of local life that emerges in varying degrees through regenerative tourism. Value can, therefore, invariably be attributed to that transformation, based on its 'quality' as an experiential service provision. However, to identify the properties of quality that would warrant such judgement requires, at that point, that we deconstruct the concept of quality in this context. The result is a spectrum with poles; on one end the ubiquitous sun-and-sand mass hedonic tourism, which requires no or very little in terms of pre-existing cultural resources, and on the other end the 'authentic' cultural tourism experience in which the tourism product(s) exude cultural and conservation meanings.

The appeal of mass tourism destinations where tourists can indulge in leisure activities has persisted for the last century (Aguiló et al., 2005; Prebensen et al., 2010). Sun-and-sand tourism constitutes a value proposition scenario in which the quality of hospitality services and amenities constitute the primary basis of value provision (Tussyadiah, 2014). This sort of stand-alone 'hospitality tourism' can be understood principally as the marketization of *space* (see Harvey, 2002; see also, Smith, 1994). Following Tuan's (1977) classical dichotomy between meaning-laden spatial entities as 'places' and such delimitations lacking meaning as 'spaces', the hospitality tourism value provision requires little more than what can be constructed *ex nihilo*, in terms of prior human activity, provided the site offers access and has amenities on it. Thus, the Juukan Gorge cave shelter in Western Australia gets a 'null' value making its evaluation a trivial exercise; naturally there was nothing present on the site that can be realistically assessed as hospitality provision, but more importantly in the understanding that propagates from markets, there was nothing 'of value' at the site in more broad terms. In fact, little about the site would have made it particularly suitable for hospitality tourism development. More interestingly, we can observe that central Venice would also perform poorly if

TOURISM AND THE MARKETIZATION OF CULTURE

Beyond being extremely vulnerable to competition, 'hospitality tourism' has little innate means to generate a sustainable competitive advantage. Its means to differentiation as opposed to its alternatives are tenuous and can evaporate almost overnight (Alegre & Garau, 2011; Crouch & Ritchie, 1999; Dwyer & Kim, 2003). The potential of such a type of tourism development to act as a realistic means to regenerate regions with positive impacts on the local economy and social and natural environments is subsequently highly limited. Rather, the resulting tourist enclaves are often anything but sustainable other than in the sense that they can constitute a profitable investment for overseas financial entities. And often, the revenues generated by these developments are extracted out of the local economy with dire environmental, social, and financial problems left in their wake. This sinister form of arrangement is in fact 'extractivism' in action, whereby most of the capital brought in by tourists never even meaningfully enters the local economy (Manuel-Navarrete, 2016; Mbaiwa, 2005; Palmer, 1994), such as is often the case with the so-called 'all inclusive' resorts. On the other end of the spectrum from this extraction and exploitation type of 'hospitality tourism' has (re)emerged what is often referred to as 'cultural tourism', which has become an increasingly dominant fixture in tourism over the last several decades (Du Cros & McKercher, 2020).

Where its sun-and-sand centred counterpart often features what can be called 'tourist reservations' that often strive to emulate, for instance, the food and drink preferences of the country-of-origin of the tourists, cultural tourism can only come to fruition as a result of an often more far-reaching interconnectedness with the local culture, nature, or 'meaning' more broadly (see Collins-Kreiner, 2010; also Di Domenico & Miller 2012). In this context, the evaluation of the experience value proposition becomes less concerned about the provision of quality services as they qualify as more or less effective provision of hospitality, instead the experience 'value' becomes contingent upon the ability to convey an experience that is understood by the tourists as *authentic*, or as a 'credible' immersion into the local culture or nature (MacCannell 1973; Richards, 1996; Wang, 1999). This type of tourism is often held up as a far more sustainable means to developing struggling economies, at least in the social and economic sense (Andéhn & L'Espoir Decosta 2021a; Barberi, 2013). The rationale underpinning such assertions is that in order to viably 'market' a culture as a central value proposition for a tourism operation, the culture and its enabling conditions for continuing preservation must also be safeguarded. At this point, it becomes evident that in all of its destructive potential, marketization also ostensibly holds the potential to crystallize into overarching structures with pockets protecting against its own warping influence.

On the basis that experiential authenticity can be understood as a central and necessary component of a cultural tourism-based value proposition, we can extrapolate a number of corollary observations. First, an 'authentic' cultural experience exhibits a situation in which there is a directly appreciable value in preserving the place that serves as the site for 'cultural tourism' (McKercher & Du Cros, 2002; Salazar, 2012). In other words, the nature of the 'value' of the activities prior to the place becoming a 'destination' for cultural tourism, and of the tourism offering per se, are convergent. Hospitality provision in this type of tourism does not extend beyond promoting access (certainly not beyond cultural carrying capacity, which can constitute a threat to the local culture). This demonstrates the primacy of the performative labour required for such a destination to effectively deliver the experience that is provided primarily through the local population's continued efforts to preserve their own culture. At face value, this cultural tourism experience approach makes it a credible path for preservation and poverty reduction (Manyara & Jones, 2007).

The local population also achieves value in that it is considered an irreplaceable asset of the destination (Kavaratzis, 2012; Braun et al., 2013). While a monetary value in this case cannot be fully appreciated on the linear and reductive basis, it does however represent, if not offer the population, a degree of protection in the sense that the value can now be shown to be pertinent in terms of its significance in markets and *inside* their nested logic. Identifying this foundational logic of cultural tourism serves as a strong rationale for the preservation of the tourism site as it is, a necessary and appreciable step of a viable and sustained value creation process. Put differently, the value provision in the case of 'cultural tourism' represents a marketization of *place* (see Tuan, 1977) as it engages primarily with the nested *meaning* of the tourism activities in situ. And for this marketization of place to be effective, it invariably requires a degree of integration with the local economic, cultural, and natural ecology. From this perspective, the destruction of Juukan Gorge cave shelter can be characterized as a short-sighted trade-off. In the destruction of a site with 46,000 years of cultural history, immeasurable historical value has been lost. The potential for heritage and cultural tourism operations, the monetary value of which is difficult to estimate, was also eradicated. And yet, we can safely advance that had the site been preserved, it would have constituted a value generating element that could be operated in perpetuity if responsibly managed (Richards, 1996; Ho & McKercher, 2004). In this case, the only gain was a mineral extraction operation with negligible public value, monetary or otherwise, thus making this event a disastrous failure even if evaluated merely from the perspective of strictly economic factors.

On the other hand, if evaluated from the perspective of foregrounded meaning, Venice, again, emerges as an oddity. While much of the city's over-tourism problems can be ascribed to its role as part of cruise tourism destination packages (González, 2018), its situation can be understood as an example in which the sign-value (see Baudrillard, 1968, 1994) of visiting the city totally eclipses prospects of any experience per se (Figure 12.1). The city's history as a 'grand tour' destination has been infused with a symbolic value which is superordinate to its value as a site of experience, or any *ability* to serve as an effective means of experience provision. A pessimistic view of the city sees it as on a trajectory from being a site where local culture and history can be experienced in a state less adulterated by tourism, to one that, at present is often being described as a 'living museum' on its way to becoming something akin to a dystopian (and possibly also underwater) theme-park version of itself. There may be a point when the city ultimately becomes even more extreme in its own commodification than are its pastiche verisimilitudes reminiscent of industrial scale casinos in places such as Las Vegas or Macau. Venice's ability to sustainably serve as a site that provides authentic experience would be null, as its marketized reality (of demand) is far separated from the romantic nostalgia of its sign (cultural symbol). Such incongruity could potentially cause dysphoric anxiety in visitors to the place, which in turn would become a second curious case of what has become colloquially known as the 'Paris syndrome'[3]: when a sign eclipses the real spatial experience of a place so much that visitors experience culture shock under the weight of the collapsing illusion.

Source: Fagarazzi, S. (2019). Venezia Autentica [Photograph]. *The Guardian*. https://www.theguardian.com/cities/2019/apr/30/sinking-city-how-venice-is-managing-europes-worst-tourism-crisis.

Figure 12.1   Crowds of tourists in Venice in 2019 prior to the outbreak of Covid-19

AN ENCOUNTER WITH LIVING CULTURE?

We have now established that a complete lack of valuation can leave a place vulnerable to competing commercial interests, or 'unchecked' market forces, as in the case of the Juukan Gorge caves. We have also shown that in the case of Venice, there was hyper-valuation far beyond what the space can accommodate. These represent variations of what is ultimately the same desolation of a place's ability to house a sustainable tourism operation, and much less a living culture. The question that subsequently arises is whether an equilibrium between these extremes can indeed be reached and maintained. In the case of agritourism, we see a particular type of tourism development that seeks to establish its experiential 'product' out of extant economic activities, often while still operating, typically an agricultural operation, simultaneously (Arroyo et al., 2013; Cai, 2002; Di Domenico & Miller, 2012). The latter facet of this practice is of crucial importance if any claim towards sustainability is to be made, as it effectively 'caps' the speed and reach of realignment efforts between agriculture and tourism. Simply put, this situation allows for the possibility that such an agritourism operation could be carried out so that it would not constitute a threat to the destination's 'spatial ability' to house and encapsulate a living culture (see Andéhn & L'Espoir Decosta, 2021a; Barbieri, 2013; Cai, 2002; Sharpley, 2002). In other words, agritourism, through its adaptability, can help navigate ecosystem dynamics without compromising long-term sustainability, a major principle of adaptability of social ecological systems (Berkes & Folke, 1998). At the micro-level, agritourism often manifests by repurposing agricultural facilities to become accommodations, and tourism offerings can manifest in simple 'living' experiences of small-scale agricultural operations that were already in place prior to the arrival of tourists (Busby & Rendle, 2000). This is facilitated by the integration of resilient 'remnants' of agricultural infrastructure and culture with adaptable tourism products re-emerging as experiential service provision. A step

170   *Handbook on tourism and conservation*

further would be refitting with relative ease the various commercial activities that once supported agriculture to become agritourism products and experiential service provision within a broader destination 'consumptionscape' (Andéhn & L'Espoir Decosta, 2021a).

The tourism operations as such might even serve to resurrect agricultural operations as tourism becomes a means to market the local specialities, produce, and even consumption practices more broadly (Andéhn & L'Espoir Decosta, 2021a). The changes brought about in an agricultural region that transitions to a tourism economy centred on visitors' experience of the culture and lifestyle around agricultural activities are likely means to regrowth thus alleviating socio-economic problems that a dwindling agriculture was causing. As agritourism can serve as a means to market products, so can the experience that the culture of the region provides around specific products (L'Espoir Decosta & Andéhn, 2018) as has long been a practice with, for instance, wines (Gomez et al., 2015; Tustin & Lockshin, 2001), and is becoming commonplace for other product-place pairings such as coffee in Cali, Colombia (Muñiz-Martínez, 2016) or chocolate confectionery in Belgium (L'Espoir Decosta & Andéhn, 2018). There appears the possibility, then, that places can be '(re)marketized' in a manner that is sustainable, or even prompts synergies, when considering at least at face value their ecosystems of signs as well. Practical examples of places that have ostensibly struck a balance between market considerations and preservation of a region's culture include Napa Valley, a district in California, USA, which gainfully combines a developed tourism economy centred around agricultural activities (see L'Espoir Decosta & Andéhn, 2018), or Tuscany, Italy, which is often hailed as a paragon of effective synergy among local agriculture, culture, and tourism (Pasquinelli, 2010; Sonnino, 2004). Unfortunately, we note that even in these cases, marketization causes reductive transformative effects on culture through the smallest common marketable denominator (see Horkheimer & Adorno, 2002), resulting in what Pasquinelli (2010) calls a trajectory towards 'monoculture', a process that operates in tandem with the 'hyper-enactment' of the properties of a place, culture, or people that are understood to be commercially gainful (Andéhn et al., 2020). This transformative process also often intersects with the tourist gaze (Urry, 1992), as the locals invariably begin to understand themselves as in proximity to an evaluation that frames their own identity as part of a commercial experience provision.

While agritourism offers some protection to the producers of local experience, it also exerts an array of performative demands on them (see also Andéhn et al., 2020). In all, what this state of affairs entails for places could be encapsulated in a reductionist tendency, inherent in the streamlining of the marketing of the place as a site of natural richness capable of providing experiences, as well as its inherent polysemic capacity (see Malpas, 2018). This is not compatible with the need to provide a clear and concise story at the lowest cost fundamental to marketing. Perhaps, from this view it may not even be meaningful to think of tourism in terms of conservation, or even sustainability as the tendency towards an industrialization of experience cannot coexist with any culture without enacting a far-reaching subversive influence upon it. The saving grace for agritourism, which could be a lesson to other contexts, is that this subversion tends to be gradual enough for the local, social, and economic ecosystems to have the chance to adjust (Andéhn & L'Espoir Decosta, 2021a). Here, the radical decrease in demand that Kotler and Levy (1971) hope to see through 'demarketing' becomes ontologically impossible given the imperative of 'growth through promotion' that is inherent in markets (Fisher, 2009) and the process that sustains it, i.e., marketing. In the case of the Juukan Gorge cave shelters in Western Australia, we are ruefully reminded that an existence 'outside' the

'Negotiating with the juggernaut': agritourism and market-driven conservation   171

market (through the landscape of laws and policies meant to protect the ancient rock shelters) not only does very little to protect a site like Juukan Gorge against destruction but also does not allay the doubt of the traditional owners of the place (the Puutu Kunti Kurrama and Pinikura people) that the laws could effectively protect their sacred sites. The broader question that arises is: does conservation have any chance against marketization?

NEGOTIATING WITH THE JUGGERNAUT

Till now, our discussion and examples have pointed to the ostensible absence of the possibility of a marketized utopia. Even if the cases of Venice and the Juukan Gorge caves are reminiscent of being between Scylla and Charybdis, we acknowledge that there is no shortage of means to 'scuttle the ship' even between these two unpleasant alternatives. An interjection at this point is that a strategy aimed at some form of 'cultural audit' would reveal the subversive influence of marketization and provide the means to a sustainable form of tourism in which space, place, and way of life can be safeguarded. Perhaps the nostalgia of a pre-marketized past creates a longing for the Utopia that never existed, or even that metathesiophobia or the fear of change, is a fitting label for those who romanticize the way of life of those who still reside largely outside the market. Ironically, the most vocal defenders of that 'authentic' pre-marketized way of life tend to, themselves, engage in such admiration from a position far removed from the realities of that existence. Perhaps, the view of conservation as counter-acting the worst excesses of tourism is positive in that it improves lives, and maybe this can even be achieved through managerial intervention and the deployment of particular strategies. However, such a proposition is up against a powerful and unpredictable opposition. As Marx and Engels observed (1974) (see also Lazzarato, 2014), capitalism operates in uncontrollable contradictions through complex financial flows and contingencies. Whoever seeks to harness the power of capitalism thus finds themselves in the role of a sorcerer's apprentice: one who cannot understand the spells they are invoking and who cannot control the power these spells unleash. Simply put, relying on market initiatives for regional development is similar to unleashing a juggernaut. In the words of ecological economist Clive Hamilton, "market-based assessments such as 'willingness to pay' favour market-based solutions", consequently leading to "a conundrum of marketising and monetising ecological values".[4] While an ecological economics perspective indeed suggests as a possible solution, the need to embed human social systems in ecological life systems, the fact remains that the long reach and considerable power of market forces present a ubiquitous set of continuing market-based problems. What we can surely advance is that in using agritourism as a means to regeneration, or to alleviate poverty in highly troubled economic ecosystems, one issue that will (continue to) require attention is the perennial power asymmetry between the wealthy visiting class and the local populace. An often-proposed solution lies in the potential reconfiguration of the political economy and con-figuration of natural resources that could enable the implementation of conservation initiatives that rely on empowering locals to preserve their natural and cultural heritage.

However, this view operates under the assumption that the local populace is somehow immune to the influence of market forces. As some have noted, a strategy that grooms the locals 'as a future proletariat' (Cammack, 2004, p. 192) to serve a visiting class can hardly appear to be one that is socially sustainable. Evidence shows that tourism is rife with the worst consequences of such power imbalances including human trafficking (Bernstein & Shih,

172 *Handbook on tourism and conservation*

2014; Paraskevas & Brookes, 2018), exploitative working conditions and sex trade (Bernstein & Shih, 2014; Carolin et al., 2015), corruption (Papathanassis et al., 2017), the widespread devastation of natural resources (Inskeep, 1987), and loss of biodiversity (UNWTO, 2010), all of which are plausible outcomes from a sudden large inflow of visitors rich in capital. More promising then is cultural tourism that draws significantly on extant heritage and way of life as components of its value provision, especially if it can be assured that such developments are coordinated at the grassroots level with a particular view to a gradual realignment of operations. Here, we might even discover synergic potentials in, for instance, exports of produce and experience of their 'productionscapes'. Such potentials require in-depth, culturally conscious and grounded methodologies for their identification, which may also invite new and interesting venues of studying tourism in the broader context of regional development, marketing of local specialities, and even place branding (see Andéhn & L'Espoir Decosta, 2021b). We cannot expect, though, markets and their logics to 'come quietly' and enable only those changes that we desire from them.

CONCLUSIONS

Even if tourism, and particularly agritourism, is considered as a cure to economic downturns, it has come with too many side-effects to be considered a viable and sustainable treatment on its own. We need to acknowledge the romantic promise of what tourism *could* be: a meeting of peoples from different cultures, a means to experience differences first hand, and a path to appreciating our shared world for all the splendour and grandness it has to offer. But we note that such meetings, seeing, and indeed even appreciation are too naïve a premise if we cannot ascertain that they do not occur in the context of the economic power imbalances that are growing unabated. Any ensuing analysis to the deployment of an axiology of markets must account for that fact because we stand to side firmly with a reductive view of citizens as producers and consumers. In plain terms, marketization cannot conserve its object, so that when we use it to 'regenerate', changes that may not necessarily help protect what was there before will invariably come. One example of an innovative approach to sustainability through regeneration is New Zealand's Bay of Plenty, where tourism professionals have been exploring regenerative ways that rely on Indigenous Māori cultural values and wisdom to interconnect with natural ecosystems that can result in the flourishing of both the planet and people. Even in light of such encouraging examples, we know that we can only ever negotiate markets, never control or contain them, and increasingly there is no true 'outside' of their logic (see Fisher, 2009). Noting that we tread precarious ground even in striving to conserve the treasures of earth and of humanity, we would do well to note that under capitalism we can easily do as much damage with an open hand as with a closed fist; consider how Marx noted the power of transmutation inherent in capitalism already a century and a half ago:

> In our days, everything seems pregnant with its contrary: machinery, gifted with the wonderful power of shortening and fructifying human labour, we behold starving and overworking it. The newfangled sources of wealth, as if by some weird spell, are turned into sources of want. (Karl Marx, 1856)[5]

NOTES

1. *The Guardian* 26/5/2020. www.theguardian.com/australia-news/2020/may/26/rio-tinto-blasts-46000 -year-old-aboriginal-site-to-expand-iron-ore-mine, accessed 10 October 2022.
2. Reuters 12/4/2019. https://uk.style.yahoo.com/squatters-occupy-venice-homes-housing-protest-tourism -surges-111954882.html, accessed 10 October 2022.
3. See www.livescience.com/what-is-paris-syndrome, accessed 10 October 2022.
4. *The Conversation* 4/11/2019. https://theconversation.com/what-is-ecological-economics-and-why -do-we-need-to-talk-about-it-123915, accessed 10 October 2022.
5. https://www.marxists.org/archive/marx/works/1856/04/14.htm#intro, accessed 10 October 2022.

REFERENCES

Aguiló, E., Alegre, J., & Sard, M. (2005). The persistence of the sun and sand tourism model. *Tourism Management 26*(2), 219–231.

Alegre, J., & Garau, J. (2011). The factor structure of tourist satisfaction at sun and sand destinations. *Journal of Travel Research 50*(1), 78–86.

Andéhn, M., Hietanen, J., & Lucarelli, A. (2020). Performing place promotion: On implaced identity in marketized geographies. *Marketing Theory 20*(3), 321–342.

Andéhn, M., & L'Espoir Decosta, J. N. P. (2021a). Authenticity and product geography in the making of the agritourism destination. *Journal of Travel Research 60*(6), 1282–1300.

Andéhn, M., & L'Espoir Decosta, J. N. P. (2021b). New perspectives welcome: A case for alternative approaches to country of origin effect research. In M. Cleveland & N. Papadopoulos (eds), *Marketing Countries, Places, and Place-Associated Brands*. Cheltenham, UK and Northampton, MA, USA: Edward Elgar Publishing, pp. 158–173.

Arroyo, C. G., Barbieri, C., & Rich, S. R. (2013). Defining agritourism: A comparative study of stakeholders' perceptions in Missouri and North Carolina. *Tourism Management 37*, 39–47.

Barbieri, C. (2013). Assessing the sustainability of agritourism in the US: A comparison between agritourism and other farm entrepreneurial ventures. *Journal of Sustainable Tourism 21*(3), 252–270.

Baudrillard, J. (1968). The system of objects. In M. Poster (ed.), *Jean Baudrillard: Selected Writings*. Stanford, CA: Stanford University Press.

Baudrillard, J. (1994). *Simulacra and Simulation*. Ann Arbor: University of Michigan Press.

Berkes, F., & Folke, C. (1998). *Linking Social and Ecological Systems: Management Practices and Social Mechanisms for Building Resilience*. Cambridge: Cambridge University Press.

Bernstein, E., & Shih, E. (2014). The erotics of authenticity: Sex trafficking and "reality tourism" in Thailand. *Social Politics 21*(3), 430–460.

Braun, E., Kavaratzis, M., & Zenker, S. (2013). My city – my brand: The different roles of residents in place branding. *Journal of Place Management and Development 6*(1), 18–28.

Busby, G., & Rendle, S. (2000). The transition from tourism on farms to farm tourism. *Tourism Management 21*(6), 635–642.

Butler, R. (ed.) (2006). *The Tourism Area Life Cycle*. Clevedon: Channel View Publications.

Cai, L. A. (2002). Cooperative branding for rural destinations. *Annals of Tourism Research 29*(3), 720–742.

Cammack, P. (2004). What the World Bank means by poverty reduction, and why it matters. *New Political Economy 9*(2), 189–211.

Carolin, L., Lindsay, A., & Victor, W. (2015). Sex trafficking in the tourism industry. *Journal of Tourism and Hospitality 4*(4), 166–171.

Collins-Kreiner, N. (2010). The geography of pilgrimage and tourism: Transformations and implications for applied geography. *Applied Geography 30*(1), 153–164.

Crouch, G. I., & Ritchie, J. R. B. (1999). Tourism, competitiveness, and social prosperity. *Journal of Business Research 44*(3), 137–52.

Di Domenico, M., & Miller, G. (2012). Farming and tourism enterprise: Experiential authenticity in the diversification of independent small-scale family farming. *Tourism Management 33*(2), 285–294.

174   *Handbook on tourism and conservation*

Du Cros, H., & McKercher, B. (2020). *Cultural Tourism*. London: Routledge.

Dunleavy, P., & Hood, C. (1994). From old public administration to new public management. *Public Money & Management, 14*(3), 9–16.

Dwyer, L., & Kim, C. (2003). Destination competitiveness: Determinants and indicators. *Current Issues in Tourism 6*(5), 369–414.

Fairclough, N. (1993). Critical discourse analysis and the marketization of public discourse: The universities. *Discourse & Society 4*(2), 133–168.

Ferguson, J. (1995). *The Anti-Politics Machine: "Development," Depoliticization and Bureaucratic Power in Lesotho*. Minneapolis: University of Minnesota Press.

Fisher, M. (2009). *Capitalist Realism: Is There No Alternative?* London: John Hunt Publishing.

Flanigan, S., Blackstock, K., & Hunter, C. (2014). Agritourism from the perspective of providers and visitors: A typology-based study. *Tourism Management 40*, 394–405.

Glusac, E. (2020). Move over, sustainable travel. Regenerative travel has arrived. *New York Times*. www.nytimes.com/2020/08/27/travel/travel-future-coronavirus-sustainable.html.

Gomez, M., Lopez, C., & Molina, A. (2015). A model of tourism destination brand equity: The case of wine tourism destinations in Spain. *Tourism Management 51*, 210–222.

González, A. T. (2018). Venice: The problem of overtourism and the impact of cruises. *Investigaciones Regionales – Journal of Regional Research 2018*(42), 35–51.

Harvey, D. (2001). *Spaces of Capital: Towards a Critical Geography*. London: Routledge.

Ho, P. S., & McKercher, B. (2004). Managing heritage resources as tourism products. *Asia Pacific Journal of Tourism Research 9*(3), 255–266.

Hood, C. (1995). The "new public management" in the 1980s: Variations on a theme. *Accounting, Organizations and Society, 20*(2–3), 93–109.

Horkheimer, M., & Adorno, T. W. (2002). *Dialectic of Enlightenment*. Stanford, CA: Stanford University Press.

Inskeep, E. (1987). Environmental planning for tourism. *Annals of Tourism Research 14*(1), 118–135.

Jameson, F. (1991). *Postmodernism, or, the Cultural Logic of Late Capitalism*. Durham, NC: Duke University Press.

Kavaratzis, M. (2012). From "necessary evil" to necessity: Stakeholders' involvement in place branding. *Journal of Place Management and Development 5*(1), 7–19.

Kotler, P., & Levy, S. (1971). De-marketing, yes, de-marketing. *Harvard Business Review*, November–December, 74–80.

Lazzarato, M. (2014). *Signs and Machines: Capitalism and the Production of Subjectivity*. Los Angeles, CA: Semiotext(e).

L'Espoir Decosta, J.-N. P., & Andéhn, M. (2018). Looking for authenticity in product geography. In J. M. Rickly & E. S. Vidon (eds), *Authenticity and Tourism: Materialities, Perceptions, Experiences*. Bingley: Emerald, pp. 15–31.

MacCannell, D. (1973). Staged authenticity: Arrangements of social space in tourist settings. *American Journal of Sociology 79*(3), 589–603.

Malpas, J. (2018). *Place and Experience: A Philosophical Topography*. London: Routledge.

Mandel, E. (1975). *Late Capitalism*. London: New Left Books.

Manuel-Navarrete, D. (2016). Boundary work and sustainability in tourism enclaves. *Journal of Sustainable Tourism 24*(4), 507–526.

Manyara, G., & Jones, E. (2007). Community-based tourism enterprises development in Kenya: An exploration of their potential as avenues of poverty reduction. *Journal of Sustainable Tourism 15*(6), 628–644.

Mbaiwa, J. E. (2005). Enclave tourism and its socio-economic impacts in the Okavango Delta, Botswana. *Tourism Management 26*(2), 157–172.

McGehee, N. G., & Kim, K. (2004). Motivation for agri-tourism entrepreneurship. *Journal of Travel Research 43*(2), 161–170.

McKercher, B., & Du Cros, H. (2002). *Cultural Tourism: The Partnership between Tourism and Cultural Heritage Management*. London: Routledge.

Muñiz-Martínez, N. (2016). Towards a network place branding through multiple stakeholders and based on cultural identities: The case of 'the coffee cultural landscape' in Colombia. *Journal of Place Management and Development 9*(1), 73–90.

'Negotiating with the juggernaut': agritourism and market-driven conservation   175

Ndaguba, E. A., & Hanyane, B. (2019). Stakeholder model for community economic development in alleviating poverty in municipalities in South Africa. *Journal of Public Affairs 19*(1), 1–11.

Palmer, C. A. (1994). Tourism and colonialism: The experience of the Bahamas. *Annals of Tourism Research 21*(4), 792–811.

Papathanassis, A., Katsios, S., & Dinu, R. N. (2017). "Yellow tourism": Crime & corruption in tourism. *Journal of Tourism Futures 3*(2), 200–202.

Paraskevas, A., & Brookes, M. (2018). Nodes, guardians and signs: Raising barriers to human trafficking in the tourism industry. *Tourism Management 67*, 147–156.

Pasquinelli, C. (2010). The limits of place branding for local development: The case of Tuscany and the Arnovalley brand. *Local Economy 25*(7), 558–572.

Petrovic M. D., Bjeljac Z., & Vuiko, A. (2016). Analysis of the life cycle of an agritourism destination: A theoretical approach. *Agricultural Bulletin of Stavropol Region 4*(2), 77–81.

Phillip, S., Hunter, C., & Blackstock, K. (2010). A typology for defining agritourism. *Tourism Management 31*, 754–758.

Prahalad, C. K. (2005). *The Fortune at the Bottom of the Pyramid: Eradicating Poverty through Profit and Enabling Dignity and Choice through Markets.* Upper Saddle River, NJ: Wharton School.

Prebensen, N., Skallerud, K., & Chen, J. S. (2010). Tourist motivation with sun and sand destinations: Satisfaction and the WOM-effect. *Journal of Travel & Tourism Marketing 27*(8), 858–873.

Richards, G. (1996). Production and consumption of European cultural tourism. *Annals of Tourism Research 23*(2), 261–283.

Salazar, N. B. (2012). Community-based cultural tourism: Issues, threats and opportunities. *Journal of Sustainable Tourism 20*(1), 9–22.

Seraphin, H., Sheeran, P., & Pilato, M. (2018). Over-tourism and the fall of Venice as a destination. *Journal of Destination Marketing & Management 9*, 374–376.

Sharpley, R. (2002). Rural tourism and the challenge of tourism diversification: The case of Cyprus. *Tourism Management 23*(3), 233–244.

Smith, S. L. (1994). The tourism product. *Annals of Tourism Research 21*(3), 582–595.

Sonnino, R. (2004). For a 'piece of bread'? Interpreting sustainable development through agritourism in southern Tuscany. *Sociologia Ruralis 44*(3), 285–300.

Sugathan, P., & Ranjan, K. R. (2019). Co-creating the tourism experience. *Journal of Business Research 100*, 207–217.

Tuan, Y. F. (1977). *Space and Place: The Perspective of Experience.* Minneapolis: University of Minnesota Press.

Tussyadiah, I. P. (2014). Toward a theoretical foundation for experience design in tourism. *Journal of Travel Research 53*(5), 543–564.

Tustin, M., & Lockshin, L. (2001). Region of origin: Does it really count? *Australia and New Zealand Wine Industry Journal* 16, 139–143.

UNWTO (2010). *Tourism and Biodiversity: Achieving Common Goals towards Sustainability.* www.e -unwto.org/doi/pdf/10.18111/9789284413713, accessed 10 October 2022.

Urry, J. (1992). The tourist gaze and the environment. *Theory, Culture & Society 9*(3), 1–26.

Wang, N. (1999). Rethinking authenticity in tourism experience. *Annals of Tourism Research 26*(2), 349–370.

13. Micro and small-scale culture-based tourism initiatives as a livelihood option for rural women in Kenya

Rita Wairimu Nthiga and Beatrice H. O. Ohutso Imbaya

INTRODUCTION

It has become necessary for rural communities to seek alternative sources of income due to insufficient income derived from agricultural practices in most sub-Saharan African countries. Tourism has thus been explored as a possible livelihood option for rural communities in Africa. Consequently, tourism has not only become a potential tool for redressing reduced agricultural income (Sharpley, 2002), but it has been lauded for its contribution to the sustainability of rural communities.

Tourism has the potential for creating wealth and engendering prosperity in rural areas. By empowering rural communities to utilize tourism resources, prospects for entrepreneurship activities that are associated with various cultures and which help to enhance the preservation of natural resources abound (Olwal & Maina, 2020). Hence, tourism can enable rural communities to uphold their unique cultural heritage and traditions. Moreover, the sector is vital for safeguarding habitat and endangered species (Ballantyne et al., 2011) as well as sustainable rural enterprises that provide employment for the youth, women and marginalized groups. Tourism also empowers women who naturally have a large presence in the agricultural workforce and who are likely to be affected more than men by insufficient agricultural incomes. Specifically, Kibaara et al. (2012) argued that tourism contributes to about 10 percent of Kenya's gross domestic product (GDP), making it the third largest contributor after agriculture and manufacturing. According to WTTC (2021), the travel and tourism sector contributed 8.1 percent to Kenya's GDP ($8,074.1 million) and 8.5 percent to total employment prior to the Covid-19 pandemic in 2019. Moreover, the Kenya Ministry of Tourism and Wildlife reported that tourism contributes to 8.3 percent of total employment translating to 1.1 million jobs (Ministry of Tourism and Wildlife, 2018).

However, despite its promotion as a viable form of economic development activity for rural areas in Africa, it has been argued that tourism development may not necessarily translate into quick benefits for rural communities (Saarinen & Lenao, 2014). Therefore, while it is generally agreed that tourism has the potential to engender community well-being and economic development (Anderson, 2013; Mitchell & Ashley, 2010; Peaty, 2012), not much is known about how rural communities can take advantage of cultural tourism to improve their livelihoods in most African countries. Moreover, the evidence existing in support of the contribution of culture-based tourism to local communities' well-being in developing countries, particularly in Africa, is still not consistent (Mitchell & Ashley, 2010).

In Kenya, micro and small-scale tourism initiatives (MSTIs) have been promoted as a supplementary source of livelihood for rural communities. In Ilesi, Kakamega County, there are a number of MSTIs aimed at providing economic incentives for communities to engage in conservation and supplement their livelihoods. These enterprises are built around the material and non-material cultural activities of the Isukha community such as pottery and artifacts, as well as rich traditional music, dance and art. Traditionally, these activities constituted the lifestyle of the Isukha community who reside close to Kakamega Forest and the "crying stone" of Ilesi. The people rely heavily on these two major tourist attractions in the county to support their livelihoods. The pottery and related activities have been packaged into commercial products and sold to tourists visiting the area for cultural experience in recent times. It is also noteworthy that policymakers and development practitioners now conceive these MSTIs as a possible pathway for vulnerable community members, including women, to achieve sustainable livelihoods. Moreover, these enterprises have elsewhere been promoted as avenues for capacity building, enhancement of local community well-being and poverty reduction (Elliot & Sumba, 2011).

Diverse studies have been undertaken in Africa to establish various contributions of tourism related initiatives to community development and specifically to livelihoods (Mbaiwa, 2004; Mbaiwa & Stronza, 2010; Mbaiwa & Sakuze, 2009; Nthiga et al., 2015; Kgathi et al., 2007). While Mbaiwa and Stronza (2010) established that benefits such as shelter, employment, income, water supply and scholarships accrued to the community as a result of tourism, Mbaiwa & Sakuze (2009) recommended the development and implementation of policies and strategies for cultural tourism and livelihood diversification.

However, limited information exists on the contribution of MSTIs to community livelihoods especially in western Kenya. To address this gap, this chapter uses the Community Capitals Framework (CCF) to evaluate the contribution of pottery initiatives to sustainable livelihoods among Ilesi Clay Potters Women Group at Kakamega County in western Kenya. The CCF builds on the Sustainable Livelihoods Framework (SLF), which focuses on how individuals, households or communities derive their living based on natural, physical, financial, social and human assets or capitals (Chambers & Conway, 1992; DFID, 1999; Ellis, 1999; Allison & Ellis, 2001; Scoones, 1998, 2009). While the SLF identifies five core asset categories or types of capital upon which livelihoods are built, CCF categorizes these assets into seven community capitals, which are natural, human, social, financial, built, cultural and political.

This chapter assesses the contribution of MSTIs to community livelihoods in Kakamega, Kenya. While the first section introduces the subject, the second section addresses culture-based tourism, and the third offers the conceptual framework of the chapter. Other sections include the methodology, findings, conclusions and recommendations.

CULTURE-BASED TOURISM

Culture-based tourism has been identified as an important development opportunity for Africa since it presents a chance to embed inclusiveness in the growth and expansion of tourism (Christie et al., 2013; Novelli, 2015; Saarinen & Rogerson, 2015). Culture-based tourism has also been identified as one of the fast-growing sectors of the tourism economy worldwide and is estimated to contribute about 40 percent of the global tourism revenue (Noonan & Rizzo, 2017). Consequently, the culture-based tourism market has been singled out as crucial to the growth of tourism globally (UNWTO, 2001; Salazar, 2012).

178   *Handbook on tourism and conservation*

According to Littrel (1997), culture is a combination of beliefs, values, attitudes and ideas (what people think); normative behavior patterns or way of life (what people do); and artworks, artifacts and cultural products (what people make). Since culture constitutes a people's way of life and the outcomes of that way of life (Richards, 2001), cultural tourism involves the consumption of a destination's material and non-material cultural heritage (McKercher & Du Cros, 2003; Leask & Yeoman, 1999). UNWTO (2018, p. 15) defines cultural tourism as "a type of tourism activity in which the visitor's essential motivation is to learn, discover, experience and consume the tangible and intangible cultural attractions/products in a tourism destination".

Additionally, cultural tourism has the potential for rural community development and poverty alleviation (Manyara & Jones, 2007), provision of markets for local products and generation of employment opportunities (Ahebwa et al., 2015). Therefore, cultural tourism may be viewed as an important and all-encompassing growth strategy for rural communities in Africa. In this chapter, we analyze a form of cultural tourism involving the sale and exhibition of culture-based products made through pottery craft. The study site comprises rural women who, through their skills, have derived economic gains from pottery making. In addition to farming, pottery is a major economic activity for most households and especially women in the area. The next section outlines the conceptual model from which the chapter derives its analysis.

THE COMMUNITY CAPITALS FRAMEWORK

The CCF was developed by Flora et al. (2004) as an approach to analyze and understand dynamics within rural communities (Stone & Nyaupane, 2018). From a systems-thinking perspective, the framework assesses the complex and dynamic relationships that exist between conservation and development (Stone & Nyaupane, 2018). The application of the CCF is appropriate in order to gain a critical understanding of the dynamics of community change (Mattos, 2015). According to Flora and Flora (2013), the CCF was first used to analyze entrepreneurial and community development efforts from a systems perspective. Flora et al. (2004, p. 1) define capital as an asset or resource that can be used, invested or exchanged to create new resources. Moreover, Emery and Flora (2006) identify seven categories of capital, which include financial, natural, cultural, human, social, political and built capitals. These capitals are further categorized into two major categories by Gutiérrez-Montes et al. (2009): human or intangible capitals (social, human, cultural and political capitals) and material or tangible capitals (financial, physical and natural capitals) (Figure 13.1). Stone and Nyaupane (2018) are of the opinion that all communities regardless of their poverty levels have resources that, if well appropriated, can be used to enhance their members' well-being as well as that of the environment in which they live.

- *Social capital* includes aspects such as the social interactions, trust and reciprocity needed to facilitate collective outcomes, networks and a sense of belonging (Bebbington, 1999; Jones, 2005; Stone & Nyaupane, 2016). According to Emery and Flora (2006), social capital also reflects the connections between people and organizations.
- *Human capital* comprises knowledge, skills and competencies embodied in individuals that are relevant to economic activity (OECD, 1998). According to Koutra and Edwards

(2012), education and training opportunities enable accumulation of knowledge and skills which in turn lead to increased productivity and earnings.
- *Natural capital* comprises the resources that exist in the natural world and include soil, lakes, natural resources, nature's beauty, rivers, forests, wildlife and local landscape (Fukuyama, 2001; Mattos, 2015). Pretty (1998) and Constanza et al. (1997) further observe that the natural capital of an area shapes the cultural capital connected to the place.
- *Financial capital* is conceived as the capacity to access funds from banks, created through investments or through micro-finance and other credit mechanisms (Koutra & Edwards, 2012). Ellis (2000) views financial capital as resources related to money and access to funding, wealth, charitable giving and grants.
- *Built capital* refers to the man-made assets used in production and includes houses, schools, businesses, libraries, water systems and communication systems, among others (Emery et al., 2006).
- *Cultural capital* reflects the way people know the world and how they act within it (Emery & Flora, 2006). Cultural resources include dances, stories, heritage, food, traditions, language and values, among others.
- *Political capital* encompasses power and community connection to people who have power (Koutra & Edwards, 2012). It also entails the ability of the community to find its own voice and engage in actions that contribute to its well-being (Aigner et al., 2001).

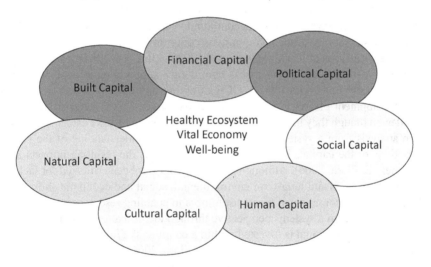

Figure 13.1   *Community capitals*

According to Stone and Nyaupane (2018), utilizing the CCF in analyzing the outcomes of participation in tourism by communities has diverse advantages. The first is that CCF incorporates cultural and political capitals, which are essential for understanding power and access to power, and highlights the relevance of local knowledge and traditions in community livelihoods. The second strength of the CCF is its alignment with systems thinking as it highlights the interdependence, interaction and synergy among the capitals because the use of the assets in one form of capital can have a positive or negative effect on the other capitals

180 *Handbook on tourism and conservation*

Table 13.1 *Conceptualization of community capitals*

Capital asset	Conceptualization
Human capital	• Knowledge
	• Skills
	• Education support
	• Training opportunities
Social capital	• Formal and informal relationships
	• Networks and associations
	• Trust
Financial capital	• Employment
	• Income
Built capital	• Buildings
	• Infrastructure
Natural capital	• Impacts on soils and natural resources
Cultural capital	• Impacts on traditions, in this case dances and pottery
Political capital	• Community voice
	• Engagement in activities that contribute to community well-being

Source: Authors (2022), adopted and modified from Emery and Flora (2006).

(Gutiérrez-Montes et al., 2009). The third strength is emphasized by Emery and Flora (2006) who argue that the loss or degradation of assets within one capital will negatively affect one or more capitals because when one of the community capitals is severely affected or depleted, the health and sustainability of the community is compromised.

Other studies (Emery et al., 2006; Matiku et al., 2020; Kline et al., 2018; Stone & Nyaupane, 2016, 2018) have used the CCF to assess community and economic development efforts. According to Emery and Flora (2006), CCF offers an approach to assess community and economic development efforts from a systems perspective. This is because the capitals are interrelated even though they have been categorized into seven distinct categories (Flora et al., 2004). An analysis from a systems perspective enables an understanding of the assets' flow and interactions of the capitals in the community as well as the resulting impacts across the capitals (Emery & Flora, 2006). Although the framework has been criticized for lack of attention on global processes and long-term environmental, social and economic shifts (Salazar et al., 2018), it has been commended for its usefulness in enabling researchers and communities to approach change from a systems perspective through identification of assets in each capital (stocks of capital), how capital is invested within a community (flow of capital) and the ways in which capitals interact (Anglin, 2015; Emery & Flora, 2006; Stone & Nyaupane, 2018).

This chapter adopts the conceptualization of capitals from diverse livelihoods scholars and researchers including Chambers and Conway (1992), Scoones (1998, 2009), DFID (1999), Ellis (1999) and Allison and Ellis (2001). Specifically, the community capitals were conceptualized based on Stone and Nyaupane (2018) who adopted Emery and Flora's (2006) conceptualization in terms of the stocks (the assets), the capital flows (the capitals invested) and the relationships among the capitals. Table 13.1 provides a summary of the aspects included for the seven capitals of the CCF.

METHODOLOGY

Study Area

The study was conducted in the Ilesi area in Shinyalu which is one of the seven sub-counties that constitute Kakamega County in western Kenya. The county covers an area of 3,051.3 km^2 and is the second most populous county in Kenya after Nairobi with the largest rural population (CGoK, 2018). According to the 2019 census (KNBS, 2019b), Kakamega County had a total population of 1,867,579 people, of which 897,133 are males, 970,406 are females and 40 are intersex[1] persons. There are 433,207 households with an average size of 4.3 persons per household and a population density of 618 people per square kilometer (GoK, 2019). The county has a predominantly agrarian economy with livestock farming taking up a small portion of the available arable land and 61 percent of households engage in crop and livestock production. According to the Kenya National Bureau of Statistics' Gross County Product Report 2019 (KNBS, 2019a), the county's GDP was estimated at KES 182.563 billion, which is 2.4 percent of the national GDP in 2017 (KES 7.524 trillion).

Kakamega County is endowed with diverse natural resources, which include water and natural forests. The main water sources are eight main rivers: Nzoia, Yala, Lusumu, Isiukhu, Sasala, Viratsi, Kipkaren and Sivilie. The county has one natural forest covering Shinyalu and Lurambi and other farm forests. The natural forest resource in the county is Kakamega forest, which is recognized as the eastern-most relic of the Guineo-Congolian lowland rainforest belt, and which once stretched from Kenya across Uganda, East and Central Africa to the West African coast (KFS, 2021). Kakamega forest is situated 35 km from Lake Victoria and is an exclusive sanctuary for an extraordinary variety of endemic flora and fauna, including insects, reptiles and birds that are not found in other parts of the country. An estimated 10–20 percent of the animal species in the forest are unique to this forest. It is also an important watershed for some of the rivers that flow into Lake Victoria. Kakamega forest was first gazetted as a Trust Forest in 1933, but a total of 4,000 ha of the northern portion of the forest, along with the adjacent 457 ha of Kisere forest, were amalgamated and gazetted as Kakamega National Park in 1986 (UNESCO, 2010).

The forest supports the local people, providing them with timber, fuel wood, herbal medicines, building materials, food and land for farming. It also attracts tourism and provides research sites for several institutions (GoK, 2014). The village of Ilesi in Shinyalu sub-county borders the Kakamega forest. The location is inhabited by the Isukha community. The Isukha people are known in western Kenya for their "isukuti" dance and bullfighting mainly performed during occasions such as initiation, funerals, weddings and other similar ceremonies. These cultural activities and ceremonies are among the main cultural tourist attractions in western Kenya. Ilesi also plays host to the "crying stone" which is not only a religious shrine for the local people, but also a major tourist attraction in Kakamega County where traditional dance groups in the area showcase their dancing skills on an occasional basis.

Research Design and Data Collection

The study utilized a qualitative approach to obtain data. The qualitative data gathering techniques used include observation (photography was used to record the status of the natural environment as shown in Figure 13.2), document analysis (minutes of meetings) and key

182   *Handbook on tourism and conservation*

informant interviews with both leaders and group members of the Ilesi Clay Potters Women Group. Data were collected between December 2020 and February 2021. Interviewees were identified through the snowball sampling method. A total of 10 interviews were undertaken. Participants were selected based on their position and membership in the group and information was obtained until saturation was achieved. Interview responses, observations and information obtained through documents were transcribed and summarized, respectively. Data were analyzed using thematic analysis in line with the CCF and supported with secondary data and literature review. The interviewees comprised four group leaders (coded as interviews 1 to 4) and six group members (coded as interviews 5 to 10).

FINDINGS

Capital Stocks and Flows

As an additional livelihood strategy for the women in the study area, pottery craft is a cultural resource, which has had a major contribution to the capital assets of the entire community. Here, we use the CCF to present the capital stocks and flows of pottery craft. In terms of *social and political capitals*, the women have organized themselves socially and politically to form the Ilesi Clay Potters Women Group. The group has a decision-making body who are elected officials (chairperson, secretary, treasurer, organizing secretary) (Interviews 1, 2). The group also has a constitution and is registered by the National Government's Ministry of Culture and Social Services. The organization of the group enhances establishment and enforcement of rules, and formation of networks, which aids devolution of power (Berkes, 2010).

Additional social capital assets are the social networks and partnerships that have emerged from the group; welfare groups have merged to enable members to meet their social needs. For example, there is the Ilesi "Tuinuane" cluster, which aims to support each other socially when faced with life challenges such as bereavement. The term "Tuinuane" is a Kiswahili word meaning "let us lift each other up" (Interviews 1, 2, 3, 4). This is in line with Sen (1997) who opines that social capital such as networks and organizations play a vital role in helping people act to improve their livelihoods, mobilize assets or pull together resources and defend themselves. The networks further enable people to have a voice, become their own agents of change and enhance members' interactions, which in turn help to improve members' quality of life. The wide acceptance of pottery products from the area has also enhanced the community's pride and unique culture. As one respondent noted, "even if you go to Uganda and Tanzania, you will find that all those doing pottery are from this village … we have this unique talent of pottery which you cannot find anywhere else" (Interviews 1, 6, 7).

However, the majority of the members are not comfortable with apportioning resources as communal investments and assets. For example, a past initiative to keep chickens and cows communally failed because some members were dishonest (Interview 1). Instead, members opted for individual initiatives due to lack of trust among members in the management of communal benefits. Similarly, some members had reservations exhibiting their products in the communal hall owned by the group. They indicated that the returns from the sale of products exhibited in this hall were far less than what they received from selling similar products outside the hall. This agrees with DFID's (1999) observation that, despite the importance of social bonds and norms in sustainable livelihood, the absence of relations of trust, reciprocity and exchanges between individuals undermines cooperation, connectedness and networking including access to wider institutions.

Results showed that the contribution to *financial capital* is the most notable benefit for the women. According to all the participants, pottery making has provided the women an opportunity for self-employment (Interviews 1–10). The women also derived income from the sale of pottery products to other regions in Kenya. The income from the sale of the products is used to cater for household expenses and children's education. Some members have also used the income to build more decent and permanent houses thus improving their quality of life. The challenge is that women opt to produce items such as pots and cookers, which they could easily sell but which tend to command low prices. The study participants were not keen to disclose the average monthly income from pottery and opted to say that "it supplements their income from farming and chicken rearing".

In terms of *built capital*, members were able to acquire modern houses from the sale of the pottery products. The women's group also received a grant and assistance by the National Government's Ministry of Culture, Sports and Social Services to set up an Exhibition Hall along the busy Kakamega–Kisumu highway, where members display and sell their earthenware. The exhibition hall, which is situated in the chief's office, serves as a focal point for group members to showcase their products and acts as a marketplace for them to sell their crafts.

Figure 13.2   Clay harvesting site in Ilesi

Regarding the contributions to *natural capital*, two main issues of concern were noted during the data collection period. The first issue is that the clay used for the pottery products is harvested from the riverbanks and wetlands causing environmental degradation (Figure 13.2). Nonetheless, though harvesting of clay may drive environmental degradation, the clay still serves as a production input for the potters who invariably derive economic benefits from product sales. Therefore, the contribution of natural capitals to livelihoods overshadows the effect of environmental degradation.

184   *Handbook on tourism and conservation*

Secondly, the potters use firewood kilns to dry their products which is a cause of deforestation and loss of tree cover in the area. Respondents revealed that the potters buy the firewood from other community members who have permits to harvest firewood from the nearby Kakamega forest. Overharvesting and overexploitation of natural resources have been identified elsewhere (Mbaiwa, 2004) as a major hindrance for the sustainability of culture-based tourism products.

From pottery making and sales, there has also been a great contribution to *human capital*. For example, since the formation of the group, there has been increased interest among members to attend workshops and training on business operations and management including financial literacy. Members have also developed an interest in attending exhibitions, especially those relating to the county activities, leading to increased opportunities for members to interact with other traders and learn more from them. Moreover, the income from the sales of pottery products is used to train children at school. Literature suggests that the ability to read and write enhances people's ability to secure better jobs and perform them efficiently. Sen (1997) views human capital holistically in terms of human capability and not purely in terms of economic benefits. Sen's (1997) standpoint is that capabilities enhance people's ability to question, challenge and usher in new ways of doing things.

The greatest contribution of *cultural capital* in the context of the study area is the ability of the community to preserve the cultural practice of pottery, which has been passed from one generation to another over several years. As argued by Bebbington (1999), cultural capital is meaningful, enabling and empowering and in most cases forms the basis for the maintenance and enhancement of other capitals. Notably, pottery making craft in the community could be rightly deemed as a source of power for members.

IMPLICATIONS

The application of the CCF to establish the contribution of pottery production and sales to the livelihoods of the women's pottery group demonstrates that the cultural assets of a place can impact on the other capital assets in that location. This is in line with the observation by Gutiérrez-Montes (2005) that an investment in one capital asset leads to a flow in other capital assets. Gutiérrez-Montes et al. (2009) also argued that there is interdependence, interaction and synergy among the capitals since the use of the assets in one capital can have a positive or negative effect over the quantity and the possibilities of other capitals. According to Emery and Flora (2006) the flow of one asset to other assets could be positive leading to an upward spiral effect or negative resulting in a downward spiral effect. In the case of the Ilesi pottery craft, there was an upward spiral effect on the financial, human, social, political and physical capitals but a downward spiral impact on the natural capitals through degradation of natural resources. This is related to the findings of Mbaiwa (2004, 2018) who also identified overharvesting and overutilization of raw materials for basket making in Botswana, which has affected the prospects for the sustainability of cultural tourism in the Okavango Delta, Botswana.

In a scenario where there is a downward spiral effect on some assets, Gutiérrez-Montes et al. (2009) advocate for a balance in the deployment of the capitals since one capital should not be favored at the expense of others. The evolution of the group into various structures and institutions represents a new form of organized social and political capital, which are instrumental for enhancing cohesion and discipline among the members, facilitating the formation of networks and helping to devolve power to the members (women). Berkes (2010), however, urges caution on the challenges of governance in a complex and uncertain world as a result of conflicting interests.

CONCLUSIONS

Based on the CCF, the chapter provides some insights into the relationship between culture and livelihood improvement among rural women potters in western Kenya. The findings reveal the interrelationships among the various community capitals. In addition, the findings indicate that an improvement in one capital may have either a positive or negative impact on another capital. For example, the exploitation of clay for pottery production and harvesting of firewood has led to the degradation of riverbanks and wetlands and deforestation in providing fuel for the kilns. The findings therefore reveal a need for a balance of the various capitals for sustainable community development. Stakeholders should therefore seek ways of mitigating the negative impacts of development initiatives. As such, the authorities, especially the county Government of Kakamega, should devise ways of ensuring the sustainability of pottery craft and at the same time protect the wetlands and the forest/tree cover in the region. This is in line with Mbaiwa's (2004) recommendations for an integrated approach to cultural tourism and participation of the potters in natural resources management initiatives. Stone and Nyaupane (2018) further recommend a systems thinking approach to linking tourism and development by initiating long-term solutions to complex socio-ecological problems as opposed to cause–effect approaches. From a methodological standpoint, there is a need to further interrogate the contexts in which communities live as well as undertake more long-term studies. More insightful details might be provided if there is an analysis of the implications of pottery production beyond the study area.

NOTE

1. Refers to people born with sex characteristics (genetic, hormonal, physical organs) that do not fit the typical definitions for male or female bodies (KNCHR, 2018).

REFERENCES

Ahebwa, W. M., Aporu, J. P., & Nyakaana, J. B. (2015). Bridging community livelihoods and cultural conservation through tourism: Case study of Kabaka Heritage Trail in Uganda. *Tourism and Hospitality Research*, 11, 1–13.

Aigner, S. M., Flora, C. B., & Hernandez, J. M. (2001). The premise and promise of citizenship and civil society for renewing democracies and empowering sustainable communities. *Sociological Inquiry*, 71(4), 493–507.

Allison, E. H., & Ellis, F. (2001). The livelihoods approach and management of small-scale fisheries. *Marine Policy*, 25(5), 377–388.

Anderson, W. (2013). Leakages in the tourism systems: Case of Zanzibar. *Tourism Review*, 68(1), 62–75.

Anglin, A. E. (2015). Facilitating community change: The Community Capitals Framework, its relevance to community psychology practice and its application to Georgia community. *Global Journal of Community Psychology Practice*, 6(2), 2–15.

186 *Handbook on tourism and conservation*

Ballantyne, R., Packer, J., & Sutherland, I. A. (2011). Visitor's memories of wildlife tourism: Implications for the design of powerful interpretive experiences. *Tourism Management, 32,* 770–779.

Bebbington, A. (1999). Capitals and capabilities: A framework for analyzing peasant viability, rural livelihoods and poverty. *World Development, 27*(12), 2021–2044.

Berkes, F. (2010). Devolution of environment and resources governance: Trends and future. *Environmental Conservation, 37*(4), 489–500.

Chambers, R., & Conway, G. R (1992). *Sustainable Rural Livelihoods: Practical Concepts for the 21st Century.* IDS Discussion Paper 296. Brighton: IDS.

Christie, I., Fernandes, E., Messerli, H., & Twining-Ward, L. (2013). *Tourism in Africa: Harnessing Tourism for Growth and Improved Livelihoods.* Washington, DC: World Bank.

Constanza, R., d'Arge, R., Farber, S., Grasso, M., Hanson, B., Limburg, K., Naeem, S., O'Neil, R. V, Parvelo, J., Raskin, R. G., Sutton, P., & Van der Belt, M. (1997). The value of the world's ecosystem services and natural capital. *Nature, 387*(6630), 253–260.

County Government of Kakamega (CGoK) (2018). *Kakamega County Integrated Development Plan 2018–2022.* Kakamega, Kenya.

DFID (1999). *Sustainable Livelihoods Guidance Sheets.* London: DFID.

Elliot, J., & Sumba, D. (2011). *Conservation Enterprise: What Works, Where and for Whom?* London: International Institute for Environment and Development.

Ellis, F. (1999). *Rural Livelihoods and Diversity in Developing Countries: Evidence and Policy Implications.* ODI Natural Resources Perspectives No. 40. London: ODI.

Ellis, F. (2000). *Rural Livelihoods and Diversity in Developing Countries.* New York: Oxford University Press.

Emery, M., Fey, S., & Flora, C. (2006). *Using Community Capitals to Develop Assets for Positive Community Change.* Workbook for the Centre for Rural Development. www.comm-dev.org/images/pdf/Issue13-2006.pdf.

Emery, M., & Flora, C. (2006). Spiraling-up: Mapping community transformation with Community Capitals Framework. *Community Development, Journal of the Community Development Society, 37*(1), 19–35.

Flora, C., & Flora, J. (2013). *Rural Communities: Legacy and Change* (4th edn). Boulder, CO: Westview Press.

Flora, C., Flora, J., & Fey, S. (2004). *Rural Communities: Legacy and Change* (2nd edn). Boulder, CO: Westview Press.

Fukuyama, F. (2001). Social capital, civil society and development. *Third World Quarterly, 22*(1), 7–22.

GoK (2014). *Agricultural Sector Development Support Programme.* Nairobi: Government of Kenya, Ministry of Agriculture, Livestock, and Fisheries.

GoK (2019). *2019 Kenya Population and Housing Census Volume II: Distribution of Population by Administrative Units.* Nairobi: Government of Kenya, Kenya National Bureau of Statistics.

Gutiérrez-Montes, I. (2005). Healthy communities equals healthy ecosystems? Evolution (and breakdown) of a participatory ecological research project towards a community natural resource management project San Miguel Chimalapa (Mexico). PhD dissertation, Iowa State University.

Gutiérrez-Montes, I., Emery, M., & Fernandez-Baca, E. (2009). The sustainable livelihoods approach and the community capitals framework: The importance of systems level approaches to community change efforts. *Community Development, 40*(2), 106–113.

Jones, S. (2005). Community-based ecotourism: The significance of social capital. *Annals of Tourism Research, 32*(2), 303–324.

Kenya Forest Service (2021). www.kenyaforestservice.org/index.php?option=com_content&view=article&id=687:kakamega-forest-station&catid=179&Itemid=686, accessed 23 January 2022.

Kenya National Commission on Human Rights (KNCHR) (2018). *Equal in Dignity and Rights: Promoting the Rights of Intersex Persons in Kenya.* Nairobi: KNCHR.

Kgathi, D. L., Ngwenya, B. N., & Wilk, J. (2007). Shocks and rural livelihoods in the Okavango Delta, Botswana. *Development Southern Africa, 24*(2), 289–308.

Kibaara, O. N., Odhiambo, N. M., & Njuguna, J. M. (2012). Tourism and economic growth in Kenya: An empirical investigation. *International Business and Economics Research Journal, 11*(5), 517–527.

Kline, C., McGehee, N., & Delconte, J. (2018). Built capital as a catalyst for community-based tourism. *Journal of Travel Research, 58*(6), 899–915.

KNBS (2019a). *Gross County Product 2019 Report*. Nairobi: Kenya National Bureau of Statistics.

KNBS (2019b). *2019 Kenya Population and Housing Census Volume 1*. Nairobi: Kenya National Bureau of Statistics.

Koutra, C., & Edwards, J. (2012). Capacity building through socially responsible tourism: A Ghanaian case study. *Journal of Travel Research*, 51(6), 779–792.

Leask, A., & Yeoman, I. (1999). *Heritage Visitor Attractions: An Operations Management Perspective*. London: Cassel.

Littrel, M. A. (1997). Shopping experiences and marketing of culture to tourists. In M. Robinson, N. Evans, & P. Carlaghan (eds), *Tourism and Culture: Image, Identity and Marketing*. Newcastle: Centre for Travel and Tourism, University of Northumbria, pp. 107–120.

Manyara, G., & Jones, E. (2007). Community-based tourism enterprises development in Kenya: An exploration of their potential as avenues of poverty reduction. *The Journal of Sustainable Tourism*, 15(6), 628–644.

Matiku, S., Zuwarimwe, J., & Tshipala, N. (2020). Community-driven tourism projects' economic contribution to community livelihoods: A case of Makuleke contractual park community tourism project. *Sustainability*, 12(8230), 1–16.

Mattos, D. (2015). Community capitals framework as a measure of community development. University of Nebraska-Lincoln, Department of Agricultural Economics. https://digitalcommons.unl.edu/agecon_cornhusker/811.

Mbaiwa, J. E. (2004). Prospects of basket production in promoting sustainable rural livelihoods in the Okavango Delta, Botswana. *Journal of Tourism Research*, 6(4), 221–253.

Mbaiwa, J. E. (2018). Effects of safari hunting tourism ban on rural livelihoods and wildlife conservation in Northern Botswana. *South African Geographical Journal*, 100(1), 41–61.

Mbaiwa, J. E., & Sakuze, L. K. (2009). Cultural tourism and livelihood diversification: The case of Gcwihaba caves and XaiXai village in the Okavango Delta, Botswana. *Journal of Tourism and Cultural Change*, 7(1), 61–71.

Mbaiwa, J. E., & Stronza, A. L. (2010). The effects of tourism development on rural livelihoods in the Okavango Delta, Botswana. *Journal of Sustainable Tourism*, 18, 635–656.

McKercher, B., & Du Cros, H. (2003). Testing a cultural tourism typology. *The International Journal of Tourism Research*, 5(1), 45–58.

Ministry of Tourism and Wildlife (2018). *Kenya Tourism Agenda 2018–2022*. Nairobi: Government of Kenya.

Mitchell, J., & Ashley, C. (2010). *Tourism and Poverty Reduction: Pathways to Prosperity*. London: Earthscan.

Noonan, D. S., & Rizzo, I. (2017). Economics of cultural tourism: Issues and perspectives. *Journal of Cultural Economics*, 41, 95–107.

Novelli, M. (2015). *Tourism and Development in Sub-Saharan Africa*. Abingdon: Routledge.

Nthiga, R. W., Van der Duim, R., Visseren-Hamakers, I. J., & Lamers, M. (2015). Tourism-conservation enterprises for community livelihoods and biodiversity conservation in Kenya. *Development Southern Africa*, 32(3), 407–423.

OECD (1998). *Human Capital Investment: An International Comparison*. Paris: OECD.

Olwal, K., & Maina, J. (2020). Tourism and rural development in Africa: Travel and tourism agenda. https://mail.google.com/mail/u/0/ - inbox/FMfcgxwLsdCrnmzxbKVhBHMrmzFlFBbg, accessed 23 January 2022.

Peaty, D. (2012). Kilimanjaro tourism and what it means for local porters and for the local environment. *Journal of Ritsumeikan Social Sciences and Humanities*, 4(1), 1–11.

Pretty, J. N. (1998). *The Living Land: Agriculture, Food and Community Regeneration in Rural Europe*. London: Earthscan.

Richards, G. (2001). *Cultural Attractions and European Tourism*. Wallingford: CAB International.

Saarinen, J., & Lenao, M. (2014). Integrating tourism to rural development and planning in the developing world. *Development Southern Africa*, 31(3), 363–372.

Saarinen, J., & Rogerson, C. M. (2015). Setting cultural tourism in Southern Africa. *Nordic Journal of African Studies*, 24(3–4), 207–220.

Salazar, N. B. (2012). Community-based cultural tourism: Issues, threats and opportunities. *Journal of Sustainable Tourism*, 20(1), 9–22.

Salazar, O. V., Ramos-Marin, J., & Lomas, P. L. (2018). Livelihood sustainability assessment of coffee and cocoa producers in Amazon region of Ecuador using household types. *Journal of Rural Studies*, 62, 1–9.

Scoones, I. (1998). *Sustainable Rural Livelihoods: A Framework for Analysis*. IDS Working Paper 72. Brighton: IDS.

Scoones, I. (2009). Livelihoods perspectives and rural development. *The Journal of Peasant Studies*, 36(1), 171–196.

Sen, A. (1997). Human capital and human capability. *World Development*, 25(12), 1959–1961.

Sharpley, R. (2002). The challenges of economic diversification through tourism: The case of Abu Dhabi. *International Journal of Tourism Research*, 4(3), 221–235.

Stone, M. T., & Nyaupane, G. P. (2016). Protected areas, tourism and community livelihoods: A comprehensive analysis approach. *Journal of Sustainable Tourism*, 24(5), 673–693.

Stone, M. T., & Nyaupane, G. P. (2018). Protected areas, wildlife-based community tourism and community livelihoods dynamics: Spiraling up and down of community capitals. *Journal of Sustainable Tourism*, 26(2), 307–324.

United Nations Educational, Scientific and Cultural Organization (UNESCO) (2010). *The Kakemega Forest*. https://whc.unesco.org/en/tentativelists/5508/.

United Nations World Tourism Organization (UNWTO) (2001). *Tourism 2020 Vision*. Madrid: United Nations World Tourism Organization.

United Nations World Tourism Organization (UNWTO) (2018). *Tourism and Culture Synergies*. Madrid: United Nations World Tourism Organization. https://doi.org/10.18111/9789284418978.

World Travel and Tourism Council (WTTC) (2021). *Kenya 2021 Annual Research: Key Highlights*.

14. Environmental impact of rural tourism
Gondo Reniko and Oluwatoyin D. Kolawole

INTRODUCTION

The concept 'rural' is contextual as rurality depends on the level of economic progress that a country has attained (Kolawole, 2014; Isserman, 2005; Hart et al., 2005; Ekong, 2003). For example, a settlement classified as rural in one country may pass for an urban settlement in another. The characterisation of rurality or urbanity is thus a social construction based on people's perceptions (Isserman, 2005). While rurality is acknowledged as nuanced, the use of certain indices (such as low population density, an abundance of farmland or remoteness, etc.) has been recognised as appropriate for delineating it from urban areas (Hart et al., 2005; Ekong, 2003). This lack of a clear definition of what is termed 'rural' has contributed to an intriguing debate about the meaning of the term rural tourism and how the industry should be operated in rural areas to meet the needs of the local people (Bramwell, 1994). Even in this case, the criteria used to distinguish rural from urban vary significantly in different contexts and not all such tourism is strictly rural. Tourism may be urban even when it is activated in rural areas. Furthermore, different forms of rural tourism have developed in different regions and hence it is hard to find characteristics that are common to all countries. Also, rural areas are in a complex process of changing due to the impact of global market communications and telecommunications that have changed market conditions for traditional production (Wilson et al., 2001).

While Alberta (2010) understands rural tourism as a countryside experience that encompasses a wide range of attractions and activities that take place in non-urban areas, Lane (1994) defines the concept as tourism that is located in rural areas. As such, tourism activities in the study area qualify as rural tourism because the settlement is a rural village based on the definition of rural and urban areas in Botswana. The term rural tourism is defined as any tourism that involves rural people and benefits them economically and socially and respects their cultural norms and values (Subhash et al., 2010). Thus, the sustainability of rural tourism needs to showcase rural life and allow local people to interact with tourists for a more enriching experience. Rural tourism encompasses activities such as eco-tourism, cultural tourism, adventure tourism and many agro-based activities. In Rotherham's (2007) opinion, the idea of a single rural tourism does not exist; it is a complex concept and difficult to define concisely. Whereas the main thrust of rural tourism in developed countries is to regenerate rural areas where traditional agrarian industries are in decline, the main thrust of the industry in developing countries like Botswana is to diversify the rural economy. Thus, the aim of rural tourism in the study area is to diversify the rural economy that is characterised by insufficient agricultural livelihoods, hence the need to search for new sources of growth and economic opportunities. The main characteristics associated with rural tourism include features such as wide-open landscapes and low levels of infrastructural development, which allow tourists or visitors to directly experience natural sceneries in their chosen destinations. Consequently, rural tourism in its purest form is situated within rural areas.

The characteristics distinguishing rural from urban tourism are shown in Figure 14.1. While rural tourism thrives on cultural, natural and historical features, urban tourism depends mainly on man-made scenic attributes. Rural tourism, which is an internationally recognised industry, is regarded as a panacea for achieving socio-economic progress in rural areas (Mohanty, 2014; Musasa and Mago, 2014; Briedenhann and Wickens, 2004). The rural tourism business is regarded as a sustainable practice for tackling economic challenges and a vital source of livelihood for the rural populace (Saarinen, 2007). Traditionally, rural communities depended on agriculture as their major economic base. However, Roberts and Hall (2004) opine that these communities are likely to be incapable of sustaining themselves without a diversified economic base in the long run. Thus, tourism is viewed as an alternative form of diversification of the rural economic base (Hwang et al., 2012).

In this chapter we discuss the application of digital image processing and its potential to automate land use and land cover change (LULCC) mapping (Chima, 2012). To realize this potential, analysts have developed image classification techniques that automatically sort pixels with similar multispectral reflectance values into clusters that correspond to functional LULCC categories (Colson, 2012). In this regard, remotely sensed data in the form of satellite images are acquired and processed in GIS software to reveal how the physical environment has been modified. The techniques show with a high degree of accuracy the changes that have taken place from a natural to a man-made environment and by that means making it possible to assess the impact of tourism on the environment.

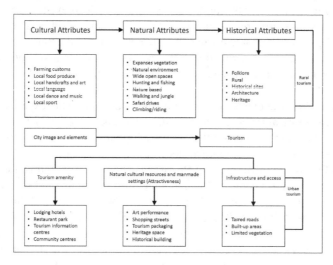

Source: Adapted from Giriwati et al. (2013) and Neto et al. (2021).

Figure 14.1 *Attributes of rural and urban tourism*

Geographic information systems (GIS) and remote sensing (RS) provide tourists with information in the form of maps and tourist attraction centres (Kushwaha et al., 2011). Therefore, GIS helps tourists to make a choice and plan their activities to match their tourism management and desired benefits. More importantly, the use of high-resolution satellite images in a GIS platform allows the identification of tourism impact on the physical environment, for example the encroachment of infrastructure on the environment. Thus, road networks, railway lines and increased construction intensity of both residential and industrial blocks can easily be identified using GIS and RS. The application of GIS in identifying location suitability also aids tourism development planning in a particular area (Slater, 2002; Arca, 2012). Identifying location suitability is perhaps the best known and most widely developed application of GIS (Farsari and Prastacos, 2002). Conflicting or complementary land uses and activities, infrastructure availability and natural resources determine the possibility and the capacity of an area as a suitable tourist destination (Bunruamkaew and Murayama, 2012). The use of GIS, therefore, enables the noting and mapping of the environmental impact of rural tourism in any given area.

In the quest for rural tourism promotion, the transformation of land use and land cover (LULC) is one of the main manifestations of its impact on the environment. Hence, the analysis of LULC dynamics is a front-burner topic in rural tourism. Estimation of temporal land cover change enables assessment of the rate at which change in LULC advances and connects them with the impact of the change, hence allowing improved prediction of future impact and trends in rural villages. It is noteworthy that a region's land cover is a constantly changing mosaic of cover types determined by both physical and anthropogenic influences. Historically, important land-use forms in Shakawe in Botswana were animal husbandry, *molapo* farming and the clearing of forest for human settlement, all of which had an insignificant impact on the LULC dynamics (Motsumi et al., 2012; Magole and Thapelo, 2005). Tourism development differs from agriculture in terms of the degree of resource utilisation and environmental impact. For instance, while agricultural enterprises like animal husbandry impact vegetation cover through grazing, the effect of tourism activities are, to a certain extent, witnessed in environmental pollution (water and air) and vegetation removal for the purpose of developing tourism infrastructures. Therefore, rural tourism development in indigenous communities mainly involves human modifications of land cover. Social and economic considerations are among the most important drivers of LULCC and yet few studies have addressed the impact of rural tourism on LULCC dynamics in Botswana and specifically in the study area. This chapter, therefore, aims to demonstrate a methodology that relies on multi-temporal RS satellite data and digital image interpretation technologies to assess the three-decade impact of rural tourism on LULCC dynamics in a selected settlement between 1990 and 2020. Thus, the spatial and temporal dynamics of LULCC in Shakawe village in northern Botswana will be analysed in this chapter.

THEORETICAL UNDERPINNINGS

The thrust of this chapter is rooted in Black and Weiler's (2015) sun-lust and wanderlust conceptual framework. For Black and Weiler (2015), sun-lust and wanderlust are two of the main reasons why people travel. Tourist motivation is regarded by Black and Weiler (2015) as one of the key elements in understanding tourist decision-making. Motivation is understood as the integration of networks of economic, social, biological and cultural forces, which gives value and direction to travel choices, behaviour and experience (Pearce, 1998). Simply put, motivation is a state of arousal of a drive that impels people to act in pursuit of certain goals. Black and Weiler's (2015) sun-lust and wanderlust conceptual framework underscores two main motives for travel: the first is the desire to go from a known to an unknown place, which is otherwise known as wanderlust. The second motive is the sun-lust experienced by certain individuals. This relates to the desire to go to a place that can provide the traveller with specific facilities that do not exist in the traveller's place of residence. Thus, the motives determining rural tourism and compelling travellers to visit Shakawe include the Tsodilo Hills (featuring culturally important rock arts) and the Okavango Delta, which are both UNESCO World Heritage Sites. Other drivers of rural tourism in this area include bird watching, fishing and boat tours. There are also guided tours into the Okavango Delta Game Park. The decision-making process is illustrated in Figure 14.2.

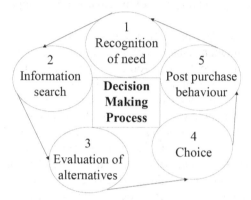

Source: Authors (2021).

Figure 14.2 A tourist's decision-making process

Before a tourist visits a destination, a decision has to be made to travel. Literature has shown that such a decision involves five clearly defined stages (Figure 14.2). The first stage in this process is the recognition of a need. The need to travel to a certain destination is stimulated by either external factors or personal desire. While external influences include mass media, friends and family members, internal desires relate to personal interests. The second stage in the decision-making process is the information searching stage. Once a place to visit has been identified and the need to visit the place established, then information about the place is sought. A wide range of sources provides the information required. These may include friends, relatives or other acquaintances who have visited the place. Thus, sources such as salespeople, displays or mass media are consulted to get information about the place to be visited. Evaluation of places to visit marks stage 3. This stage involves the potential traveller making a choice on which destination among many is to be visited first. The traveller assesses the importance of the alternatives through a cost–benefit analysis. This is then followed by stage four, which involves choosing the destination. The decision to choose the place to visit is determined by personal interests as well as an assessment of the importance of visiting the destination. Friends and relatives also play a part in this stage to influence the tourist to choose the destination. Post-purchase behaviour is the stage where a person has visited a destination and is evaluating the place based on the services at the destination. Thus, the tourist is either satisfied or dissatisfied. This stage depends on the tourist's prior expectations. Thus, if the tourist's expectation is met or not met, the tourist is, therefore, satisfied or dissatisfied as the case may be.

METHODOLOGY

In the assessment of LULCC of the study area, this study utilised both geospatial technologies including GIS and RS and literature review. Geospatial technology is related to the processing or assemblage of data linked with the location (van Manen et al., 2009).

Data Collection

The freely available National Aeronautics and Space Administration (NASA) archives of Landsat images provides an opportunity for cost-effective use of RS for monitoring LULCC anywhere on Earth (Campbell and Wynne, 2011; Quattrochi et al., 2003). Our research uses Landsat 5 Thematic Mapper (TM) and Landsat 8, Operational Land Imager and Thermal Infrared Sensor (OLI/TIRS), Path 170 and Row 074 of 25 and 28 October 1990, 2000, 2010 and 2020. These images were chosen because of their clarity and for being three decades apart. ArcGIS 10.7.1 and system for automated geoscientific analysis (SAGA) software were used in the preparation, spatial analysis and mapping. Ancillary data used in the research include Google Earth images, which were used as reference data during the classification and validation phase of the analysis. The Landsat programme is a very prolonged venture for acquiring satellite imageries of the Earth. It was launched as Earth Resources Technology Satellite in 1972 and later renamed Landsat (Irons et al., 2016). Fieldwork was conducted between June 2018 and December 2019 to determine ambiguous land cover classification and to visit the area of major change to observe and determine the possible causes of the LULCC. This also provides a secondary validation of the classification accuracy for the most current image date. A Garmin handheld global positioning system (GPS) unit with an accuracy of 5 metres was used to map and collect the coordinates of important land use features during pre- and post-classification field visits to the study area to prepare LULC maps.

RESULTS AND DISCUSSION

The driver of LULCC over the past thirty years in the Okavango Delta in general and particularly in the study area may be attributed to rural tourism. The transformation of LULC is one of the most important manifestations of anthropogenic influences on the environment. Thus, LULC analysis is an important issue if policymakers are to understand the impact of tourism in the Okavango Delta. Such an analysis is vital as it enables both local authorities and policymakers to make an informed decision on the impact of tourism within the entire Okavango Delta and specifically in Shakawe. The rate at which change in LULC advances helps to link the alteration to the development of rural tourism, hence allowing improved predictions on the future impact of tourism. This section of the chapter, therefore, describes LULC classifications from 1990 to 2020.

LULC analysis is a technique that deals with the determination of different classes of LULC over a given period about a given area (Dey et al., 2018). In the case of this study, it refers

194   *Handbook on tourism and conservation*

to the identification of LULC from satellite images for Shakawe village from 1990 to 2020. The study used both unsupervised and supervised classification techniques. The classification of images is based on the pixel value, that is, a digital number that each cell carries, and the different types of classes have a set of different digital numbers (Talukdar et al., 2020). Unsupervised and supervised classifications are performed by partitioning the multispectral feature space of the satellite image and creating multiple numbers of classes for the given input (Chatterjee et al., 2020). While the unsupervised classification technique does not require training data, the supervised technique requires training data. To determine the impact of rural tourism on LULC this chapter analyses land use changes for the periods covering 1990–2000, 2000–2010 and 2010–2020.

Accuracy Assessment

Accuracy assessment is an important step in the process of analysing remotely sensed data because the products serve as the basis for economic decision-making. The potential user of the product should know about the reliability of the data when confronted with thematic maps derived from RS data. In the context of this chapter, the probability of built-up area was 0.94, indicating that the LULC classification on the thematic map is 94 per cent similar to the built-up area in the real-life situation in Shakawe. Similarly, the probable occurrence of the shrubs/trees category was 0.88, which implies that 88 per cent of the area covered in shrubs/trees on the ground also appears as shrubs/trees in the classified images. The average accuracy in the study is 88 per cent, which implies that what appears on the classified maps is an 88 per cent resemblance of LULC on the actual ground in the study area. This figure is above the recommended minimum probability of 75 per cent (Jenness and Wynne, 2005). The reliability, which is also known as user's accuracy, refers to the probability that a pixel classified on the image represents that category on the ground (Story and Congalton, 1986). A Kappa coefficient[1] of 0.95, which is higher than the minimum recommended Kappa coefficient of 0.75, was obtained in this study (Jenness and Wynne, 2005).

Land Use Land Cover Change Analysis

The improvement of local infrastructure enhances tourism activities while the development of the tourism sector, in turn, provides financial support for the improvement of tourism features. The development of tourism has enhanced Shakawe's economic power to improve its local infrastructure. Prior to 2014, Botswana had realised that tourism arrival figures were declining especially from the lucrative overseas tourism markets. The decline was attributed to the global economic recession, which contributed to the weakening of the US dollar. The focus on low-volume, high-cost tourism further contributed to the perception that Botswana is an expensive tourism destination. Given these factors, there was an urgent need for a clear national market-driven tourism strategy to direct the future marketing and development of key tourism attractions towards greater diversification and market expansion. Such strategies would imply substantial government investments in infrastructure like roads and airstrip improvements as well as accommodation provision in national parks. All these were made to ensure that tourism areas become attraction centres, which are accessible and appealing to a wider range of potential tourists. Demand for Botswana wildlife tourism products increased substantially as a result of greater awareness and government investment into the sector.

Consequently, tourism in and around Shakawe grew rapidly. For instance, the available beds for visitors per night in Shakawe grew by over 66 per cent between 2000 and 2014 (Mbaiwa and Hambira, 2020). The developments placed major pressure on the park's infrastructure as visitors to Shakawe and other satellite villages in the area increased.

The study monitored the spatial and temporal LULCC in the study area situated within the Okavango Delta over 30 years. The classification of LULC produced five distinct classes of land use and these are shrubs/trees, open spaces, water, built-up and hydrophytes. Overall, a fast LULCC was observed during the periods 2000–2010 and 2010–2020 as compared to 1990–2000. Significant changes occurred in the open spaces and the built-up areas. While open spaces decreased in 2000, built-up areas significantly increased between 2010 and 2020. This period coincided with the listing of the Okavango Delta as the 1000th UNESCO World Heritage Site in 2014 (Matswiri, 2017). Designation as a World Heritage Site makes it more famous (Robinson, 2015), such that its previous obscure status (if that was the case) suddenly transforms and is immediately brought into the international limelight. This implies that Shakawe village, which is strategically situated along the Okavango Delta panhandle, may have become more famous in 2014 as soon as the delta was inscribed as a UNESCO World Heritage Site. Shakawe's acquisition of a new status naturally would warrant the attraction of more tourists than in the past, which in turn would impact significantly on its infrastructure. It has been noted by Yang (2010) that the listing of a place as a heritage site is one of the most powerful driving forces in the attraction of tourists to a destination. A study by Robinson (2015) reveals that the designation of a place as a World Heritage Site leads to the construction of modern facilities in the area. One of the areas in which tourism has impacted Shakawe village is through the development of infrastructure. Based on Mbaiwa's (2003) findings, there has been an expansion of infrastructure since the 1990s in a bid to meet the needs of the growing tourism industry in the village. Thus, the construction of infrastructure results in a significant impact on the physical environment. Some infrastructural development includes road networks in northern Botswana where Shakawe is situated. Mbaiwa (2003) notes that northern Botswana was inaccessible and tarred roads were virtually non-existent before 1990. During the same period, the built-up areas in Shakawe covered a small area and there were more open spaces than built-up areas. Vegetation was abundant especially in the north-western part of the village. The scenario gradually changed towards the end of 1990 when the government of Botswana realised the importance of tourism in infrastructure development including tarred roads, which facilitates tourism activities in the Okavango Delta (Lenao and Saarinen, 2015). This development process led to the building of the 500 km Maun–Mohembo Road, which was completed in 1995 (Mbaiwa, 2003; Robinson, 2005). The construction of this road generated a host of construction activities in the study area. Infrastructural development significantly improved over time and by 2000, a significant change in the built-up area could be noticed (Mmopelwa et al., 2005). Robinson's (2005) study in the Okavango Delta and that of Folorunso et al. (2020) in Lagos had similar results in which a significant change in infrastructure was noticed between 2005 and 2020 after an increase in tourist arrivals. The sudden change in infrastructure implies that hospitality businesses were constructing accommodation and tourism-related facilities to meet tourists' demands. The impact of tourism in the village may have been implicated in the noticeable changes in LULC from 1990 to 2020 in the study area.

The changes in LULC between 1990 and 2000 reveal that shrubs/trees covered an area of 96.7 hectares (ha) in 1990 and this was increased to 144.6 ha in 2000. In the same vein, the

196   *Handbook on tourism and conservation*

built-up area was 3.7 ha in 1990 and it almost doubled to 6.9 ha in 2000. It is noteworthy that during the same period, open spaces covering only 64.3 ha in 1990 slightly decreased to cover 56.1 ha of land in 2000. This implies that there was a decrease in open spaces due to an increase in shrubs/trees as well as the expansion of built-up areas. Hydrophytes decreased from covering 57.6 ha in 1990 to 37.5 ha in 2000. While this decrease in hydrophytes may be attributed to the general decrease in water in the Okavango River perhaps due to low rainfall in the Angolan catchment area, the possibility of siltation of the river is, however, high given that the area is flat, and more vegetation cover was cleared. Also, the reduction of hydrophytes could be as a result of plant succession. Plant succession refers to the process in which there is a gradual replacement of one plant community by another plant community that is of a steady type (Khorchani et al., 2021). Thus, hydrophytes were replaced by shrubs/trees.

Changes in LULC between 2000 and 2010 reveal that while shrubs/trees occupied 144.6 ha of land in 2000, there was a decrease in the area covered in vegetation to 100.7 ha in 2010. During the same period, the built-up area increased from 6.9 ha in 2000 to 11.1 ha in 2010. While the decrease in vegetation is attributed to the clearing of trees to pave the way for the construction of tourism-related infrastructure (such as roads and accommodation), the increase in built-up area was due to the construction of tourism-related infrastructure. However, the results of the accuracy assessment with a kappa coefficient of 0.75 as well as the producer and user's accuracy of 93 and 94 refute the inaccuracies in classification. There was an increase in the volume of water in the river from 6.4 ha in 2000 to 53.7 ha in 2010. Indeed, an increase in precipitation in the Angolan highlands may have been implicated in the increase in water in the Okavango River as shown in the spatial extent of water in the river. A slight change in open spaces occurred between 2000 and 2010 during which 56.1 ha of open spaces increased to 58.1 ha. The area covered in hydrophytes declined from 37.5 ha in 2000 to 27.9 ha in 2010 due to an increase in water, which inundated the hydrophytes owing probably to the increase in rainfall in the Angola catchment area.

Tourism is the second-largest government revenue earner in Botswana and has contributed to the gross domestic product (GDP) after diamonds (Mbaiwa, 2003). In the 1996/97 financial year, a total of P800 million amounting to 4.5 per cent of the country's GDP was realised. Literature has shown that tourism has a propensity to create employment for the people of Ngamiland and elsewhere in the world (Richardson, 2021; Stone et al., 2020; Saarinen et al., 2020). Arising from tourism activities, there was an establishment of facilities such as camps and lodges in the entire Okavango Delta including in Shakawe. Wilson et al.'s (2001) study shows that 735 people were employed in 20 camps and lodges in the Okavango Delta in 2001. Mbaiwa (2003) also establishes that 923 people were employed in 30 other camps in the same area in 2003 not reported in Wilson et al.'s (2001) research. The government of Botswana estimated that the number of people employed in tourism-related activities such as hotels, airlines, safari companies and transport was 27,000 in 1990 and the figure increased four-fold between 1990 and 2019 (Vumbunu, 2020). However, the Covid-19 pandemic is now a threat to the tourism industry. According to a report by the Hospitality and Tourism Association of Botswana (Botswana Tourism, 2020), Covid-19 resulted in the closure of international borders and restricted movements of people. Hence, the tourism industry lost some operations as a result of travel bans and restrictions. Between March 2020 and March 2021, hotels in the study area were operating at 10 per cent of their full capacity (United Nations World Tourism Organization, 2020). By April 2020, only 300 of the 26,000 workers remained employed in the tourism sector. Besides impacting people's health, the effect of Covid-19 has also affected people's social lives, behaviour, livelihoods and economics (Stone et al., 2021). Botswana Tourism (2020) estimates that the tourism industry lost approximately P1,283,422.74, which is the equivalent of US$111,734.81. The Botswana Presidential Covid-19 Task Force Bulletin (2020) notes that some major annual sporting events including the Toyota Desert Race, Khawa Dune Challenge and cultural festivals were cancelled due to Covid-19. The events provided entertainment and are of social and economic significance to the locals and traders (Botswana Tourism, 2020). Their cancellation had major impacts on local people who depend on selling goods and services during such events.

The changes in LULC that occurred between 2010 and 2020 reveal that whereas shrubs/ trees covered 57.2 ha in 2010, the LULC category has decreased to 40.7 ha in 2020. The built-up area significantly increased from 22 ha in 2010 to 72.5 ha in 2020. There was a slight increase in the area covered with water between 2010 and 2020, from 3.3 ha to 5 ha, respectively. While 70.7 ha were open spaces in 2010, there has been a change in land use, leading to the reduction in open spaces to 14.8 ha in 2020. However, hydrophytes increased from 18.2 ha in 2010 to 33.2 ha in 2020, probably due to increased washing of organic matter into the water bodies through surface run-off as a result of population growth and construction activities. This is in line with Crossley et al. (2002) who observe that an increased supply of nutrients, particularly those present in manure in sufficient quantity, has the propensity to boost the growth and population of aquatic plants.

This might imply that people's purchasing power increased as they became employed in tourism-related industries. This in turn may have consequently led to a plethora of activities in the construction industry, which took place between 2010 and 2020. The rate of deforestation due to the construction of accommodation and other tourism-related facilities was high during the period. Important land resources, which are impacted heavily in this area, include land water and vegetation. Thus, the increased construction of recreational facilities exerted pressure on these resources leading to a change in the scenic landscape of the village. There were, therefore, several construction activities that occurred in the study area between 2010 and 2020 as shown by widespread open spaces and expansion of the built-up area. The direct impact was noticeable in the provision of tourist facilities such as accommodation and feeder road networks. Forests are often at the receiving end of the impact of tourism due to deforestation caused by fuelwood collection and land clearance. Results of this study indicated that over 50 per cent of vegetation cover were cleared and over 75 per cent of the space in the village was at the same time under construction between 2010 and 2020 alone. The constructions include accommodation, lodges, hotels as well as pavements such as tarred roads and recreational facilities like stadia, among others. Just as any industrial concern, tourism can also result in environmental pollution including air, noise and water pollution. A study by Mmualefe et al. (2011) reveal that water in the main channels of the Okavango River is slightly acidic to neutral (pH 6.5–7.0) and becomes alkaline downstream in the floodplain especially around cattle farms.

The change in built-up area over the three decades in the study area deserves scrutiny. The increase in built-up area concurs with Mbaiwa and Stronza's (2010) study, which reveals that 72 per cent of the rural households in the Okavango Delta benefited from tourism through income, employment and improved infrastructure such as roads, accommodation, etc., all of which had an impact on LULC. Also, water supply and roads facilities had an impact on the physical environment of the study area. A study conducted by Kgathi et al. (2006), which was designed to determine the proportion of cleared land in the upper Okavango Basin, reveals

198   *Handbook on tourism and conservation*

that there was a general increase in cleared area and a reduction in vegetation cover in the Upper Okavango Delta. Ramberg et al. (2006) attributed the clearance of vegetation to the establishment of vegetable gardens and agricultural fields, all of which are tourism-related entities. However, results in the current study show that the reduction in vegetation cover was not a result of the removal of vegetation for agricultural purposes but the result of the need to build more tourist-related facilities. This concurs with Mbaiwa and Stronza's (2010) findings, which reveal that 82 per cent of the people studied in three selected villages in the Okavango Delta no longer practised crop farming but relied on tourism for their livelihood. The study further reveals that many traditional livelihood activities like subsistence hunting and collection of rangelands products, livestock and crop farming have been replaced by tourism activities in the Okavango Delta (Mbaiwa and Stronza, 2010). However, similar studies elsewhere in Botswana reveal that increase in tourism arrivals has led to a general increase in the construction of facilities. A similar study, which was conducted by Akinyemi and Mashame (2018) in Palapye village, attributed the rapid developmental change in LULC within the period 2000–2004 to the social and economic transformation of Palapye since the beginning of the twenty-first century. In the same vein, the pattern of change in LULC in Shakawe village over the years is partly attributable to the listing of the Okavango Delta as the 1000th World Heritage Site in 2014, which was part of the strategies to conserve the wetland. At a national level there were issues surrounding the dredging of the Boro River in order to supply water to Orapa mining and prospecting which dates back to the 1990s (Armstrong, 1991). There has also been the possibility of future oil prospecting and expanded agricultural activities especially livestock farming in the Okavango Delta. As such, the listing of the Okavango Delta as a World Heritage Site was meant to check any possibility of degrading the environment and it has thus opened a host of tourism-related activities in the area. As noted by Darkoh and Mbaiwa (2014), the listing of the delta led to road and airstrip constructions in and around the protected areas in Shakawe. However, the movement of vehicles and boats continues to impact on wildlife. Also, the construction of the Mohembo Bridge will most likely contribute to the infrastructure and socio-economic development in Shakawe village because of the potential improvement in human mobility and accessibility to other remote villages (Mogomotsi et al., 2020). When commissioned, the Mohembo Bridge has the propensity of boosting regional tourism, which may eventually have an impact on the LULCC in Shakawe village.

LULC Losses and Gains in 30 Years

Analysis of LULC categories between 1990 and 2020 revealed that different land categories were converted from one class of land use to another. There were gains in some categories and losses in others. Among the five LULC classes, findings reveal that there was a significant gain in the built-up area category and a loss in the shrubs/trees category. Depending on the fluxes of tourist activities in the study area, changes in bare open spaces fluctuated within the period. Results show that the built-up area increased five-fold ($16.1 km^2$ to $72.5 km^2$) between 1990 and 2020. The increase in the built-up area is attributed to the increased population. In other words, the expansion witnessed in the built-up area is a result of more people migrating to Shakawe village, which in turn overspills into the neighbouring ungazetted settlements such as Mohembo and Skondomboro among others. Safarov and Sh's (2020) study, which aimed at understanding the impact of tourism on Uzbekistan's economy, also found that a boom in the tourism industry has a spillover effect on the human population, which in turn engenders the expansion of the built-up area. Literature has also shown that the increase in population boosts activities in the construction industry and if left unchecked might lead to unprecedented environmental challenges like incessant water and land degradation (Lin et al., 2021). Studies conducted by Cavric and Keiner (2004) and Ringrose and Matheson (1995) reveal that population growth increases built-up area, which in turn impacts negatively on vegetation cover. Thus, the conversion of shrubs/trees to residential land uses does increase impervious surfaces, which have a high propensity of accelerating surface run-off. This has a high probability of degrading the water in the Okavango River.

Natural vegetation (shrubs/trees) is another category of LULC that shows a significant change over 30 years in the study area. Findings revealed that the shrubs/trees category covered 57.2 km² in 1990 and this figure got reduced to 49.7 km² in 2000 before yet another record reduction, which was as low as 5 km² in 2010. However, an analysis of the 2020 image shows an increase in vegetation cover in the form of the shrubs/trees category, which slightly increased from 5 km² to 15 km² in the 2010 to 2020 period. This result is valid because the period is too long for vegetation to grow into mature vegetation. A similar study by Chiutsi and Mudzengi (2017) on the impact of tourism seasonality and destination management in the Mana Pools National Park in Zimbabwe also showed an overall fluctuation in the vegetation cover due to increases in the number of tourism businesses.

CONCLUSION

The chapter utilises multi-temporal RS satellite data and digital image interpretation technologies to assess the environmental impact of rural tourism on the LULC dynamics in Shakawe village over three decades. Unsupervised image classification was conducted and five LULC classes were identified. The five LULC classes were built-up area, shrubs/trees, open spaces (bare land), hydrophytes and water. Findings showed that there was an increase in the built-up area while the shrubs/trees category witnessed a spatial shrink within 30 years. The open space category was the least in 1990 and became the highest in 2010 but witnessed a slight increase in 2020. The gains and losses of LULC were attributed to increased population growth with its accompanying boost in tourist-related activities. This implies that rural tourism (in the form of ecotourism in this case) has a tremendous negative environmental impact if not properly regulated.

NOTE

1. Kappa coefficient is generated from a statistical test to evaluate the accuracy of classification. It ranges between −1 and 1. Whereas a Kappa index of −1 means there were gross errors in the classification process, a Kappa coefficient closer to one implies perfect classification and the map portrays almost exactly the real-life situation.

REFERENCES

Akinyemi, F. O., & Mashame, G. (2018). Analysis of land change in the dryland agricultural landscapes of eastern Botswana. *Land Use Policy, 76*, 798–811.

Alberta, Department of Agriculture and Rural Development, Rural Development Division, & Irshad, H. (2010). *Rural Tourism: An Overview*. Edmonton, Canada: Government of Alberta.

Arca, C. (2012). Social media marketing benefits for businesses: Why and how should every business create and develop its social media sites. MSc thesis, Aalborg University, Denmark.

Armstrong, S. (1991). Botswana's water plan hits the rocks. *New Scientist, 1752*. www.newscientist.com/article/mg12917521-700-botswanas-water-plan-hits-the-rocks/#ixzz77quoIUKk, accessed 20 December 2021.

Black, R., & Weiler, B. (2015). *Theoretical Perspectives on Tour Guiding. Demystifying Theories in Tourism Research*. Wallingford: CABI.

Botswana Tourism (2020). *Okavango Delta*. www.botswanatourism.co.bw/explore/okavango-delta, accessed 23 July 2021.

Bramwell, B. (1994). Rural tourism and sustainable rural tourism. *Journal of Sustainable Tourism, 2*(1–2), 1–6.

Briedenhann, J., & Wickens, E. (2004). Tourism routes as a tool for the economic development of rural areas: Vibrant hope or impossible dream? *Tourism Management, 25*(1), 71–79.

Bunruamkaew, K., & Murayama, Y. (2012). Land use and natural resources planning for sustainable ecotourism using GIS in Surat Thani, Thailand. *Sustainability, 4*(3), 412–429.

Campbell, J. B., & Wynne, R. H. (2011). *Introduction to Remote Sensing*, 4th edition. New York: Guilford Press.

Cavric, B. I., & Keiner, M. (2004). *Toward Sustainable Rural Land Use Planning Practices in Botswana: The Concept for an Integrated Rural Plan*. ETH Zurich.

Chatterjee, A., Saha, J., Mukherjee, J., Aikat, S., & Misra, A. (2020). Unsupervised land cover classification of hybrid and dual-polarized images using deep convolutional neural network. *IEEE Geoscience and Remote Sensing Letters, 18*(6), 969-973.

Chima, C. I. (2012). Monitoring and modelling of urban land use in Abuja Nigeria, using geospatial information technologies. PhD dissertation, Coventry University.

Chiutsi, S., & Mudzengi, B. K. (2017). Tourism seasonality and destination management implications for Mana Pools tourist destination in Zimbabwe. *African Journal of Hospitality, Tourism and Leisure, 6*(2), 1–13.

Colson, L. (2012). Using high resolution imagery to assess local urban growth patterns and its relationship with population density in Accra, Ghana. PhD dissertation, The George Washington University.

Crossley, M. N., Dennison, W. C., Williams, R. R., & Wearing, A. H. (2002). The interaction of water flow and nutrients on aquatic plant growth. *Hydrobiologia, 489*(1), 63–70.

Darkoh, M. B., & Mbaiwa, J. E. (2014). Okavango Delta: A Kalahari oasis under environmental threats. *Journal of Biodiversity & Endangered Species, 2*(4).

Dey, J., Sakhre, S., Gupta, V., Vijay, R., Pathak, S., Biniwale, R., & Kumar, R. (2018). Geospatial assessment of tourism impact on the land environment of Dehradun, Uttarakhand, India. *Environmental Monitoring and Assessment, 190*(4), 1–10.

Ekong, E. E. (2003). *An Introduction to Rural Sociology* (pp. 198–207). Uyo, Nigeria: Dove Educational Publishers.

Farsari, Y., & Prastacos, P. (2002). GIS contribution for the evaluation and planning of tourism: A sustainable tourism perspective. Presentation at Scientific Meeting: Social Practices and Spatial Information: European and Greek Experiences in GIS, Thessaloniki. Aristotelian University of Thessaloniki and Hellas.

Folorunso, C. O., Ayeni Dorcas, A., & Ayeni, T. O. (2020). Impact of tourism oriented architectural features on sales in shopping malls of metropolitan Lagos, Nigeria. *American Journal of Tourism Management, 9*(1), 19–23.

Giriwati, N., Homma, R., & Iki, K. (2013). Urban tourism: Designing a tourism space in a city context for social sustainability. *The Sustainable City VIII (2 Volume Set): Urban Regeneration and Sustainability, 179*, 165–176.

Hart, L. G., Larson, E. H., & Lishner, D. M. (2005). Rural definitions for health policy and research. *American Journal of Public Health, 95*(7), 1149–1155.

Hwang, D., Stewart, W. P., & Ko, D. W. (2012). Community behavior and sustainable rural tourism development. *Journal of Travel Research, 51*(3), 328–341.

Irons, J. R., Taylor, M. P., & Laura, R. (2016). *Landsat1*. Landsat Science. NASA.

Isserman, A. M. (2005). In the national interest: Defining rural and urban correctly in research and public policy. *International Regional Science Review, 28*(4), 465–499.

Jenness, J., & Wynne, J. J. (2005). Cohen's Kappa and classification table metrics 2.0: An ArcView 3. x extension for accuracy assessment of spatially explicit models. Open-File Report OF 2005-1363. Flagstaff, AZ: US Geological Survey, Southwest Biological Science Centre.

Kgathi, D. L., Kniveton, D., Ringrose, S., Turton, A. R., Vanderpost, C. H., Lundqvist, J., & Seely, M. (2006). The Okavango: A river supporting its people, environment and economic development. *Journal of Hydrology, 331*(1–2), 3–17.

Khorchani, M., Nadal-Romero, E., Lasanta, T., & Tague, C. (2021). Effects of vegetation succession and shrub clearing after land abandonment on the hydrological dynamics in the Central Spanish Pyrenees. *CATENA, 204*, 105374.

Kolawole, O. D. (2014). Whither sustainable rural development? A critical exploration of remote communities in and around the Okavango Delta, Botswana. *Spanish Journal of Rural Development, 5*(3), 99–114.

Kushwaha, A., Chatterjee, D., & Mandal, P. (2011). Potentials of GIS in heritage & tourism. *Proceedings of the Geospatial World Forum* (pp. 18–21). Hyderabad, India.

Lane, B. (1994). What is rural tourism? *Journal of Sustainable Tourism, 2*(1–2), 7–21.

Lenao, M., & Saarinen, J. (2015). Integrated rural tourism as a tool for community tourism development: Exploring culture and heritage projects in the North-East District of Botswana. *South African Geographical Journal, 97*(2), 203–216.

Lin, H. H., Ling, Y., Lin, J. C., & Liang, Z. F. (2021). Research on the development of religious tourism and the sustainable development of rural environment and health. *International Journal of Environmental Research and Public Health, 18*(5), 2731.

Magole, L., & Thapelo, K. (2005). The impact of extreme flooding of the Okavango River on the livelihood of the molapo farming community of Tubu village, Ngamiland sub-district, Botswana. *Botswana Notes & Records, 37*(1), 125–137.

Matswiri, G. M. (2017). Two in one: Explaining the management of the Okavango Delta World Heritage Site, Botswana. Master's thesis, University of Cape Town.

Mbaiwa, J. E. (2003). The socio-economic and environmental impacts of tourism development on the Okavango Delta, north-western Botswana. *Journal of Arid Environments, 54*(2), 447–467.

Mbaiwa, J. E., & Hambira, W. L. (2020). Enclaves and shadow state tourism in the Okavango Delta, Botswana. *South African Geographical Journal, 102*(1), 1–21.

Mbaiwa, J. E., & Stronza, A. L. (2010). The effects of tourism development on rural livelihoods in the Okavango Delta, Botswana. *Journal of Sustainable Tourism, 18*(5), 635–656.

Mmopelwa, G., Raletsatsi, S., & Mosepele, K. (2005). Cost benefit analysis of commercial fishing in Shakawe, Ngamiland. *Botswana Notes & Records, 37*(1), 11–21.

Mmualefe, L. C., Mpofu, C., & Torto, N. (2011). Modern sample preparation techniques for pesticide analysis. In M. Stoytcheva (ed.), *Pesticides in the Modern World: Trends in Pesticides Analysis* (pp. 199–220). Rijeka, Croatia: Intech.

Mogomotsi, P. K., Stone, L. S., Mogomotsi, G. E. J., & Dube, N. (2020). Factors influencing community participation in wildlife conservation. *Human Dimensions of Wildlife, 25*(4), 372–386.

Mohanty, P. P. (2014). Rural tourism in Odisha: A panacea for alternative tourism. A case study of Odisha with special reference to Pipli village in Puri. *American International Journal of Research in Humanities, Arts and Social Sciences, 14*, 557.

Motsumi, S., Magole, L., & Kgathi, D. (2012). Indigenous knowledge and land-use policy: Implications for livelihoods of flood recession farming communities in the Okavango Delta, Botswana. *Physics and Chemistry of the Earth, Parts A/B/C, 50*, 185–195.

Musasa, G., & Mago, S. (2014). Challenges of rural tourism development in Zimbabwe: A case of the Great Zimbabwe-Masvingo district. *African Journal of Hospitality, Tourism and Leisure, 3*(2), 1–12.

Neto, V. R. de Souza, & Marques, O. (2021). Rural tourism fostering welfare through sustainable development: A conceptual approach. In A. R. C. Perinotto, V. F. Mayer, & J. R. R. Soares (eds), *Rebuilding and Restructuring the Tourism Industry: Infusion of Happiness and Quality of Life* (pp. 38–57). Hershey, PA: IGI Global.

Pearce, D. G. (1998). Tourist districts in Paris: Structure and functions. *Tourism Management, 19*(1), 49–65.

202 *Handbook on tourism and conservation*

Presidential (COVID-19) Task Force Bulletin (2020, September 3). Tourism sector threatened by COVID-19, Issue 111.

Quattrochi, D. A., Walsh, S. J., Jensen, J. R., & Ridd, M. K. (2003). Remote sensing and its relationship to geography. In G. L. Gaile & C. T. Willmott (eds), *Geography in America at the Dawn of the 21st Century* (pp. 376–418). New York: Oxford University Press.

Ramberg, L., Hancock, P., Lindholm, M., Meyer, T., Ringrose, S., Sliva, J., ... & Vander Post, C. (2006). Species diversity of the Okavango Delta, Botswana. *Aquatic Sciences, 68,* 310-337.

Richardson, R. B. (2021). The role of tourism in sustainable development. In *Oxford Research Encyclopedia of Environmental Science*. Oxford University Press. https://doi.org/10.1093/acrefore/9780199389414.013.387.

Ringrose, S., & Matheson, W. (1995). An update on the use of remotely sensed data for range monitoring in south-east Botswana, 1984–1994. *Botswana Notes & Records, 27*(1), 257–270.

Roberts, L., & Hall, D. (2004). Consuming the countryside: Marketing for 'rural tourism'. *Journal of Vacation Marketing, 10*(3), 253–263.

Robinson, P. (2015). Conceptualizing urban exploration as beyond tourism and as anti-tourism. *Advances in Hospitality and Tourism Research, 3*(2), 141–164.

Robinson, R. N. (2005). Tourism, health and the pharmacy: Towards a critical understanding of health and wellness tourism. *Tourism, 53*(4), 13327461.

Rotherham, I. D. (2007). Sustaining tourism infrastructures for religious tourists and pilgrims within the UK. In R. Raj & N. D. Morpeth (eds), *Religious Tourism and Pilgrimage Festivals Management: An International Perspective* (pp. 64–77). Wallingford: CAB International.

Saarinen, J. (2007). Contradictions of rural tourism initiatives in rural development contexts: Finnish rural tourism strategy case study. *Current Issues in Tourism, 10*(1), 96–105.

Saarinen, J., Moswete, N., Atlhopheng, J. R., & Hambira, W. L. (2020). Changing socio-ecologies of Kalahari: Local perceptions towards environmental change and tourism in Kgalagadi, Botswana. *Development Southern Africa, 37*(5), 855–858.

Safarov, B., & Sh, D. O. (2020). Tourism industry development is increasing, new jobs will increase incomes and living standards. *Архивариус, 2*(47).

Slater, A. (2002). Specification for a dynamic vehicle routing and scheduling system. *International Journal of Transport Management, 1*(1), 29–40.

Stone, L. S., Mogomotsi, P. K., Stone, M. T., Mogomotsi, G. E., Malesu, R., & Somolekae, M. (2020). Sustainable tourism and the SDGs in Botswana: Prospects, opportunities and challenges towards 2030. In S. O. Keitumetse, L. Hens, & D. Norris (eds), *Sustainability in Developing Countries* (pp. 153–181). Cham: Springer.

Stone, L. S., Stone, M., Mogomotsi, P., & Mogomotsi, G. (2021). The impacts of Covid-19 on nature-based tourism in Botswana: Implications for community development. *Tourism Review International, 25*(2–3), 263–278.

Story, M., & Congalton, R. G. (1986). Accuracy assessment: A user's perspective. *Photogrammetric Engineering and Remote Sensing, 52*(3), 397–399.

Subhash, K. B., Weiermair, K. C., Lee, C., & Scaglione, M. (2010). What constitutes health tourism? An Ayurvedic viewpoint. Conference: 9th Biennial Asia Tourism Forum (ATF 2010) Conference on Tourism and Hospitality Industry in AsiaAt: Taiwan Hospitality & Tourism College, Hualien, Taiwan.

Talukdar, S., Singha, P., Mahato, S., Pal, S., Liou, Y. A., & Rahman, A. (2020). Land-use land-cover classification by machine learning classifiers for satellite observations: A review. *Remote Sensing, 12*(7), 1135–1155.

United Nations World Tourism Organization (2020). UNWTO highlights potential of domestic tourism to help drive economic recovery in destinations worldwide. News release. https://webunwto.s3.eu-west-1.amazonaws.com/s3fs-public/2020-09/200911-domestictourism-en.pdf, accessed 1 May 2020.

van Manen, N., Scholten, H. J., & van de Velde, R. (2009). Geospatial technology and the role of location in science. In H. J. Scholten, R. Velde, & N. van Manen (eds), *Geospatial Technology and the Role of Location in Science* (pp. 1–13). Dordrecht: Springer.

Vumbunu, T. (2020). A diversification framework for eco-tourism products of Botswana. PhD dissertation, North-West University, South Africa.

Wilson, S., Fesenmaier, D. R., Fesenmaier, J., & Van Es, J. C. (2001). Factors for success in rural tourism development. *Journal of Travel Research*, *40*(2), 132–138.

Yang, J. T. (2010). Antecedents and consequences of knowledge sharing in international tourist hotels. *International Journal of Hospitality Management*, *29*(1), 42–52.

15. Towards agritourism development in Zimbabwe: growth potential, benefits and challenges

Rudorwashe Baipai, Oliver Chikuta, Edson Gandiwa and Chiedza N. Mutanga

INTRODUCTION

Agritourism is generally defined as visiting a working agricultural setting, usually a farm, ranch or any agribusiness enterprise for relaxation, entertainment, educational purposes or involvement in the farm operations (Santeramo & Barbieri, 2016; Joyner et al., 2017). Rogerson and Rogerson (2014) defined agritourism as a form of rural tourism that involves tourist services being offered in an agricultural setting providing the tourists with the opportunity to experience and learn about the daily farm operations.

On the other hand, Chase et al. (2018) developed a framework for defining agritourism activities. In their framework Chase et al. (2018) defined agritourism as the act of visiting a working farm for entertainment, education, recreation, hospitality and purchase of farm products. The framework categorised agritourism activities into five categories that can exist on farm and deeply connected to agriculture activities (core agritourism activities) and off farm or on farm but lacking a deep connection to agriculture production (peripheral agritourism activities). There seems to be general consensus among researchers about the core agritourism activities (Lamie et al., 2021; Quella et al., 2021; Rogerson & Rogerson, 2014; Streifeneder, 2016). However, there is less agreement on the periphery, as some researchers consider these activities to be part of agritourism while others are of a different view. For example, Streifeneder (2016) distinguished authentic agritourism, which occurs on working farms where agricultural activities predominate over the agritourism ones, and countryside tourism which occurs on a non-working farm or on a working farm with touristic activities not linked to agriculture practices. Lamie et al. (2021) and Chase et al. (2018) noted the variations in definitions of agritourism that exist in different countries and highlighted that efforts are being made to harmonise these definitions through ongoing international dialogue.

These definitions reflect the broadness and diversity of the agritourism offer. The notion of 'visiting' highlighted in the first definition depicts the tourism aspect of this concept. Relaxation and entertainment depict the hospitality services such as farm stays, farm bed and breakfast and the wide variety of agricultural attractions (farm crops, farm animals, natural features, wildlife, farm machinery and buildings) and voluntary activities (feeding farm animals, horse riding, culinary activities, fishing, wildlife photography) found at the farms. Other researchers, for example, Van Zyl (2019), noted that agritourism overlaps with other types of tourism and whenever each of these types of tourism includes farm-based experiences it can be referred to as agritourism.

Despite the variations in definitions, tremendous growth in agritourism has been witnessed in developed countries (Chase, 2020) and extensive research on agritourism has been carried out in these regions (Baipai et al., 2021). The trend is settling in the developing countries and the industry is growing as well though most of the agritourism enterprises in these regions are still at the developmental stage. This growth is attributed to many of the socio-economic benefits associated with agritourism which include additional income, improved standards of living, creation of new job opportunities, conservation of biodiversity resources, preservation of local culture, respect for marginal rural culture and lifestyle (Leh et al., 2017; Tulla et al., 2018; Kunasekaran et al., 2018; Lago, 2017; Chaiphan, 2016).

AGRITOURISM DEVELOPMENT IN ZIMBABWE

Agriculture has been the backbone of Zimbabwe's economy, both pre- and post-colonial (Maiyaki, 2010) and it has been reported that it will continue to be one of the economy's pillars for the foreseeable future (Mutami, 2015). Agriculture underpins the social, political and economic lives of the general populace in the country. The agricultural sector employs over 60 per cent of the country's population and contributes 15 per cent of the country's gross domestic product (GDP). Smallholder farmers living in the rural farming areas carry out most of the farming with commercial farmers also producing for export (Brazier, 2018; Poulton et al., 2002). The country is endowed with rich agricultural resources in the form of vast areas of arable, fertile land and a good climate (Maiyaki, 2010). During the colonial era, land policies in Rhodesia[1] favoured the then minority white farmers. Hence, the racial distribution of land was done in an unequal manner between the minority white farmers and the black farmers. The colonial government put in place policies that segregated the indigenous farmers in marginal lands (communal areas) whilst the European farmers took the most fertile lands. Some commercial white farmers had farm lodges and kept game animals at the farms that they used to attract tourists (Poulton et al., 2002).

Overpopulation in the communal areas caused a lot of pressure resulting in disputes over land among the local people (Poulton et al., 2002; Scoones et al., 2011). The fast land reform programme, which began in 2000, saw the redistribution of land from the commercial white farmers to the black Zimbabwean majority. Large scale commercial farms were acquired from a few commercial white farmers and redistributed to black farmers in the form of A1, A2 and commercial farm models. This gave the black farmers the opportunity to take up both agricultural and agritourism activities in former white-owned commercial farms. This change in land ownership has been reported to have temporarily resulted in a collapse of commercial agriculture as well as agritourism in Zimbabwe (Mukwereza, 2013). The seizure of land from the minority white farmers without compensation did not go down well with the international community, leading the West to impose sanctions on the country (Sibanda & Makwata, 2017), which resulted in Zimbabwe's isolation with its attendant negative impacts on tourism.

This triggered a wave of poverty among black people (as most did not have inputs, equipment and financial resources to carry out farming activities) as well as a decline in the economic performance of the country. Consequently, the country lost its status as the bread basket of southern Africa (Chitsike, 2003). The situation was exacerbated by erratic rainfalls, high temperatures and recurrent droughts, all of which aggravated the suffering of the majority of rural communities (Kanyenze et al., 2017). The Government of Zimbabwe intervened by

206 *Handbook on tourism and conservation*

launching several government input schemes from 2005 with the most recent being the Special Agriculture (Maize) Production Programme popularly known as the command agriculture (Mutami, 2015).

In order to address the lopsided land tenure issues, the government introduced the 99-year lease as way of formalising the redistributed land and giving confidence to beneficiaries (Government of Zimbabwe, 2018). The consolidation of the legal standing of the 99-year lease and offer letters paved the way for farm investments and production through improved land utilisation (Ministry of Finance and Economic Development, 2017). These are key issues that may also enable the development of agritourism. In 2016, the Government of Zimbabwe launched the land audit programme with the key goal of identifying multiple farm owners as well as underutilised and idle land. Command agriculture, provision of bankable leases, irrigation, farm mechanisation and infrastructural development programmes reflect the efforts by the government to make the land reform programme a success (Ministry of Finance and Economic Development, 2017). These efforts, in a way, create an enabling environment for agritourism.

Although Zimbabwe is agro-based and has several farms that can be developed into agritourism destinations, agritourism is still in its infancy and still underdeveloped (Baipai et al., 2021). Moreover, there is limited literature on agritourism in Zimbabwe (Chiromo, 2016). The little research that has been conducted on agritourism in Zimbabwe does not reflect the significance that the subsector has gained in the international arena (Baipai, 2022). The limited secondary data available reveal that agritourism was more pronounced before the land reform programme of 2000 (Guvamombe, 2019). Guvamombe (2019) noted that the commercial white farmers had invested in farm tourism. They had farm lodges, snake parks, wildlife sanctuaries and monuments as well as many other attractions that they exploited in order to attract tourists. Since the land reform programme of 2000, there has not been any deliberate effort to promote agritourism amongst the local black farmers (Guvamombe, 2019). The objectives of this study, therefore, were to (i) assess the growth potential for agritourism development on Zimbabwean farms; (ii) analyse the impacts of agritourism development in Zimbabwe; and (iii) evaluate challenges faced in developing agritourism on agricultural farms.

THEORETICAL FRAMEWORK

This study was guided by the Triple Bottom Line (TBL) approach to sustainability. The framework was developed by John Elkington in 1994 after the realisation that the extent of business sustainability was difficulty to measure. Moreover, Elkington argued that the success of a business must not be based on a measure of profit or loss only but also on the well-being of people and the health of the planet (Alhaddi, 2015). Thus, Elkington developed the TBL framework to enable effective measurement of sustainability (Slaper, 2011; Purvis, 2018) through the integration of the 3Ps (profit, people and planet) dimensions (Alhaddi, 2015). Incorporating sustainability considerations into any development in general, and agritourism in particular, will allow the enhancement of local communities' livelihoods while also strengthening biodiversity conservation and the reduction of land degradation as well as building and developing economies without straining resources (Addinsall et al., 2016).

The economic aspect of the TBL framework focuses on the impact that the operations of an organisation have on the economic system of a country within which it is operating

(Elkington, 1994). These include but are not limited to employment creation, personal income, revenue contributing to GDP, establishment sizes, taxes and job growth (Slaper, 2011). The social dimension of TBL represents the degree to which an organisation conducts business in a manner that is beneficial and fair to the employees and to the community at large (Elkington, 1994). The social dimension focuses mainly on the interaction between the local community and the organisation, and it addresses issues that are related to community participation, fair wages, employee relations, education, quality of life, access to social resources, equity, social capital and health care package provisions (Azevedo & Barros, 2017; Slaper, 2011; Alhaddi, 2015; Woodcraft, 2015). The environmental measure of TBL encourages organisations to engage in practices that lead to conservation of the environmental resources for use by future generations (Elkington, 1994). It relates to the cost-effective use of energy resources, reduction of greenhouse gas emissions, air and water quality, solid waste management, and minimisation of the ecological footprint (Kearney, 2009).

The agritourism sector is projected to grow at an exponential rate globally. Sustainable agritourism therefore entails development of tourism products within a farm setting in a manner that leads to conservation of agro-biodiversity resources, equitable social development and an all-encompassing economic growth for the betterment of local livelihoods (Ciolac et al., 2019). The subsector brings numerous benefits to the environment, tourists, farmers, local people and the economy if the entire TBL dimensions are adhered to. Slaper (2011) emphasised that for a community-based project, the TBL dimensions that can be used to measure the success of the project can be best determined by the local community. This makes the TBL framework very applicable for sustainable agritourism development especially since agritourism mostly benefits rural farming communities by improving their socio-economic well-being. Moreover, TBL encourages stakeholder participation as most of the data required for measuring the 3Ps are obtained from stakeholders (Elkington, 1994). This makes the framework most applicable to agritourism development which relies more on participation of local people and conservation of natural environments for its success (Ciolac et al., 2020).

METHODOLOGY

The Study Context

Zimbabwe is a landlocked country in southern Africa, bordered by Botswana, Mozambique, South Africa and Zambia. The country covers a total area of 390,760 km^2 and has a population of approximately 14.6 million people (Zimstat, 2012) with 41 per cent of the population under the age of 15, and 60 per cent living in the rural areas (Food and Agriculture Organization, 2016). The country has ten provinces and is divided into five agroecological (natural) regions based on climate, vegetation and soil types.

Region I is located in the eastern part of the country. The region receives rainfall of over 1,000 mm per year and farmers in this region specialise in growing high value crops such as bananas, tea, potatoes and coffee. Farmers in natural region II engage in cotton, tobacco, maize, wheat, horticulture and intensive livestock production. The region receives medium rainfall of 750–1,000 mm per year. Natural region III receives low rainfall of 500–750 mm per year and is suitable for cattle ranching, maize, wheat and tobacco. Natural region IV is suitable for maize short season varieties, drought resistant crops such as millet, rapoko and sorghum

Handbook on tourism and conservation

Table 15.1   Distribution of respondents' age and gender

Stakeholder	Age (years)						Gender	
	35–40	41–45	46–50	51–55	56–60	60–65	Male	Female
Farmers	5	0	7	0	7	15	22	12
Government officials	5	5	4	2	0	0	9	6
Tour operators	2	3	4	1	0	0	8	2
Total	12	8	15	3	7	15	39	20

and extensive livestock production. This region receives low rainfall of 450–650 mm per year, with frequent seasonal droughts and severe dry spells during the rainy season. Natural region V is suitable for livestock, wildlife management, beekeeping and non-timber forest products. The region experiences very low rainfall of less than 650 mm per year (Brazier, 2018). The characteristics of the agroecological regions make all regions suitable for agritourism development as they all have agricultural attractions that are unique to them.

Data Collection and Analysis

This study adopted a qualitative approach and a multiple case study design. A multiple case study design enabled the researchers to gather the perceptions of stakeholders using more than one case study (Gustafsson, 2017). Further, a multi-stakeholder approach was also adopted, which allowed the researchers to gather multiple perceptions of the individuals (Baipai et al., 2021). In-depth interviews were conducted with a total of 59 stakeholders relevant to agritourism development in Zimbabwe who were purposively selected from Manicaland and Mashonaland West provinces of Zimbabwe. The two provinces are relevant case studies because they are situated in regions of productive agricultural land, which increases their potential for growth in agritourism thus making them a true representation of the study population. The stakeholders included 34 farmers whose farms have potential for agritourism development, 10 key informants from the Ministry of Lands, Agriculture, Fisheries, Water and Rural Development, 5 from the Ministry of Environment, Climate, Tourism and Hospitality Industry and 10 tour operators operating in the study provinces. Of the 10 key informants from the Ministry of Lands, Agriculture, Fisheries, Water and Rural Development, 2 were from the Department of Agricultural, Technical and Extension Services (AGRITEX) provincial level, 2 were from the Lands Department and 6 were District AGRITEX officers. Of the 5 key informants from the Ministry of Environment, Climate, Tourism and Hospitality Industry, 2 were Zimbabwe Tourism Authority (ZTA) Area Managers, 1 was a ZTA Research and Product Development Manager, 1 was an Executive Director International Marketing and 1 was from the Ministry of Environment, Climate, Tourism and Hospitality Industry. The interviews with farmers were conducted until data saturation was reached (Saunders et al., 2018). From the government departments, only those who were relevant to agritourism development were selected. The distribution of their ages and gender is shown in Table 15.1.

The authors asked the respondents about their perceptions on the growth potential of agritourism in Zimbabwe, its impacts and challenges faced in developing the subsector in the country. The interviews were carried out between October 2020 and June 2021. Consent to conduct the interviews was sought from the research participants through a consent form before carrying out the interviews. Approval to conduct the research was obtained from Chinhoyi University of Technology, the Ministry of Lands, Agriculture, Fisheries, Water and

Rural Development and from the Ministry of Environment, Climate, Tourism and Hospitality Industry. The data obtained from the interviews were analysed using thematic content analysis. NVIVO 12 was used to conduct a word frequency count. The word frequency count was used to show the most frequently mentioned themes and their weighted percentages.

FINDINGS AND DISCUSSION

Growth Potential for Agritourism on Zimbabwe's Farms

The study highlighted two important factors, which indicate that agritourism development has potential in Zimbabwe. These factors include (i) the presence of many agricultural farms in the country, and (ii) the availability of a market for agritourism, which, however, is mainly international. All the 59 participants (100 per cent) perceived that there was a huge potential for agritourism development in the farms albeit not being fully tapped. For example, a provincial AGRITEX officer from Manicaland pointed out that the few agricultural farms that were offering agritourism needed refocusing and boosting. Another provincial AGRITEX officer from Mashonaland West indicated that those who were trying to embrace agritourism were commercial farms and very few small farms, which still had to work on infrastructure. A key informant from the government ministry responsible for tourism supported the above assertions and highlighted that:

> Before independence there were quite a number of agritourism farms, but these have since faded away because of the disturbances that were caused by the land reform. However, there is great potential. Every corner of Zimbabwe has unique agricultural offerings; each region has some icons. (Key informant, government ministry responsible for tourism)

Our results thus indicate that there is great potential for agritourism in Zimbabwe. These findings concur with Khairabadi et al. (2020) who also identified great potential for agritourism development in farms located in the Simin region of Iran. They noted that combining tours to the region with agritourism activities would enable maximum utilisation of such opportunities available on farms. Similarly, Priyanka and Kumah (2016) revealed that there was agritourism potential in India and highlighted that this potential could only be fully utilised through developing and implementing strategies that address the challenges that developing countries face in trying to develop agritourism. Elsewhere, in the Kisimu Region in Kenya, Bwana et al. (2015) also identified untapped potential for agritourism. As in this present study, the reasons for the failure to utilise this potential to the maximum included lack of concerted effort of all relevant stakeholders, the fear of disrupting agricultural production and lack of knowledge on how to fully utilise this potential (Sawe et al., 2018; Van Zyl & Merwe, 2021).

On the other hand, all 10 tour operators constituting 17 per cent of the respondents affirmed that there was a market for agritourism in the country. For example, while one tour operator commented that 'there is a big market that needs to be tapped into through vigorous marketing and promotion of the concept', another tour operator observed that 'tea plantations used to bring in vast numbers of tourists before the COVID-19 pandemic. We also have Honde Valley which is popular for their banana plantations. There is need to continue marketing these areas.' Although they revealed their positive perceptions about agritourism development in the

210 Handbook on tourism and conservation

country, tour operators indicated that the concept is still not popular with local tourists. One of the tourism operators remarked that:

> The local tourists would not want to engage a tour operator because they know their way around, so they do self-drive. Our clienteles are usually the international senior travellers who prefer wine tourism, culinary tourism, u-pick your own fruit and vegetables but these are usually not available in most of our farms. (Tour Operator 8, Manicaland province)

The tour operators opined that the future of agritourism is bright but encouraged farmers to improve on production efficiency so as to increase agricultural attractions. They revealed that their clients were mostly international senior travellers (60 years and above) who, therefore, prefer accessible tourism and a totally green establishment.

The findings indicate that there is a potential for agritourism market in Zimbabwe. It has, however, been pointed out that emphasis should be placed on understanding the preferences and motivations of this market so as to ensure the design of appropriate marketing strategies (Lucha et al., 2016; Arru et al., 2019). The views of tour operators who participated in this study are similar to Karampela et al. (2019), who emphasised the need for market segmentation and devising marketing strategies that suit the characteristics of each market segment. Results from tour operators highlighted that the agritourism market in Zimbabwe mainly consists of senior travellers. Chen et al. (2019) and Lupi et al. (2017) assert that agritourism caters for a niche market due to its exclusiveness. Our findings thus corroborate those of Tugade (2020), who posit that the growth of the agritourism sector depends on the ability to respond to new market demands through new product development. Failure by farmers to jump into this new mode of income diversification may result in them failing to keep their businesses afloat (Tugade, 2020).

BENEFITS OF AGRITOURISM DEVELOPMENT

The stakeholders that were interviewed had the perceptions that agritourism, if fully developed, would bring benefits to the farming communities and to the country at large as shown in Figure 15.1.

Only three of the farmers (1 from Manicaland and 2 from Mashonaland West) emphasised that they had no idea at all of what they would benefit from the subsector; the rest of the participants (56) constituting 95 per cent of the respondents pointed out that agritourism was indeed an income generating option that could significantly provide employment for the local people and ultimately improve their standard of living. The results agree with Ariffin et al. (2014) and Zacal et al. (2019) who concluded that agritourism largely contributed to the economic well-being of participating farmers. Also, Van Zyl and Merwe (2021) concluded that economic benefits (job creation, income generation) are some of the main motives for engaging in agritourism in South African farms.

On the contrary, Maetzold (2002) questioned the assumption that agritourism can deliver sustainable and inclusive development in the Philippines. The author argued that agritourism farmers in the Philippines are represented not only by wealthy farming classes but also by elite networks that involve the state and other private entities. In such a scenario the underprivileged small farmers are excluded from benefiting from agritourism development because of the conditions that favour the elite farming class (Montefrio & Sin, 2019). Leh et al. (2017)

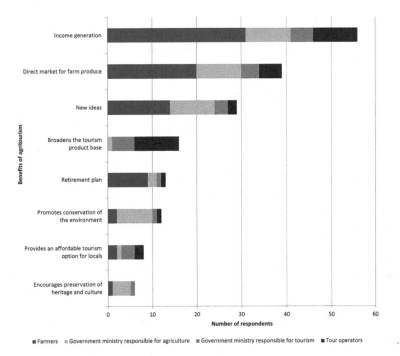

Figure 15.1   Benefits of agritourism in Zimbabwe

argue that although the economic contribution of agritourism is apparent and extensively agreed, its magnitude has been regularly questioned by many.

About 32 participants constituting 54 per cent of the respondents viewed agritourism as a subsector that would provide a market for horticultural produce closer to the farm. For example, a Manicaland farmer stated that: 'I sometimes face problems in getting a market for tomatoes. Tomatoes are perishable; once you fail to get a market on time you lose out. It becomes easy if people come to buy here at the farm.'

These findings support Tugade (2020) who argued that the presence of tourists at the farm provides opportunities for farm-direct sales. Chase et al. (2018) also noted that on-farm direct sales are an important category of agritourism activities. They added that agricultural products that are not sold through the traditional markets can be used to prepare food to table meals at the farm restaurant.

About 29 participants constituting 49 per cent of the respondents perceived that visitors to farms will bring in more ideas, which could assist in improving their farming methods. Supporting this standpoint, a Manicaland farmer commented that 'Visitors can also bring new ideas that can help us in growing our business'. Also, a key informant, a government official from the ministry responsible for agriculture in Manicaland, commented that 'as tourists come, they bring in a lot, the farmer may get some knowledge and ideas through interactions'. Another key informant from the government ministry responsible for agriculture in Mashonaland West explained that agritourism enables the farmers to learn more about cultural diversity and he agreed that visitors bring in new agricultural practices that farmers can adopt. These results are in agreement with Asmelash and Kumar (2019) who highlight that agritourism creates an opportunity for local people to communicate and interact with tourists. Through this interaction, the locals can learn new ideas and foreign languages. Besides being able to learn a foreign language, interaction with tourists is usually in English, thus the more they interact with tourists, the more they practise speaking in English and improve their communication skills (Sawe et al., 2018).

About 16 participants constituting 27 per cent of the respondents were of the view that agritourism broadens the tourism product base. A key informant from the government ministry responsible for tourism stated that 'agritourism can help to broaden the tourism product base; it will attract a new niche market different from our traditional market which consists of leisure tourists.' Another key informant from the same ministry pointed out that the concept can result in longer stays because of the availability of more products. She stated that: 'instead of the tourist just visiting the parks, they may also want to visit the agritourism facilities in the area thus prolonging their stay and spending more.' These assertions by respondents are in agreement with Ammirato et al. (2020) who pointed out that agritourism provides a niche market in the tourism industry.

About 13 of the participants, constituting 22 per cent of the respondents, regarded agritourism as a good retirement plan. A farmer in Manicaland explained that 'As I can no longer work in the field because of old age, agritourism will provide me with a less stressful retirement job whilst my children take care of the more strenuous activities in the fields.' Tugade (2020) referred to this as an 'intrinsic drive' to transform the farm into an agritourism venture. The Manicaland farmer's views reveal that the need to prepare for retirement can motivate farmers to venture into agritourism as a means of sustenance later in life. Our results are in support of Yildirim and Kilinc (2018) who mentioned retirement plans as a factor that drives farmers to engage in agritourism.

About 12 of the respondents, constituting 20 per cent, mentioned that agritourism promoted sustainability of the environment in various ways. A provincial AGRITEX officer from Manicaland stated that agritourism would encourage farmers to be environmentally friendly. Some notable sustainability strategies were being implemented at some of the farms, for example, at one of the Mashonaland West farms where animal waste was being recycled into biogas and food waste was minimised through using leftovers and unsold farm produce to feed farm animals. Ammirato et al. (2020) posited that many agritourism farmers are developing more sensitive strategies that exploit natural resources in order to produce energy with minimal environmental impact. Elsewhere, Bajgier-Kowalska et al. (2017) pointed out that the natural environment is the raw material for agritourism activities and protecting it, therefore, becomes the farmers' priority. The protection of the environment is a prerequisite condition for the successful development of agritourism.

Agritourism benefits the locals by providing them with an affordable tourism option as highlighted by 8 (14 per cent) of the respondents. A key informant from the government ministry responsible for tourism explained that agritourism provides 'an alternative accommodation which is more affordable'. This was supported by a farmer from Mashonaland West who added that 'having agritourism here means the locals will have somewhere near and affordable to visit during their holidays'. A farmer from Manicaland emphasised that 'some of the locals cannot afford to travel to far and expensive resort areas and an agritourism farm gives them an affordable option'. This is in agreement with Dumitras et al. (2021) who assert that urban dwellers want destinations that are affordable and easy to find and agritourism presents that option. However, Guliyev and Nuriyeva (2018) argue that although some agritourism packages may be affordable, some may be too costly for consumers who do not understand the value of such packages.

Agritourism was also applauded by 6 participants, constituting 10 per cent of the respondents, for its ability to preserve values and cultural lifestyles of the farming community. For example, a key informant from the government ministry responsible for tourism opined that 'agritourism offers cultural experiences and promotes eating of local foods'. Van Zyl and Merwe (2021) also pointed out that agritourism preserves local heritage and culture. Agritourism exploits rural heritage and culture in order to entertain visitors while educating them and generating income at the same time. The realisation that rural heritage and culture can attract tourists to their farms motivates farmers to preserve cultural heritage and resources (Mykletun, 2018). Through agritourism, local people can now share their culture and heritage with visitors. This gives the local people a sense of identity and pride. Knowing that their resources can provide entertainment and educational activities encourages the locals to guard their heritage and culture jealously (Guliyev & Nuriyeva, 2018).

CHALLENGES FACED IN DEVELOPING AGRITOURISM IN AGRICULTURAL FARMS

This chapter noted that the development of agritourism was hindered by several challenges as indicated in Figure 15.2.

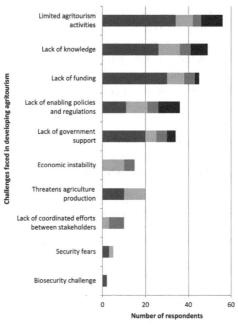

Figure 15.2   *Challenges faced in developing agritourism*

About 56 participants, constituting 95 per cent of the respondents, highlighted that limited agritourism activities at the farms were the main challenge hindering them from offering agritourism. One farmer from Manicaland highlighted that they had not yet developed activities for tourists, and they were working on it. Another farmer from Mashonaland West indicated that he had some knowledge of what agritourism was all about by commenting that 'We are still at a planning stage and we are working on putting up structures and diversifying our agricultural activities so that when visitors come, they have something interesting to do.' These results are in line with findings by Moraru et al. (2016) and Adamov et al. (2020), who revealed that moderately few agritourism ventures are market-ready. Tugade (2020) posited that farmers venturing into agritourism needed to be able to identify the agritourism product that they are presenting to their consumers and be in a position to communicate their offering to the potential visitors.

About 49 of the participants, constituting 83 per cent of the respondents, noted that lack of knowledge of what agritourism is, how to develop the subsector and how they can get income from it was the major challenge for farmers. A farmer from Mashonaland West commented that: 'I don't really know how I can generate income from agritourism' and another farmer from the same province commented 'I don't have enough information about its profitability'. Lack of information regarding agritourism markets and trends are some of the challenges that were highlighted by Moraru et al. (2016). A key informant from the government ministry responsible for tourism agreed that: 'Lack of knowledge is the main hindrance, so it's a question of education. Awareness is the key. If farmers get awareness, they will seek more information on how they can develop agritourism on their farms.' This is in line with Adamov et al. (2020) who pointed out that more structured agricultural farms have a high potential of being able to meet the demands of tourists if they are provided with adequate education and training. The researchers noted that the tour operators were not familiar with the term 'agritourism' but used the term 'safari-based holiday packages' and the package usually encompasses activities such as farm tours, scenic drives, photographic safaris, game drives, game viewing and horse riding. Tugade (2020) pointed out the importance of being able to identify agritourism products, present them to consumers and be in a position to communicate the offering to potential visitors.

Unavailability of capital to support the development of agritourism was cited by 45 participants, constituting 76 per cent, as another major reason why they had not yet developed agritourism at their farms. One of the farmers commented thus:

> I used to have international visitors who came to my farm for camping and sightseeing, but they have since stopped coming. I am willing to have visitors here, but my main challenge is lack of water and accommodation. I tried to acquire a loan from the bank to fund my water project and construction of chalets for visitors, but the bank has not been forthcoming. (Mashonaland West farmer)

Yamagishi et al. (2021) also pointed out that farmers lack the capital required to develop agritourism activities. A key informant from the government ministry responsible for tourism felt that 'high interest rates deter farmers from borrowing from banks'. Liu et al. (2017) also noted that farmers often face difficulties in securing technical and financial support.

Lack of enabling policies and regulations was mentioned by about 36 of the participants, constituting 61 per cent of the respondents, as another challenge hindering agritourism development in the country. One farmer explained that:

I have done a lot of research on agritourism, and I have travelled a lot. I got some of the ideas from Tanzania. I have interviewed headmasters of schools to get their views on agritourism and they think it's a good idea. I have consulted National Parks and ZTA officials to hear their views. According to these officials there are a lot of regulations that one should follow in order to have agritourism at a farm. (Manicaland farmer)

A provincial AGRITEX officer from Mashonaland West opined that there was inconsistent policy implementation in the country. Elsewhere, Kubickova and Campbell (2020) and Leh et al. (2017) emphasised that policy formulation is one key area that governments should address as policies and proper planning are crucial for agritourism development. On another note, a Manicaland farmer mentioned that the 99-year lease agreement instituted by the government does not give him confidence to invest in agritourism.

About 34 participants, constituting 54 per cent of the respondents, commented that government support towards agritourism development was only limited to advisory services through AGRITEX and little promotional effort through ZTA. A district AGRITEX officer from Manicaland highlighted that: 'Unavailability of government support may hinder agritourism development.' This is buttressed by Yamagishi et al. (2021) who noted the importance of government support in agritourism development. They had opined in their study on the future of farm tourism in the Philippines that model agritourism sites were critical in changing mindsets and farmers' perceptions in including an agritourism aspect in their activities. Yamagishi and team however, noted that a lot of improvements still needed to be done in most farms and government support was imperative in ensuring that the improvements are done. Moraru et al. (2016) also pointed out that government support towards agritourism development is usually poor.

Economic instability characterised by hyperinflation and very high prices for agricultural inputs was cited as a hindrance to agritourism development by 25 per cent of the respondents. They affirmed that economic stability, government economic policies and government support through grants and incentives will determine the successful development of agritourism. A provincial AGRITEX officer from Mashonaland West observed that the situation was exacerbated by the prevailing macroeconomic environment which was a hindrance to agritourism development. However, despite the negative perceptions about the macro environment prevailing in Zimbabwe, a key informant from the government ministry responsible for tourism acknowledged that there were agritourism ventures that were thriving in the country though there are no published statistics on the number of agritourism ventures in Zimbabwe. Cristina et al. (2017) highlighted that in order to enjoy all the benefits of agritourism ventures, there are specific conditions and incentives that should prevail in the environment at macro level. Also, Bajgier-Kowalska et al. (2017) noted that agritourism development has more often been affected by prevailing social and economic conditions.

About 10 of the participants, constituting 17 per cent of the respondents, expressed their concerns about lack of coordinated efforts between the Ministries of Agriculture and Tourism, and many other organisations in the country that have an important role to play in the development of agritourism. A district AGRITEX officer from Mashonaland West commented that 'there is conflict between the need to increase agricultural production and the need to channel resources to tourism'. Yamagishi et al. (2021) also noted the poor coordination between relevant government agencies in the promotion of farm tourism in the Philippines. According to Adamov et al. (2020) and Barbieri and Streifeneder (2019), the conflict of interest between agriculture and tourism industries and the lack of coordinated efforts lead to the absence of a structure that supports agritourism development.

All 10 government ministry participants responsible for agriculture, who constituted 17 per cent of the respondents, revealed their fears that agritourism might disturb agriculture activities and lead to low production, which may in turn threaten food security in the country. A provincial AGRITEX officer from Manicaland envisaged the possibility of a reduced area for extensive commercial food production as staff and material resources may be diverted to agritourism development especially during the start-up stages. A key informant from the Lands Department commented that: 'Competition for resources in the forms of capital, land and labour in the development of tourism sites versus farming, and mixed priorities resulting from diversification may result in low standards.' To add to that, the provincial AGRITEX officer from Mashonaland West highlighted that: 'Farmers might be lazy if more and more tourists are attracted to farms or nearby farms, and some would prefer providing hospitality services other than production.' One district AGRITEX officer explained that: 'If international tourists are the main target, farmers may end up prioritising agricultural products needed by the foreigners at the expense of the local community even the nation.' Another district AGRITEX officer felt that: 'Agritourism needs farms to be big enough, a measure that would disadvantage the agriculture industry considering the high demand for land at present.'

Our results are similar to those of Yildirim and Kilinc (2018) who argue that agritourism has some disadvantages and these need to be considered seriously. They emphasised that planned development of agritourism should be ensured in order to counter some of these negative impacts. However, the views of respondents that farmers may choose agritourism over agricultural production contradict Ciolac et al.'s (2020) viewpoint who indicated that the tourism aspect of agritourism should be complementary to agriculture but not a substitute to it. Streifeneder (2016) also emphasised 'agriculture first', through agritourism activities that are deeply connected to agriculture. Similarly, Chase et al. (2018) emphasised that a working farm is a prerequisite for core agritourism activities. Any other activities that are not deeply connected to agriculture are peripheral agritourism activities (Chase et al., 2018). This implies that agritourism cannot thrive without agriculture. Staged agritourism holds sway in cases where touristic activities override agricultural activities (Chatterjee & Prasad, 2019) or rural tourism activities (Streifeneder, 2016).

About 5 participants, constituting 8 per cent of the respondents, mentioned security issues as the main challenge that can arise from having visitors coming to the farm. A farmer from Manicaland commented that: 'some visitors may come to spy on you or your farm'. Another farmer from Mashonaland West province stated that 'some visitors may come back to steal from you'. Leh et al.'s (2017) study confirmed that farmers may feel unsafe as a result of growth in agritourism. In Leh et al.'s (2017) study respondents agreed that the growth of agritourism in Malaysia has resulted in an increase in crimes in the form of theft and vandalism.

About 2 participants, constituting 3 per cent of the farmers, mentioned biosecurity risk as a challenge hindering the development of agritourism. A farmer from Manicaland who owns a dairy farm feared that having visitors at the farm posed the risk of spreading of viruses and bacteria, which are harmful to animals, plants and humans with the latest being the Covid-19 virus. The farmer highlighted that they used to have agritourism at the farm, but they have since stopped.

Towards agritourism development in Zimbabwe 217

[W]e used to have agritourism, but we have since stopped because of fear of disease outbreaks. We used to have primary and secondary school children, university students, small-scale and large-scale farmers coming to learn how we do things here, but we have since stopped receiving visitors long before the Covid-19 outbreak. The main reason why we stopped is because of fears of foot and mouth disease (FMD). We have a friend who lost three quarters of his dairy cows because of FMD. How can you be interested in having visitors and risk losing your dairy cows? It's better to look after your cows and produce more milk than having visitors. (Manicaland farmer)

The farmer's fears are in agreement with Roman and Golnik's (2019) study in Italy where participants had perceptions that visitors may have negative effects in terms of spread of diseases as well as on security issues. Zacal et al. (2019) and Schilling et al. (2006) also recorded fears of biosecurity challenges that arise as a result of having visitors at an agricultural farm. However, the concern of agritourism operators in Schilling et al.'s study was more on the security of visitors than animals. They claimed that agritourism operators had fears that visitors may get bitten by zoo animals or contract diseases from animals and not the other way round, as highlighted by the Manicaland farmer above.

CONCLUSION

Based on the two provinces studied, the chapter revealed that there is great potential for agritourism that has not been fully utilised in the Zimbabwean farms. However, there are many issues to be resolved if the country is to fully utilise its agritourism potential. The chapter noted that the benefits of agritourism development in Zimbabwe would include income generation, opportunities for on-farm direct sales, bringing in of new ideas, broadening of the tourism product base, a good retirement plan, promotion of conservation of the environment, provision of affordable tourism options for the locals and encouragement of preservation of local heritage and culture. Challenges that hindered the development of agritourism in Zimbabwe were also identified and they include lack of agritourism activities, knowledge, funding, enabling policies and regulations, limited government support, economic instability and uncoordinated efforts between key stakeholders. Perceptions by some stakeholders that agritourism could be a threat to agricultural production, security and biosecurity in farms also act as barriers to agritourism development in the country.

The chapter concludes that stakeholders have positive perceptions towards agritourism development in the country in terms of the potential and the prospects for growth. Further, the chapter concludes that agritourism is a sustainable diversification strategy which has the potential to improve the livelihoods of local people, conservation of the environment as well as contributing to the growth of the economy of Zimbabwe. This is in line with the TBL approach to sustainability which highlights that a successful and sustainable enterprise is one that generates profit, contributes to the well-being of the local people and the overall health of the planet.

In order to promote the development of sustainable agritourism, the study recommends that there is a need for farmers to expand their ventures and set aside specific areas of their farms where tourism activities could be operationalised. By doing so, integration of agricultural activities and tourism can be realised, hence enhancing the resilience of business ventures on the farms. Awareness programmes could also help clear the misconceptions that agritourism may replace agricultural activities and this may assist farmers realise that the subsector indeed serves as a sustainable diversification strategy that could improve their livelihoods. Provision

218   *Handbook on tourism and conservation*

of start-up capital funding through long-term loans or grants, which may stimulate agritourism growth associated with the formulation of enabling policies related to agritourism ventures, is also suggested.

NOTE

1.   Rhodesia is the former name of Zimbabwe.

REFERENCES

Adamov, T., Ciolac, R., Iancu, T., Brad, I., Pet, E., Popescu, G., & Smuleac, L. (2020). Sustainability of agritourism activity: Initiatives and challenges in Romanian mountain rural regions. *Sustainability*, *12*(6), 1–23.

Addinsall, C., Weiler, B., Scherrer, P., & Glencross, K. (2016). Agroecological tourism: Bridging conservation, food security and tourism goals to enhance smallholders' livelihoods on South Pentecost, Vanuatu. *Journal of Sustainable Tourism*, *25*(8), 1100–1116.

Alhaddi, H. (2015). Triple bottom line and sustainability: A literature review. *Business and Management Studies*, *1*(2), 6–10.

Ammirato, S., Felicetti, A. M., Raso, C., Pansera, B. A., & Violi, A. (2020). Agritourism and Sustainability: What We Can Learn from a Systematic Literature Review. *Sustainability*, *12*(22), 1–18.

Ariffin, A. R. M., Ali, Z. M., Zainol, R., Rahman, S., Hua, A. K., & Sabran, N. (2014). Sustainable highland development through stakeholders' perceptions on agro ecotourism in Cameron highlands: A preliminary finding. *SHS Web of Conferences 12*, 01086.

Arru, B., Furesi, R., Madau, F. A., & Pulina, P. (2019). Recreational services provision and farm diversification: A technical efficiency analysis on Italian agritourism. *Agriculture*, *9*(2), 42, 1–15.

Asmelash, A. G., & Kumar, S. (2019). The structural relationship between tourist satisfaction and sustainable heritage tourism development in Tigrai, Ethiopia. *Heliyon*, *5*(3), e01335. https://doi.org/10.1016/j.heliyon.2019.e01335.

Azevedo, S., & Barros, M. (2017). The application of the triple bottom line approach to sustainability assessment: The case study of the UK automotive supply chain. *Journal of Industrial Engineering and Management*, *10*(2), 286–322.

Baipai, R. (2022). Critical success factors for agritourism development in Zimbabwe. *African Journal of Hospitality, Tourism and Leisure*, *11*(SEI), 617–632.

Baipai, R., Chikuta, O., Gandiwa, E., & Mutanga, C. (2021). A critical review of success factors for sustainable agritourism development. *African Journal of Hospitality, Tourism and Leisure*, *10*(6), 1778–1793.

Bajgier-Kowalska, M., Tracz, M., & Uliszak, R. (2017). Modeling the state of agritourism in the Malopolska region of Poland. *International Journal of Tourism Space, Place and Environment*, *19*(3), 502–524.

Barbieri, C., & Streifeneder, T. (2019). Agritourism advances around the globe: A commentary from the editors. *Open Agriculture*, *4*(1), 712–714.

Brazier, A. (2018). *Climate Change in Zimbabwe: A Guide for Planners and Decision Makers*. Harare: Konrad-Adenauer-Stiftung.

Bwana, M. A., Olima, W. H. A., Andika, D., Agong, S. G., & Hayombe, P. (2015). Agritourism: Potential socio-economic impacts in Kisumu County. *Journal of Humanities and Social Science*, *20*(3), 78–88.

Chaiphan, N. (2016). *Study of Agritourism in Northern Thailand: Key Success Factors and the Future of the Industry*. Northern Thailand: Thammasat University.

Chase, L. (2020). *Agritourism in Vermont*. https://accd.vermont.gov/sites/accdnew/files/document/VDTM/benchReseaech-2017BenchmarkStudyFullReport.pdf.2.1/2020, accessed 18 January 2022.

Chase, L., Stewart, M., Schilling, B., Smith, B., & Walk, M. (2018). Agritourism: Toward a conceptual framework for industry analysis. *Journal of Agriculture, Food, Systems, and Community Development*, 8(1), 13–19.

Chatterjee, S., & Prasad, M. V. D. (2019). The evolution of agri-tourism practices in India: Some success stories. *Madridge Journal of Agriculture and Environmental Sciences*, 1(1), 19–25.

Chen, Y., Dax, T., & Zhang, D. (2019). Complementary effects of agricultural tourism and tourist destination brands in preserved scenic areas in mountain areas of China and Europe. *Open Agriculture*, 4(1), 517–529.

Chiromo, P. (2016). *Social Innovation and Agrotourism Growth in Rural Communities of Zimbabwe*. Chinhoyi, Zimbabwe: Chinhoyi University of Technology.

Chitsike, F. (2003). A critical analysis of the land reform programme in Zimbabwe. 2nd FIG Regional Conference, 1–12.

Ciolac, R., Adamov, T., Iancu, T., Popescu, G., Lile, R., Rujescu, C., & Marin, D. (2019). Agritourism: A sustainable development factor for improving the 'health' of rural settlements. Case study Apuseni Mountains area. *Sustainability*, 11(5), 1–24.

Ciolac, R., Iancu, T., Brad, I., Popescu, G., Marin, D., & Adamov, T. (2020). Agritourism activity: A "smart chance" for mountain rural environment's sustainability. *Sustainability*, 12(6237), 1–24.

Cristina, M., Iamandi, I., & Munteanu, S. M. (2017). Incentives for developing resilient agritourism entrepreneurship in rural communities in Romania in a European context. *Sustainability Science*, 9, 1–30. https://doi.org/10.3390/su9122205.

Dumitras, D. E., Mihai, V. C., Jitea, I. M., Donici, D., & Muresan, I. C. (2021). Adventure tourism: Insight from experienced visitors of Romanian national and natural parks. *Societies*, 11(2), 1–11. https://doi.org/10.3390/soc11020041.

Elkington, J. (1994). Enter the triple bottom line. In J. Elkington (ed.), *The Triple Bottom Line: Does it All Add Up?* London: Routledge, pp. 1–16.

Food and Agriculture Organization (2016). *Policy Brief 9: Top-Down and Bottom-Up Approaches*. www .fao.org/climatechange/55804/en, accessed 11 January 2022.

Government of Zimbabwe (2018). *Towards an Upper Middle Income by 2030: "New Dispensation Core Values"*.

Guliyev, S., & Nuriyeva, K. (2018). Adventure tourism marketing: A research on the tourists' behaviours regarding adventure tourism in Azerbaijan. *SSRN Electronic Journal*, 9(1), 64–79.

Gustafsson, J. (2017). *Single Case Studies vs. Multiple Case Studies: A Comparative Study*. Halmstad, Sweden: Academy of Business, Engineering and Science.

Guvamombe, I. (2019). Farm tourism trendy, refreshing. *Herald*, 3 August.

Joyner, L., Kline, C., Oliver, J., & Kariko, D. (2017). Exploring emotional response to images used in agritourism destination marketing. *Journal of Destination Marketing & Management*, 9, 44–55.

Kanyenze, G., Chitambara, P., & Tyson, J. (2017). *The Outlook for the Zimbabwean Economy*. USA: ODI Report.

Karampela, S., Papapanos, G., & Kizos, T. (2019). Perceptions of agritourism and cooperation: Comparisons between an island and a mountain region in Greece. *Sustainability*, 11(3), 1–19. https://doi.org/10.3390/su11030680.

Kearney, A. T. (2009). *Green Winners: The Performance of Sustainability-Focused Companies during the Financial Crisis*. www.kearney.com/documents/291362523/291365012/green_winners.pdf/9f0ab012-8678-35c9-15c1-a7f02b2ec0ed?t=1493922129000, accessed 23 November 2021.

Khairabadi, O., Sajadzadeh, H., & Mohammadianmansoor, S. (2020). Assessment and evaluation of tourism activities with emphasis on agritourism: The case of Simin Region in Hamedan City. *Land Use Policy*, 99, 105045.

Kubickova, M., & Campbell, J. M. (2020). The role of government in agro-tourism development: A top-down bottom-up approach. *Current Issues in Tourism*, 23(5), 587–604.

Kunasekaran, P., Fuza, N., Hassan, E., & Ramachandran, S. (2018). Factors influencing perceptions of local community on 'Kelulut' honey as agrotourism. *International Journal of Business and Society*, 19(SI), 66–78.

Lago, N. A. A. (2017). Tourism demand and agriculture supply: Basis for agritourism development in Quezon Province. *Asia Pacific Journal of Multidisciplinary Research*, 5(3), 1–9.

220 *Handbook on tourism and conservation*

Lamie, R. D., Chase, L., Chiodo, E., Schmidt, C., Flanigan, S., Dickes, L., & Streifeneder, T. (2021). Agritourism around the globe: Definitions, authenticity, and potential controversy. *Journal of Agriculture, Food Systems, and Community Development, 10*(2), 573–577.

Leh, O. L. H., Mohd Noor, M. H. C., Marzukhi, M. A., & Mohamed Musthafa, S. N. A. (2017). Social impact of agro-tourism on local urban residents. Case study: Cameron Highlands, Malaysia. *Journal of the Malaysian Institute of Planners, 15*(6), 51–66.

Liu, S., Yen, C., Tsai, K., & Lo, W. (2017). A conceptual framework for agri-food tourism as an eco-innovation strategy in small farms. *Sustainability, 9*(1693), 1–11.

Lucha, C., Ferreira, G., Walker, M., & Groover, G. (2016). Profitability of Virginia's agritourism industry: A regression analysis. *Agricultural and Resource Economics Review, 45*(1), 173–207.

Lupi, C., Giaccio, V., Mastronardi, L., Giannelli, A., & Scardera, A. (2017). Exploring the features of agritourism and its contribution to rural development in Italy. *Land Use Policy, 64*, 383–390.

Maetzold, J. A. (2002). *Nature-Based Tourism & Agritourism Trends: Unlimited Opportunities*. Washington DC: Kerr Center for Sustainable Agriculture.

Maiyaki, A. A. (2010). Zimbabwe's agricultural industry. *African Journal of Business Management, 4*(19), 4159–4166.

Ministry of Finance and Economic Development (2017). *National Budget Statement for 2018*. www .zwrcn.org.zw/images/Budget_corner/2018_Budget_Statement_Final.pdf, accessed 18 October 2021.

Montefrio, M. J. F., & Sin, H. L. (2019). Elite governance of agritourism in the Philippines. *Journal of Sustainable Tourism, 27*(9), 1338–1354.

Moraru, R.-A., Ungureanu, G., Bodescu, D., & Donosa, D. (2016). Motivations and challenges for entrepreneurs in agritourism. *Agronomy Series of Scientific Research, 59*(1), 267–272.

Mukwereza, L. (2013). Reviving Zimbabwe's agriculture: The role of China and Brazil. *IDS Bulletin, 44*(4), 116–126.

Mutami, C. (2015). Smallholder agriculture production in Zimbabwe: A survey. *The Journal of Sustainable Development, 14*(2), 140–157.

Mykletun, R. J. (2018). Adventure tourism in the North: Six illustrative cases. *Scandinavian Journal of Hospitality and Tourism, 18*(4), 319–329.

Poulton, C., Davies, R., Matshe, I., & Urey, I. (2002). *A Review of Zimbabwe's Agricultural Economic Policies: 1980–2000*. ADU Working Papers 10922, Imperial College at Wye, Department of Agricultural Sciences.

Priyanka, S., & Kumah, M. (2016). Identifying the potential of agri-tourism in India: Overriding challenges and recommend strategies. *International Journal of Core Engineering & Management, 3*(3), 7–14.

Purvis, B. (2018). Three pillars of sustainability: In search of conceptual origins. *Sustainability Science, 5*. https://doi.org/10.1007/s11625-018-0627-5.

Quella, L., Chase, L., Wang, W., Conner, D., Hollas, C., Leff, P., Feenstra, G., Sindh-Knights, D., Virginia, W., & Stewart, M. (2021). *Agritourism and On-Farm Direct Sales Interviews: Report of Qualitative Findings*. www.uvm.edu/vtrc/agritourism-survey, accessed 13 October 2021.

Rogerson, C. M., & Rogerson, J. M. (2014). Agritourism and local economic development in South Africa. *Bulletin of Geography, 26*, 93–106.

Roman, M., & Golnik, B. (2019). Current status and conditions for agritourism development in the Lombardy region. *Bulgarian Journal of Agricultural Science, 25*(1), 18–25.

Santeramo, F. G., & Barbieri, C. (2016). On the demand for agritourism: A cursory review of methodologies and practice. *Tourism Planning and Development, 14*(1), 139–148. https://doi.org/10.1080/21568316.2015.1137968.

Saunders, B., Sim, J., Kingstone, T., Baker, S., Waterfield, J., Bartlam, B., Burroughs, H., & Jinks, C. (2018). Saturation in qualitative research: Exploring its conceptualization and operationalization. *Quality and Quantity, 52*(4), 1893–1907.

Sawe, B. J., Kieti, D., & Wishitemi, B. (2018). A conceptual model of heritage dimensions and agrotourism: Perspective of Nandi County in Kenya. *Research in Hospitality Management, 8*(2), 101–105.

Schilling, B. J., Marxen, L. J., Heinrich, H. H., & Brooks, F. J. A. (2006). *The Opportunity for Agritourism Development in New Jersey*. New Jersey: Rutgers Cook College.

Scoones, I., Marongwe, N., Mavedzenge, B., Murimbarimba, F., Mahenehene, J., & Sukume, C. (2011). *Zimbabwe's Land Reform: A Summary of Findings*. Brighton: IDS.

Sibanda, V., & Makwata, R. (2017). *Zimbabwe Post Independence Economic Policies: A Critical Review*. London: LAP Lambert Academic Publishing.

Slaper, T. F. (2011). The triple bottom line: What is it and how does it work? The triple bottom line defined. *Indiana Business Review, 86*(1), 4–8.

Streifeneder, T. (2016). Agriculture first: Assessing European policies and scientific typologies to define authentic agritourism and differentiate it from countryside tourism. *Tourism Management Perspectives, 20*, 251–264. https://doi.org/10.1016/j.tmp.2016.10.003.

Tugade, L. O. (2020). Re-creating farms into agritourism: Cases of selected micro-entrepreneurs in the Philippines. *African Journal of Hospitality, Tourism and Leisure, 9*(1), 1–13.

Tulla, A. F., Vera, A., Valldeperas, N., & Guirado, C. (2018). Social return and economic viability of social farming in Catalonia: A case-study analysis. *Sciendo, 10*(3), 398–428.

Van Zyl, C. (2019). The size and scope of agri-tourism in South Africa. Master's thesis, North-West University, South Africa.

Van Zyl, C. C., & Merwe, P. Van Der (2021). The motives of South African farmers for offering agri-tourism. *Open Agriculture, 6*, 537–548.

Woodcraft, S. (2015). Understanding and measuring social sustainability. *Journal of Urban Regeneration and Renewal, 8*(2), 133–144.

Yamagishi, K., Gantalao, C., & Ocampo, L. (2021). The future of farm tourism in the Philippines: Challenges, strategies and insights. *Journal of Tourism Futures*. https://doi.org/10.1108/JTF-06-2020 -0101.

Yildirim, G., & Kilinc, C. C. (2018). A study on the farm tourism experience of farmers and visitors in Turkey. *Journal of Economic & Management Perspectives, 12*(3), 5–12.

Zacal, R. G., Virador, L. B., & Canedo, L. P. (2019). State of selected agritourism ventures in Bohol, Philippines. *International Journal of Sustainability, Education, and Global Creative Economic, 2*(1), 9–14.

Zimstat (2012). *Census 2012: Preliminary Report*.

PART IV

DESTINATION COMMUNITIES AND NATURAL RESOURCES CONSERVATION

16. Commodification of nature and territorialization: conservation, local communities and Botswana's international cooperation

Kekgaoditse Suping

INTRODUCTION

Botswana is endowed with natural resources such as diamonds, cattle and wildlife (Thumberg-Hartland, 1978; Leith, 2005; Harvey & Lewis, 1990; Picard, 1985; Acemoglu et al., 2001). Botswana's natural resource endowment and good governance have been credited for the country's socio-economic development from one of the poorest at independence in 1966 to a middle-income country two decades later (1986), before rising to be an upper middle-income economy in 2005 (Edge, 1998; Acemoglu et al., 2001; Leith, 2005; UNDP, 2009; Leftwich, 1995; Samatar, 1999). Botswana has a lot of wildlife resources, mostly concentrated in the northern and northwestern parts of the country such as Kwando-Linyanti, Chobe and Okavango Delta (DEA, 2008; DWNP, 2013). Of particular interest to this chapter is certain charismatic megafauna such as elephants and others that are classified as endangered species, and the focus of international actors as they are perceived to be of global importance.

To protect its wildlife resources, Botswana developed legal, institutional and policy frameworks for wildlife conservation. Some of Botswana's wildlife resources are transboundary, and as such, the country is party to both regional and international agreements aimed at wildlife conservation such as the Convention on International Trade in Endangered Species of Wild Fauna and Flora (CITES) and the Southern African Development Community (SADC) Protocol on Wildlife Conservation and Law Enforcement. Participation in these wildlife conservation agreements has influenced Botswana's relations with neighbouring states with which it shares transboundary natural resources and several international environmental organizations. As such, effective wildlife conservation has facilitated Botswana's international cooperation. This is because Botswana commands a good international reputation on wildlife conservation, attracts international tourists and tourism enterprises, and has grown its tourism sector to be the second highest foreign revenue earner after diamonds (Mbaiwa, 2003; WTTC, 2018).

Botswana's wildlife conservation and tourism are well documented. However, the existing literature does not analyse the connection between wildlife conservation, tourism and Botswana's international cooperation. The existing literature covers Wildlife Management Areas (WMAs) and their associated challenges (Parry & Campbell, 1990; Wilmsen, 1989; Suzman, 2001). Some literature focuses on the involvement of local communities in wildlife conservation in Botswana (Mbaiwa & Hambira, 2020; Mbaiwa, 2011, 2017a). Injustices such as the exclusion, marginalization, evictions and exploitation of local people, except for a few elites, in Botswana's tourism sector is also documented (Good, 2008; Stadler, 2005; Wilmsen,

224 *Handbook on tourism and conservation*

1989; Bolaane, 2004; Kiema, 2010; Mbaiwa, 2017b). Botswana's wildlife laws, policies and practices have also been scrutinized by some writers who identified some exogenous influence (Suzman, 2001; Marobela, 2010; Wilmsen, 1989; Kiema, 2010). Some literature covers the development of Botswana's wildlife conservation and tourism to generate income through eco-tourism (DEA, 2016; DWNP, 2013; Mbaiwa, 2017a). Collectively, the existing literature does not interrogate the connection between the country's wildlife conservation and its international cooperation, which is the focus of this chapter. International cooperation refers to social, political, environmental, and economic relations or interactions between actors in the international system such as states, regional organizations, international organizations, commercial entities, non-governmental organizations and citizens (Young, 1994; Milner, 1992).

This chapter argues that wildlife conservation has facilitated Botswana's international cooperation, but also created several challenges. First, wildlife conservation areas in Botswana are commodified, serve private interests and benefit exogenous tourism enterprises at the expense of local people. Second, conservation areas in Botswana are externalized and managed by international treaties and regional protocols, the effect of which is Botswana's loss of sovereignty over those spaces. Third, Botswana is caught up in the complexity of satisfying both international conservation and tourism demands and the needs of local communities in WMAs. In conclusion, this chapter contends that there is an urgent need to enact an effective legal framework for sustainable wildlife management that will avert human–wildlife conflict while sustaining Botswana's tourism sector, and its international cooperation agreements and reputation.

This chapter uses secondary sources of data that include journal articles, books, conservation treaties and protocols, government reports on wildlife conservation, Botswana wildlife conservation laws and policies, and news articles. The chapter also makes use of online articles from international organizations. These sources of data provide a pool of information intended for various purposes and audiences, thus useful to avoid an analysis of a one-sided view. Interpretive analysis is employed to conflate the data from the secondary sources and draw informed conclusions. This method of content analysis is best suited for this chapter as it allows for thick description and use of narrative explanation, meanings and preferences of people involved in a particular setting or actions (Bevir & Rhodes, 2005).

TERRITORIALIZATION, LOCAL COMMUNITIES AND WILDLIFE CONSERVATION IN BOTSWANA

Botswana's legal and policy framework on wildlife conservation has institutionalized territorialization. According to Igoe and Brockington (2007, p. 437), territorialization refers to "the demarcation of territories within states for purposes of controlling people and resources". The created WMAs or 'territories' in Botswana seem to be mere abstract geographical demarcations illustrated on maps, but such territories are experienced and/or lived realities. A 'territory' is a lived space that directly impacts on and/or alters people's lives as it "ignores and contradicts people's lived social relationships and the histories of their interactions with the land" (Vandergeest & Peluso, 1995, p. 389). Soon before Botswana's independence and thereafter, some land areas were demarcated for wildlife conservation as national parks and game reserves, such as the Central Kalahari Game Reserve (CKGR) and Moremi Game Reserve (Parry & Campbell, 1990; Wilmsen, 1989; Bolaane, 2004). In 1986, Botswana adopted the Wildlife Conservation Policy (WCP) which gave effect to the creation of exclusive territories for wildlife conservation. The WCP resulted in the establishment of WMAs for efficient management and control of wildlife resources and people movements in such areas. The Botswana Wildlife Conservation and National Parks Act was later enacted in 1992 to give legal effect to the WCP of 1986. By 2013, the Wildlife Act and policies had resulted in the demarcation of wildlife estate of approximately 214,600 km², which is about 39 per cent of Botswana's total land area (DWNP, 2013). Of this wildlife estate, 45.7 per cent (98,000 km²) is national parks and game reserves, while the remaining 116,000 km² (54.3 per cent) is WMAs (DWNP, 2013). In addition to the aforementioned wildlife estate, there is allocation of private game reserves and ranches to individuals that have interest in game farming and conservation, and such allocations are done by the land authorities with the approval of the Department of Wildlife and National Parks (DWNP) (DWNP, 2013, 2016; Parry & Campbell, 1990). Furthermore, there are forest reserves in Botswana which are part of the protected areas and serve as wildlife habitats and biodiversity conservation areas.

The creation and increase of wildlife territories ensured the conservation of wildlife species, but also negatively impacted local communities in wildlife-endowed areas as is characteristic of territorialization (Igoe & Brockington, 2007). According to Igoe and Brockington (2007), one of the defining features of territorialization is an increase in the number of environmental organizations and forceful eviction of local residents from protected areas to make way for tourism enterprises. This best describes the case of Botswana where the San or Basarwa were evicted from the CKGR by the Government of Botswana (Kiema, 2010; Suzman, 2001). Basarwa are the indigenous people of Southern Africa residing in Botswana, South Africa and Namibia (Kiema, 2010; Hitchcock, 2002; Stadler, 2005; Suzman, 2001). In Botswana, some of them inhabited the CKGR where they were banished to in the 1960s but were forcibly removed between 1997 and 2002 to make way for exogenous mining and tourism companies, although the government claimed the relocation was intended for their development and wildlife conservation (Stadler, 2005; Marobela, 2010; Kiema, 2010; Gupta, 2013; Solway, 2009; Hitchcock, 2002). The resistance of the Basarwa to relocation has often been met with state brutality to pave the way for conservation organizations and transnational tourism companies (Good, 2008; Gupta, 2013; Hitchcock, 2002; Marobela, 2010). This resonates well with territorialization which is premised on the exertion of social pressure, coercion, threat and the use of violence (Vandergeest & Peluso, 1995). The Basarwa communities are the most widely affected victims of territorialization for wildlife conservation, tourism and other economic activities which saw them evicted from their ancestral lands more than any tribe in Botswana even prior to the country's independence (Anaya, 2010; Hitchcock, 2002; Marobela, 2010; Winters, 2019). Bolaane (2004) notes that when Moremi Game Reserve, located on the eastern part of the Okavango Delta, was established in 1963, the Basarwa of Khwai River known as the Bugakhwe were relocated from their ancestral land to make way for wildlife conservation, resulting in the loss of their traditional habitats within the area and alienation from their customary and land rights.

Besides Basarwa, various communities in wildlife endowed areas have also been negatively affected by Botswana's adoption of enclave tourism in other areas like Chobe and Ngamiland (Mbaiwa, 2003). As Mbaiwa (2003) correctly observed, enclave tourism has denied local people benefits and the government a lot of revenue, because it is structured in a manner that excludes and discriminates against local people by setting up facilities and prices that are unaffordable to locals and ensures that tourism services are paid for in foreign currency at

226 *Handbook on tourism and conservation*

the offices of exogenous tour operators outside Botswana. While private tourism enterprises remain the main beneficiaries in WMAs, there are no proper cost-sharing mechanisms as the government bears the costs and responsibility of protecting such areas while local communities face threats to their livelihoods (DWNP, 2016; DEA, 2016; Mbaiwa, 2003; Bolaane, 2004; Hitchcock, 2002). The Government of Botswana accrues revenue from WMAs through rents and concessions, and further earns international recognition for conservation and the thriving tourism industry. However, income from WMAs is not shared with communities that have been denied the opportunity to utilize the land demarcated as WMAs and negatively affected by living in close proximity to the habitat of wild animals, thus making them more vulnerable to threats to their lives and livelihoods than other people in non-WMAs (DWNP, 2016; Mbaiwa, 2003; Kiema, 2010; Winters, 2019; Wilmsen, 1989; Good, 1992).

COMMODIFICATION OF WILDLIFE RESOURCES, EXOGENOUS ACTORS AND LOCAL COMMUNITIES

Botswana's wildlife resources have been transformed into revenue accruing commodities. The Department of Environmental Affairs (DEA) estimates the total value of game stock at about BWP 3 billion, of which the Zambezian flooded grassland ecoregion contributes 60 per cent, and the dryland ecosystems 34 per cent (DEA, 2016). Botswana's wildlife conservation laws and policies provide for and promote the participation of multiple stakeholders in conservation by advocating for decentralization, privatization and re-regulation of wildlife resources (DWNP, 2013). Re-regulation refers to the transformation of previously untradable things into tradable commodities using the state (Langholz, 2003; Igoe & Brockington, 2007; Castree, 2008). The introduction of multiple stakeholders to wildlife conservation has attracted external tourism enterprises in the name of eco-tourism and they endeared themselves to the Government of Botswana and some elites to develop and benefit from the country's wildlife conservation and tourism (Good, 2008; Mbaiwa, 2017a). For example, after diamonds, the wildlife-based tourism sector is the second largest foreign revenue earner for Botswana, contributing about 17.6 per cent to the GDP by 2011, before dropping down to 12 per cent in 2017 (WTTC, 2018; DEA, 2016). Prior to the advent of Covid-19, Botswana's tourism sector was thriving with a lot of exogenous tourism investors who exerted a lot of influence on wildlife conservation laws and policies (DWNP, 2013; Mbaiwa, 2017a; Good, 2008). The dominance of exogenous actors in tourism is facilitated, amongst others, by their partnership with the local business and political elites at the expense of local people or communities (Mbaiwa & Hambira, 2020). As Mbaiwa (2003), and Mbaiwa and Hambira (2020) rightly observed, there are relations between the local elites and exogenous investors that enable high capital investment and monopoly of the tourism sector by the latter.

Botswana resonates well with the justifications for the commodification of nature, that regard "corporate sponsorship of conservation organizations, increased management of protected areas for profit companies and increased emphasis on eco-tourism as a means of achieving economic growth, community prosperity and bio-diversity conservation" (Igoe & Brockington, 2007, p. 433). As such, the rhetoric on wildlife conservation in Botswana has been, and is, based on the view that "bio-diversity also provides opportunities for communities to generate income through utilisation and management of biological resources in their proximity" (DEA, 2016, p. 12). Therefore, the Government of Botswana has delegated the management of some wildlife resources to private investors and local communities through the Community-Based Natural Resource Management (CBNRM) programme (DWNP, 2013; Mbaiwa, 2017a; DEA, 2016). The CBNRM programme was intended for wildlife conservation and to derive benefits for local communities from wildlife resources (DWNP, 2013). However, CBNRM initiatives in Botswana have yielded mixed results since their inception, with successes in a few instances (Mbaiwa & Stronza, 2010), and failures in many others owing to a lack of clearly defined ways for local communities to access and enjoy the benefits of managing specific resources (Mbaiwa, 2015).

The level of local people's participation in wildlife conservation and tourism activities in Botswana remains very low due to their lack of capital, skills and expertise in the tourism sector (Mbaiwa, 2003; Bolaane, 2004; Solway, 2009). The power relations and systemic inequities that characterize global wildlife conservation (McAfee, 1999; Svarstad et al., 2018; Igoe & Brockington, 2007) have defeated the goals of CBNRM programmes to yield tangible results in favour of local communities in Botswana (DEA, 2016; Mbaiwa, 2003; Mbaiwa & Hambira, 2020). For example, most of the tourism enterprises in Botswana are owned by non-citizens, mostly white South Africans, and the profits from the tourism operations are rarely injected into the local or national economies since almost all the payments are made outside Botswana (Mbaiwa, 2003; Good, 2008; Marobela, 2010). The eco-tourism promotes private enterprises as the solution to Botswana's lack of revenue challenges (DWNP, 2016; WTTC, 2018). In the transaction of wildlife resources, governments serve as brokers for external tourism investors and justify the commodification of nature as a necessity to sustain its existence, accrue some revenue for the government and avert possible over-exploitation of natural resources arising from the fact that no value is attached to them by the local communities (Ferguson, 2006; Liverman, 2004). The foregoing best describes the case of Botswana where exogenous tourism enterprises are dominant actors and have forged close ties with the local business and political elites to exclusively benefit from the country's wildlife resources (Good, 1993; Kiema, 2010; Mbaiwa & Hambira, 2020). Therefore, eco-tourism as it is currently conceptualized and practised in Botswana marginalizes local communities, thus rendering CBNRM an ineffective foreign concept that is beyond the reach of most local people who aspire to succeed in and benefit from the tourism sector (Stadler, 2005). Furthermore, the outward orientation of Botswana's tourism sector makes it susceptible to exogenous shocks such as pandemics like Covid-19 and its associated travel restrictions (Stone et al., 2021). The emergence of Covid-19 in 2020 crippled the tourism sector in Botswana as the flow of tourists into the country drastically decreased.

Local communities in Botswana feel marginalized or excluded in wildlife conservation and tourism decision-making processes. Historically, wildlife was perceived as a common property of the community and controlled by the same (Hitchcock, 2002; Good, 2008; Wilmsen, 1989). However, authority over management and utilization of wildlife shifted from the local communities to the state in post-independence Botswana as wildlife resources became commercialized and exploited for profit (Sylvain, 2003; Hitchcock, 2002; Winters, 2019; Solway, 2009). The shift from community ownership and control of wildlife resources to the state, and the attachment of monetary value to wildlife resulted in the communities abandoning their responsibility to protect wildlife resources and a surge in poaching, as the people felt that they no longer benefited from the animals (Biggs et al., 2017; Bwalya & Suping, 2020). The power relations in the wildlife conservation and tourism industry have left local communities in particular, and Batswana in general, on the periphery of decision-making processes on

228 *Handbook on tourism and conservation*

matters that directly affect them (Good, 1992; Solway, 2011; Sylvain, 2003; Gupta, 2013). For example, when the hunting ban was introduced by the Government of Botswana in 2014, the local communities directly affected by the soaring numbers of wildlife were not consulted, nor were their interests taken into consideration as the decision was externally driven by exogenous safari companies and international environmental organizations to have their demands met (Sylvain, 2005; Mbaiwa, 2017a; Nyoni, 2018; Christy & Hartley, 2013; Winters, 2019; Bwalya & Suping, 2020; Crookes & Blignaut, 2016). The shift from hunting safaris by the Government of Botswana to photographic and game-viewing safaris required skill that local communities did not have and licences they could not afford (Tebele, 2018; 't Sas-Rolfes et al., 2014). Consequently, the commodification of nature and re-regulation have transformed Botswana's wildlife resources and WMAs into tradable commodities that are often beyond the affordability of local communities (Stadler, 2005; Mbaiwa, 2003; DWNP, 2016).

STATE-MAKING, LOSS OF CITIZENSHIP AND WILDLIFE MANAGEMENT AREAS

Wildlife conservation and its dependent tourism have promoted 'state-making' in Botswana. This involves activities that promote dependence on foreign direct investment, external funding and exogenous expertise, thus making the state vulnerable to undue influence by exogenous actors and institutions (Igoe & Brockington, 2007). State-making is inherent in wildlife conservation in most developing countries as it gives transnational actors, and some local elites, leverage over the local communities through its preference for exogenous actors (Igoe & Croucher, 2007; Jones, 2006; Goldman, 2001; McAfee, 1999; Lemos & Agrawal, 2006). The features of state-making are prevalent in Botswana's wildlife conservation and tourism. For decades, Botswana has attracted and depended on exogenous expertise and transnational enterprises to drive the development of the country's tourism sector (Stadler, 2005; Marobela, 2010; Kiema, 2010; Suzman, 2001). The local communities in Botswana are sidelined and treated as second-class citizens whose rights, especially those related to land and culture, are trampled upon by the powerful tourism and political elites who wield power and unduly influence the state concerning WMAs (Sylvain, 2003; Good, 1992, 1993; Solway, 2009; Suzman, 2001; Mbaiwa & Hambira, 2020). In state-making, ecosystems and economies, such as incentives for wildlife conservation and CBNRM, that are not understood by local communities are introduced by exogenous actors, ultimately disadvantaging the former (Fletcher, 2010; Goldman, 2001; Igoe & Fortwangler, 2007). In this new form of state, the lives, and sometimes movements, of local people in WMAs are increasingly criminalized, and their access to and activities in and around such areas are highly restricted and even banned by their own governments to appease exogenous actors (Igoe & Croucher, 2007; Castree, 2008; Büscher & Dressler, 2007). Ultimately, state-making leads to alienation of local communities, leaving them feeling like non-citizens in their own country due to their marginalization and perpetual threat of relocation or eviction from their ancestral lands (Igoe & Croucher, 2007; Liverman, 2004).

The prevailing practices in Botswana's WMAs with respect to the treatment of local communities adjacent to such areas, most of which had their land rights revoked in favour of wildlife conservation and tourism, are consistent with the concept of state-making. Some of Botswana's WMAs and protected areas are rented or operated as concessions to private exoge-

Commodification of nature and territorialization 229

nous tourism investors, thus sidelining the local communities without helping them build their capacity in the tourism sector (Good, 2008; Kiema, 2010; Mbaiwa, 2017a). There is a systemic exploitation and exclusion of local communities that ought to have a stake in Botswana's WMAs through tourism activities. The local people and communities in the WMAs are paid exploitative wages and racially discriminated against by some 'foreign nationals' who own tourism enterprises in Botswana and the government has done nothing much to intervene because of elite connections to exogenous actors (Good, 2008; Mbaiwa, 2003; Gupta, 2013; Wilmsen, 1989; Kiema, 2010; Stadler, 2005; Mbaiwa & Hambira, 2020).

Moreover, in state-making, where the interests and activities of the local communities are deemed to interfere with the operations of exogenous enterprises, the former are evicted, disposed of or banished to areas with less or no economic value without regard for their cultural and socio-economic rights (Langholz, 2003; Lemos & Agrawal, 2006). This has been the situation of most Basarwa communities in Botswana for decades (Stadler, 2005; Wilmsen, 1989; Vidal, 2014; Saugestad, 2001). The Basarwa of CKGR were banished to the New Xade settlement by the Government of Botswana in the early 2000s (Marobela, 2010; Gupta, 2013; Sylvain, 2003; Good, 2008; Kiema, 2010; Hitchcock, 2002; Stadler, 2005). The local communities were regarded as obstacles to the commodification of nature and spaces deemed to be necessary for sustainable conservation and profit-making (McAfee, 1999; Lemos & Agrawal, 2006). For much of Botswana's history of wildlife conservation, the exclusion of local communities around WMAs is well documented and has been a defining feature of wildlife conservation (Gupta, 2013; Solway, 2009, 2011; Suzman, 2001; Sylvain, 2003; Kiema, 2010; Good, 1992; Hitchcock, 2002; Saugestad, 2001; Anaya, 2010; Wilmsen, 1989; Mbaiwa & Hambira, 2020; Stadler, 2005). There are only a few cases, such as the Okavango area, where CBNRM has proved beneficial to some local communities (Mbaiwa & Stronza, 2010). The removal of local communities from their ancestral lands and loss of their rights to claim the same in Botswana is often cunning and brutal (Marobela, 2010; Suzman, 2001; Saugestad, 2001; Anaya, 2010; Solway, 2011; Sylvain, 2005). Local communities are made to be receptive to 'outsiders' interested in their landscapes and wildlife resources, both of which are valuable for exogenous tourism enterprises (McAfee, 1999; Büscher & Dressler, 2007; Berlanga & Faust, 2007; Igoe & Fortwangler, 2007; Langholz, 2003; Saugestad, 2001; Anaya, 2010; Stadler, 2005). However, consistent with state-making, the local communities are often lured into solemnly ceding their rights and claims to their ancestral lands hoping to be brought into the tourism industry through initiatives such as CBNRM (Berlanga & Faust, 2007). More often that turns out not to be the case as has been seen with most communities in WMAs in Botswana (Saugestad, 2001; Suzman, 2001; Solway, 2009; Stadler, 2005).

The dignity, well-being and citizenship of local communities in WMAs is forfeited due to loss of their cultural and land rights (Mbaiwa, 2003; Bolaane, 2004; Good, 1993; Solway, 2011; Suzman, 2001; Robins, 2000). The pleas of some local communities in WMAs, such as the Basarwa of CKGR and those of the Khwai River in the eastern part of the Okavango, to have their cultural and land rights recognized, protected and guaranteed by the government were not heeded (Vidal, 2014; Anaya, 2010; Robins, 2000; Bolaane, 2004; Solway, 2011; Suzman, 2001; Saugestad, 2001; Mbaiwa, 2017b). Throughout their history, "the Basarwa have had to deal with domination and discrimination from other groups and the nation-state" (Hitchcock, 2002, p. 800). Consequently, some individuals and international organizations such as the British-based Survival International and other activists on the rights of minorities criticized the Government of Botswana for serving the interests of exogenous tourism elites

230   *Handbook on tourism and conservation*

in what BBC's John Simpson once referred to as "the hunt of Basarwa by their government" (Simpson, 2013; Ontebetse, 2013; Saugestad, 2001; Solway, 2009; Good, 1992).

TRANSNATIONALIZATION OF SPACES AND LOSS OF STATE SOVEREIGNTY: TRANS-FRONTIER CONSERVATION AREAS AND BOTSWANA'S INTERNATIONAL ENVIRONMENTAL COOPERATION

Wildlife resources facilitate Botswana's international cooperation. International cooperation can be state-led or driven by non-state actors such as interest groups, elites, regional and/ or international organizations or commercial entities (Krasner, 1983). International cooperation is often negotiated in a bargaining process that is explicit but can also be imposed by strong actors forcing others to adjust their policies and behaviour (Young, 1994; Gowa, 1986). Furthermore, international cooperation can occur without communication or explicit agreement (Young, 1994), because "cooperative behaviour emerges where expectations of actors converge" (Milner, 1992, p. 469). Botswana has established multilateral environmental relations and agreements with several international and regional institutions. At the international level, Botswana ratified CITES, which aims to conserve and ban trade of endangered plants and animals, such as elephants and rhinos. Botswana is also party to the United Nations Environmental Programme (UNEP) and has actively participated in international environmental cooperation summits and conferences such as the Earth Summit and its Conference of Parties (COP). Botswana is also party to the United Nations Convention on Biological Diversity, the Ramsar Convention on Wetlands of International Importance and the World Heritage Convention under which the Okavango Delta – the largest inland wetlands – is listed as a World Heritage Site (DEA, 2008; Department of Tourism, 2000; Mladenov et al., 2007; Hamandawana & Chanda, 2013). Additionally, Botswana has been receptive to other international wildlife conservation entities such as the World Wildlife Foundation (WWF) and Conservation International. Botswana also has some local conservation organizations with external connections, like Elephants Without Borders based in Kasane. Botswana's participation in international agreements has earned the country a positive international reputation for wildlife conservation and made the country attractive to international actors with interest in wildlife resources and tourism. The international environmental agreements have also fostered new networks that purport to help states like Botswana to become 'champions of conservation' (Frank et al., 2000; Igoe & Brockington, 2007; Langholz, 2003).

Regionally, Botswana is located in Southern Africa which is rich in biodiversity. The region also plays a critical role in international wildlife conservation and attracts intra-regional and international tourists, trophy hunters, as well as poachers (Bwalya & Suping, 2020). To ensure and sustain wildlife conservation and generate revenue from it through eco-tourism, the SADC member states, including Botswana, adopted and enacted several wildlife conservation initiatives and protocols, such as the Protocol on Wildlife Conservation and Law Enforcement of 1999, SADC Law Enforcement and Anti-Poaching Strategy, and Charter on the Regional Tourism Organizations of Southern Africa of 1997. The SADC Protocol on Wildlife Conservation and Law Enforcement seeks, amongst others, to build capacity for wildlife management at both national and regional levels. The protocol has also led to regional cooperation on wildlife conservation through the establishment of trans-frontier conservation areas (TFCAs) such as Kavango-Zambezi Trans-frontier Conservation Area (KAZA), jointly connecting Angola, Botswana, Namibia, Zambia and Zimbabwe. Additionally, the Kgalagadi Trans-frontier Park shared by Botswana and South Africa, and the only peace park in the SADC region, was also established because of the SADC Protocol on Wildlife Conservation and Law Enforcement (Peace Parks Foundation, 2000). The aforementioned SADC wildlife conservation initiatives have facilitated and fostered cordial relations between Botswana and other states in the region with which it shares transboundary natural resources. Moreover, Botswana's participation in regional conservation agreements has boosted the country's tourism sector, because its "comparative advantage in terms of tourism has been its wilderness state in most of its protected areas that offer tranquillity and relaxation that most of the SADC regions lack" (DWNP, 2016, p. 5).

Botswana's demarcation of transnational spaces for wildlife conservation has promoted its international cooperation. Vandergeest and Peluso (1995, p. 388) rightly observed that "the territory of a national park is nested in national territory, which is nested in a global territorial grid". More importantly, territories are not just demarcated land spaces, but instruments of control that are indicative of power relations. For example, the international environmental agreements and TFCAs have ensured effective management and conservation of transborder natural resources, especially wildlife, but have also led to Botswana's loss of sovereignty over such areas. This is the case as conservation areas are governed by international and regional agreements, and spaces are managed as transnational networks and enterprises dictate (Ferguson, 2006; Berlanga & Faust, 2007). The transnationalization of Botswana's spaces in the name of conservation has dismantled restrictive state structures and practices that could protect local people from transnational private enterprises and international environmental organizations or agreements (Good, 2008; Kiema, 2010; Nieuwoudt, 2008; Langholz, 2003).

Botswana's TFCAs have been transformed into closely guarded enclaves due to their economic value, especially to exogenous tourism operators (Mbaiwa, 2003; Nieuwoudt, 2008; Saugestad, 2001; Suzman, 2001). Any attempt by local communities to be included in the tourism value chain is highly contested by exogenous players who have invested in these areas by establishing tourism resorts (Stadler, 2005; Hitchcock, 2002; Mbaiwa & Hambira, 2020). The transnationalization of spaces serves transnational interests and national elites that have formed tourism joint ventures with exogenous actors (Ferguson, 2006; Berlanga & Faust, 2007). It also comes as a major challenge that constrains Botswana's ability to make conservation decisions contrary to international agreements without international backlash. This is the case even where local people are adversely affected such as in human–elephant conflict that has been prevalent in the country for a number of years (Ontebetse, 2019; Chase et al., 2016; Bwalya & Suping, 2020). The transnationalization of wildlife areas forges and strengthens the relations between Botswana, neighbouring states and transnational actors, but often disregards the interests of local people living near WMAs and TFCAs. Furthermore, the ratification of international wildlife conservation agreements and protocols earns Botswana some international reputation, but sometimes denies the country the liberty to utilize its wildlife resources as it deems necessary ('t Sas-Rolfes et al., 2014; Ontebetse, 2019).

To a large extent there is privatization of sovereignty on Botswana's wildlife resources. According to Ferguson (2006) and Igoe and Brockington (2007), the privatization of state sovereignty refers to a system where sovereignty of the state is highly fragmented and decentralized or controlled by different state and private actors in different contexts for varying purposes. The sovereignty of the state in Botswana's WMAs and TFCAs is under the control

232 *Handbook on tourism and conservation*

of different transnational actors for varying purposes, as some prescribe conservation treaties like CITES, others promote the establishment of TFCAs, while others form joint tourism enterprises with the local elites (Mbaiwa, 2017b; Suzman, 2001; Solway, 2011; Saugestad, 2001; Marobela, 2010; Kiema, 2010; Mbaiwa & Hambira, 2020). Often, the Government of Botswana is forced to choose between its people's livelihoods and the interests of exogenous actors in wildlife conservation and tourism (Good, 2008; Chase et al., 2016; Saugestad, 2001; Ontebetse, 2019; Mbaiwa & Hambira, 2020). For instance, in 2019, Botswana found itself restrained by CITES from reducing the ever-increasing elephant population that threatened the people's livelihoods and claimed some lives (Ontebetse, 2019; Bwalya & Suping, 2020). The exogenous actors sometimes use their dominant position to persuade and/or lobby other developing countries to support their wildlife conservation initiatives at the expense of other states. An example is Kenya, a signatory of CITES, opposed Botswana and Zimbabwe's bid for the lifting of the ban on the sale of ivory, and Botswana was powerless to deal with the soaring elephant population and increasing number of human–wildlife conflict cases ('t Sas-Rolfes et al., 2014; Ontebetse, 2019; Daily Nation, 2013). As observed by some scholars, international agreements, summits and protocols are also an expansion of some Western ideologies to developing countries, using various means such as donor assistance, to legitimize their actions and erode the sovereignty of small or weak host states over wildlife resources (Howson & Smith, 2008; Svarstad et al., 2018; Castree, 2008; Berlanga & Faust, 2007), and Botswana is no exception (Ontebetse, 2019; Bwalya & Suping, 2020; Stadler, 2005; Marobela, 2010; Good, 1993).

CONCLUSION

Wildlife conservation has earned Botswana international recognition. The participation of Botswana in international and regional environmental agreements has facilitated Botswana's cordial relations with international environmental organizations, institutions and other states with which it shares transboundary natural resources. The value of wildlife conservation and the resultant thriving tourism industry in Botswana cannot be denied. It has managed to sustain Botswana's tourism sector by attracting international tourists and transnational tourism enterprises (Mbaiwa, 2003; WTTC, 2018). However, Botswana's wildlife conservation and the tourism industry remain externally controlled by international environmental organizations and dominated by exogenous enterprises at the expense of locals (Saugestad, 2001; Mbaiwa & Hambira, 2020; Mbaiwa, 2017a; Stadler, 2005). Botswana's conservation management areas, though close to local communities in proximity, remain out of their reach due to extreme and systemic marginalization of such communities by exogenous actors who dictate access conditions and prices of wildlife resources in such areas (Kiema, 2010; Mbaiwa, 2003; Saugestad, 2001; Vidal, 2014). The CBNRM initiatives have remained largely mere rhetoric as the local communities often lack capacity, skill and resources to compete with exogenous tourism actors, and the government interventions are seemingly limited by the international and regional agreements (Vidal, 2014; Solway, 2009; Suzman, 2001). As the Department of Environmental Affairs correctly stated, "CBNRM contributes little to poverty reduction" (DEA, 2016, p. 12). For example, as mentioned, the Government of Botswana, in 2014, was compelled to appease international conservationists by imposing a ban on hunting even though elephant populations were increasingly encroaching on people's land and claiming human lives (Chase et al., 2016; Mbaiwa, 2017b; Miller, 2014; Nieuwoudt, 2008).

The prevailing conservation challenges and inequities in the tourism sector disadvantage local communities and make their aspirations to benefit from the wildlife resources in their areas an illusion. There is a need for Botswana to introduce inclusive wildlife legal and policy frameworks that: offer practical solutions to the local communities' uncertainties; solve the government's challenge to balance international cooperation with the needs of local communities; address the feeling of loss of 'citizenship' amongst victims of evictions for wildlife conservation; and deal with the complex inequities in Botswana's wildlife conservation and tourism. Such frameworks ought to include, amongst others, improved incentives for conservation amongst local communities to reduce human–wildlife conflicts; creation and reservation of some opportunities for local people in the tourism sector; and localization of the tourism value-chain by ensuring that operations and bookings made in Botswana can also benefit local people through employment. Furthermore, the Government of Botswana should aim to widen participation of marginalized communities in wildlife conservation and tourism decision-making processes, including prior to committing to international wildlife conservation treaties. Including local voices in Botswana's wildlife conservation and tourism narratives can also help paint a real picture of the situation on the ground and reduce the dominance of conservation and tourism narratives told by exogenous actors.

REFERENCES

Acemoglu, D., Johnson, S., & Robinson, J. (2001). An African success story: Botswana. In D. Rodrik (ed.), *In Search of Prosperity: Analytical Narrative on Economic Growth* (pp. 80–119). Princeton, NJ: Princeton University Press.

Anaya, J. (2010). Addendum – The situation of indigenous peoples in Botswana. United Nations Human Rights Council. www.refworld.org/country,COI,UNHRC,,BWA,,4ac47f652,0.html, accessed 2 May 2021.

Berlanga, M., & Faust, B. (2007). We thought we wanted a reserve: One community's disillusionment with government conservation management. *Conservation and Society*, 5(4), 450–477.

Bevir, M., & Rhodes, R. A. W. (2005). Interpretation and its others. *Australian Journal of Political Science*, 40(2), 169–187.

Biggs, D., Cooney, R., Roe, D., Dublin, H. T., Allan, J. R., Challender, D. W. S., & Skinner, D. (2017). Developing a theory of change for a community-based response to illegal wildlife trade. *Conservation Biology*, 31(1), 5–12.

Bolaane, M. (2004). The impact of game reserve policy on the River BaSarwa/Bushmen of Botswana. *Social Policy & Administration*, 38(4), 399–417.

Büscher, B., & Dressler, W. (2007). Linking neo-protectionism and environmental governance: On the rapidly increasing tensions between actors in the environment-development nexus. *Conservation and Society*, 5(4), 586–611.

Bwalya, E., & Suping, K. (2020). Africa's rhino poaching crisis: The role of Vietnam. *IPADA*, 5, 561–570.

Castree, N. (2008). Neoliberalising nature: Processes, effects, and evaluations. *Environment and Planning A: Economy and Space*, 40(1), 153–173.

Chase, M. J., Schlossberg, S., Griffin, C. R., Bouché, P. J. C., Djene, S. W., Elkan, P. W., Ferreira, S., Grossman, F., Kohi, E. M., Landen, K., Omondi, P., Peltier, A., Selier, S. A. J., & Sutcliffe, R. (2016). Continent-wide survey reveals massive decline in African savannah elephants. *PeerJ*, 4, e2354. https://doi.org/10.7717/peerj.2354, accessed 18 May 2021.

234 *Handbook on tourism and conservation*

Christy, B., & Hartley, A. (2013). Battle for the elephants. *New National Geographic Special*, 27 February. www.nationalgeographic.org/education/channel/battle-for-elephants/, accessed 22 May 2021.

Crookes, D. J., & Blignaut, J. N. (2016). A categorisation and evaluation of rhino management policies. *Development Southern Africa*, 33(4), 459–469.

Daily Nation (2013). Kenya gets China support to fight poaching – President Kenyatta. https://nation.africa/kenya/news/kenya-gets-china-support-to-fight-poaching-president-kenyatta--887138, accessed 12 April 2021.

DEA [Department of Environmental Affairs] (2008). *Okavango Delta Management Plan*. Gaborone: Government Printer.

DEA [Department of Environmental Affairs] (2016). *National Biodiversity Strategy and Action Plan*. Gaborone: Government Printer.

Department of Tourism (2000). *Botswana Tourism Master Plan*. Gaborone: Government Printer.

DWNP [Department of Wildlife and National Parks] (2013). *Wildlife Policy*. Gaborone: Government Printer.

DWNP [Department of Wildlife and National Parks] (2016). *Action Plan for Implementing the Convention on Biological Diversity's Programme of Work on Protected Areas*. Gaborone: Government Printer.

Edge, W. (1998). Botswana: A developmental state. In W. Edge and M. Lekorwe (eds), *Botswana: Politics and Society* (pp. 333–348). Pretoria: Van Schaik Publishers.

Ferguson, J. (2006). *Global Shadows: Africa in the Neoliberal World Order*. Durham, NC: Duke University Press.

Fletcher, R. (2010). Neoliberal environmentality: Towards a poststructuralist political ecology of the conservation debate. *Conservation and Society*, 8(3), 171–181.

Frank, D. J., Hironaka, A., & Schofer, E. (2000). The nation-state and the natural environment over the twentieth century. *American Sociological Review*, 65, 96–116.

Goldman, M. (2001). Constructing an environmental state: Eco-governmentality and other trans-national practices of a 'green' World Bank. *Social Problems*, 48(4), 499–523.

Good, K. (1992). Interpreting the exceptionality of Botswana. *The Journal of Modern African Studies*, 30(1), 69–95.

Good, K. (1993). At the ends of the ladder: Radical inequalities in Botswana. *The Journal of Modern African Studies*, 31(2), 203–230.

Good, K. (2008). *Diamonds, Dispossession and Democracy in Botswana*. Johannesburg: Jacana Media.

Gowa, J. (1986). Anarchy, egoism and third images: The evolution of cooperation and international relations. *International Organizations*, 40(1), 167–186.

Gupta, C. (2013). A genealogy of conservation in Botswana. *PULA: Botswana Journal of African Studies*, 27(1), 45–67.

Hamandawana, H., & Chanda, R. (2013). Environmental change: In and around the Okavango Delta during the nineteenth and twentieth centuries. *Regional Environmental Change*, 12(3), 681–694.

Harvey, C., & Lewis, S. (1990). *Policy Choice and Development Performance in Botswana*. New York: St. Martin's Press.

Hitchcock, R. K. (2002). 'We are the first people': Land, natural resources and identity in the Central Kalahari, Botswana. *Journal of Southern African Studies*, 28(4), 797–824.

Howson, R., & Smith, K. M. (2008). Hegemony and the operation of consensus and coercion. In R. Howson & K. M. Smith (eds), *Hegemony: Studies in Consensus and Coercion* (pp. 107–124). New York: Routledge.

Igoe, J., & Brockington, D. (2007). Neoliberal conservation: A brief introduction. *Conservation Sociology*, 5(4), 432–449.

Igoe, J., & Croucher, B. (2007). Conservation, commerce, and communities: The story of community-based wildlife management areas in Tanzania's northern tourist circuit. *Conservation and Society*, 5(4), 534–561.

Igoe, J., & Fortwangler, C. (2007). Whither communities and conservation? *International Journal of Biodiversity Science and Management*, 3(2), 65–76.

Jones, S. (2006). A political ecology of wildlife conservation in Africa. *Review of African Political Economy*, 33(109), 483–495.

Kiema, K. (2010). *Tears for My Land: A Social History of the Kua of the Central Kalahari Game Reserve, Tc'amnqoo*. Gaborone: Mmegi Publishing House.

Krasner, S. D. (1983). *International Regimes*. Ithaca, NY: Cornell University Press.

Langholz, J. (2003). Privatizing conservation. In S. R. Brechin, P. R. Wilshusen, C. L. Fortwangler and P. C. West (eds), *Contested Nature: Promoting International Biodiversity Conservation with Social Justice in the Twenty-First Century* (pp. 117–135). New York: State University of New York Press.

Leftwich, A. (1995). Bringing politics back in: Towards a model of the developmental state. *Journal of Development Studies*, 31(3), 400–427.

Leith, J. (2005). *Why Botswana Prospered*. Montreal: McGill-Queen's University Press.

Lemos, C., & Agrawal, A. (2006). Environmental governance. *Annual Review of Environment and Resources*, 31, 297–325.

Liverman, D. (2004). Who governs, at what scale, and at what price? Geography, environmental governance, and the commodification of nature. *Annals of the Association of American Geographers*, 94(9), 734–738.

Marobela, M. N. (2010). The state, mining and the community: The case of Basarwa of the Central Kalahari Game Reserve in Botswana. *Labour, Capital and Society / Travail, Capital et Société*, 43(1), 137–154.

Mbaiwa, J. E. (2003). Enclave tourism and its socio-economic impacts in the Okavango Delta, Botswana. *Tourism Management*, 26(2), 157–172.

Mbaiwa, J. E. (2011). The effects of tourism development on the sustainable utilisation of natural resources in the Okavango Delta, Botswana. *Current Issues in Tourism*, 14(3), 251–273.

Mbaiwa, J. E. (2015). Community-based natural resource management in Botswana. In R. van der Duim, M. Lamers, & J. van Wijk (eds), *Institutional Arrangements for Conservation, Development and Tourism in Eastern and Southern Africa* (pp. 59–80). Dordrecht: Springer.

Mbaiwa, J. E. (2017a). Poverty or riches: Who benefits from the booming tourism industry in Botswana? *Journal of Contemporary African Studies*, 35(1), 93–112.

Mbaiwa, J. E. (2017b). Effects of the safari hunting tourism ban on rural livelihoods and wildlife conservation in Northern Botswana. *South African Geographical Journal*, 100(1), 1–21.

Mbaiwa, J. E., & Hambira, W. L. (2020). Enclaves and shadow state tourism in the Okavango Delta, Botswana. *South African Geographical Journal*, 102(1), 1–21.

Mbaiwa, J. E., & Stronza, A. L. (2010). The effects of tourism development on rural livelihoods in the Okavango Delta, Botswana. *Journal of Sustainable Tourism*, 18(5), 635–656.

McAfee, K. (1999). Selling nature to save it? Biodiversity and green developmentalism. *Environment and Planning D: Society and Space*, 17(2), 133–154.

Miller, J. (2014). Botswana to inaugurate diamond mine on bushmen ancestral land. Mumbai: Rapaport.

Milner, H. (1992). International theories of cooperation among nations: Strengths and weaknesses. *World Politics*, 44(3), 466–496.

Mladenov, N., Gardner, R. J., Flores, E. N., Mbaiwa, E. J., Mmopelwa, G., & Strzepek, M. K. (2007). The value of wildlife-viewing tourism as an incentive for conservation of biodiversity in the Okavango Delta, Botswana. *Development Southern Africa*, 24(3), 409–423.

Nieuwoudt, S. (2008). *Development–Botswana: Of tourists, bushmen – and a borehole*. Washinton, DC: Inter-Press Service. www.ipsnews.net/2008/05/development-botswana-of-tourists-bushmen-and -a-borehole/, accessed 15 April 2021.

Nyoni, D. T. (2018). Botswana – shoot-to-kill anti-poaching policy and summary executions. https:// africasustainableconservation.com/2018/06/20/botswana-shoot-to-kill-anti-poaching-policy-and -summary-executions/, accessed 17 April 2021.

Ontebetse, K. (2013). Survival International threatens to take up new Basarwa case. *Sunday Standard*, Gaborone, 30 May.

Ontebetse, K. (2019). Elephant overpopulation tramples on Botswana-Kenya relations. *Sunday Standard*, Gaborone, 8 May.

Parry, D., & Campbell, B. (1990). Wildlife management areas of Botswana. *Botswana Notes and Records*, 22, 65–77.

Peace Parks Foundation (2000). *Kgalagadi Trans-frontier Park*. Stellenbosch: Peace Parks. www .peaceparks.org/tfcas/kgalagadi/, accessed 10 May 2021.

236 *Handbook on tourism and conservation*

Picard, L. A. (1985). *Politics and Rural Development in Southern Africa: The Evolution of Modern Botswana*. London: Rex Collings.

Robins, S. (2000). Land struggles and the politics and ethics of representing 'Bushman' history and identity. *Kronos*, 26(1), 56–75.

Samatar, A. I. (1999). *An African Miracle: State and Class Leadership and Colonial Legacy in Botswana Development*. Portsmouth, NH: Heinemann Publishing.

Saugestad, S. (2001). *The Inconvenient Indigenous: Remote Area Development in Botswana, Donor Assistance and the First People of the Kalahari*. Uppsala: Nordic Africa Institute.

Simpson, J. (2013). Hunted by their own government – the fight to save Kalahari Bushmen. *The Independent*, 25 October.

Solway, J. (2009). Human rights and NGO 'wrongs': Conflict diamonds, culture wars and the 'bushman question'. *Africa: Journal of the International African Institute*, 79(3), 321–346.

Solway, J. (2011). Culture fatigue: The state and minority rights in Botswana. *Indiana Journal of Global Legal Studies*, 18(1), 211–240.

Stadler, A. (2005). *Conservation for Whom? Telling Good Lies in the Development of Central Kalahari*. Linkoping: Linkoping University.

Stone, L., Mogomotsi, P., & Mogomotsi, G. (2021). The impacts of Covid-19 on nature-based tourism in Botswana: Implications for community development. *Tourism Review International*, 25(2–3), 263–278.

Suzman, J. (2001). *An Introduction to the Regional Assessment of the Status of the San in Southern Africa*. Windhoek: Legal Assistance Centre.

Svarstad, H., Benjaminsen, T. A., & Overå, R. (2018). Power theories in political ecology. *Journal of Political Ecology*, 25(1), 350–363.

Sylvain, R. (2003). Class, culture and recognition: San farm workers and indigenous identities. *Anthropologica*, 45(1), 111–119.

Sylvain, R. (2005). Disorderly development: Globalization and the idea of "culture" in the Kalahari. *American Ethnologist*, 32(3), 354–370.

't Sas-Rolfes, M., Moyle, B. J., & Stiles, D. (2014). The complex policy issue of elephant ivory stockpile management. *Pachyderm*, 55, 62–77.

Tebele, M. (2018). Masisi revokes 'shoot-to-kill' policy. *Southern Times*, 25 May. https://southerntimesafrica.com/site/news/masisi-revokes-shoot-to-kill-policy#, accessed 23 April 2021.

Thumberg-Hartland, P. (1978). *Botswana: An African Growth Economy*. Boulder, CO: Westview.

United Nations Development Programme (UNDP) (2009). *Assessment of Development Results: Evaluation of UNDP Contribution – Botswana*. New York: UNDP.

Vandergeest, P., & Peluso, N. (1995). Territorialization and state power in Thailand. *Theory and Society*, 24(3), 385–426.

Vidal, J. (2014). Botswana bushmen: 'If you deny us the right to hunt, you are killing us'. *The Guardian*, 18 April.

Wilmsen, E. N. (1989). *Land Filled with Flies: The Political Economy of the Kalahari*. Chicago: University of Chicago Press.

Winters, O. J. (2019). The Botswana bushmen's fight for water & land rights in the Central Kalahari Game Reserve. *Consilience*, 21(1), 172–186.

World Travel and Tourism Council (WTTC) (2018). *Travel and Tourism Power and Performance*. London: WTTC.

Young, O. (1994). *International Governance: Protecting the Environment in a Stateless Society*. Ithaca, NY: Cornell University Press.

17. Community-based natural resources management and poverty reduction

Israel R. Blackie

INTRODUCTION

The community-based natural resources management (CBNRM) approach is based on the notion that local communities are likely to engage in sustainable utilisation of natural resources found in their localities because their livelihoods depend on them (Twyman, 2000; Mbaiwa, 2013). This supposition is based on the view that outsiders, including government and private conservation entities, may lack understanding and genuine interest in local environments. This may lead to negative impacts on the natural resource base (Government of Botswana, 2007; Blaikie, 2006). Ostensibly, the CBNRM approach in Botswana entailed the decentralisation of management of natural resources with redistribution of power and devolution of responsibilities from central government to rural communities (Mbaiwa, 2013; Stone and Nyaupane, 2015). Through CBNRM, central government sought to extend devolution of natural resources usufructs or user rights to local communities.

The aim of this chapter is to examine how local communities as intended beneficiaries of the CBNRM programme became excluded from accessing and utilising natural resources in their localities in Botswana. This chapter provides a context for discussing how local communities as intended beneficiaries of CBNRM find themselves excluded from accessing and utilising communally held natural resources. The chapter concludes by providing policy options in the form of recommendations that should facilitate changes to structures that directly or indirectly perpetuate marginalisation and/or exclusion of certain sectors of local communities.

STUDY AREA AND APPROACH

The study was carried out in Ngamiland and Chobe Districts of Botswana. These two districts have more active CBNRM trusts than anywhere else in the country because of the availability of natural resources (wildlife) that facilitate tourism as the main attribute of CBNRM operations in Botswana. The study was concurrently timed following a convergent parallel design in which both qualitative and quantitative methods were prioritised equally (Creswell, 2014). Convergent parallel mixed-methods design involved qualitative research methods (documentary analysis, focus group discussions (FGDs) and in-depth interviews with key stakeholders) and quantitative research methods (heads of household questionnaire and an online survey). Systematic probability sampling was adopted for heads of households as the main survey to ensure that each household head in the village had an equal probability of being interviewed (n=101). The systematic probability sampling method adopted also ensured administrative convenience for the location of sampling units. Communities are made up of aggregation of households (*malwapa*). As a result, households were used as sampling units. Participants in

238   *Handbook on tourism and conservation*

Table 17.1   Socioeconomic characteristics of study communities

Village	Pop size	Gender make-up		Ethnic composition	CHAs[a] & area covered (km²)	Types of tourism activities engaged in	Households	Average Household Survey
		M	F					
Gudigwa	725	351	374	San people	NG12	Non-consumptive or photographic tourism	145	5.0
Sankuyo	410	208	202	Bayeyi, Basubiya and Basarwa	NG 33 and 34 (870 km²)	Non-consumptive or photographic tourism	77	5.2
Khwai	360	118	113	San, Batawana and Basubiya	NG 18 and NG 19 (1, 995km²)	Non-consumptive or photographic tourism	–[b]	–
Kachikau	1,356	669	687	Basubiya and Batawana	CH1 and CH2 (117 00 km²)	Consumptive hunting tourism and	989	6.6
Parakarungu	899	397	502	Basubiya		non-consumptive photographic tourism	–	6.6
Kasane	9,084	4,512	4,572	Basubiya, Batawana, and others	CH1 and CH2	Provision of tourism accommodation (hotels, lodges, campsites), MICE (meetings, incentives, conferences and exhibitions) as well as safari drives	6,830	2.7

Notes: [a] CHA refers to a controlled hunting area which has often been delineated for community use in their lease. [b] Not depicted from the national census conducted by Statistics Botswana (2011).

this study were interviewed at a single point in time, and hence the study was a cross-sectional survey. Data were analysed using both Atlas.ti and the Statistical Package for Social Sciences (SPSS). Atlas.ti was used to capture the main themes arising from qualitative data while SPSS was used to analyse quantitative data and also to generate tables and frequencies where appropriate.

Following the guidelines on conducting FGDs (Glaser and Strauss 1967), where an estimate of 6 to 8 participants is recommended in each session, this study included 4–6 participants per session in each of the six FGDs conducted in Gudigwa, Sankuyo, Khwai, Kasane, Kachikau and Parakarungu villages to represent the two districts of Ngamiland and Chobe. The majority of the respondents were aged between 30 and 60 years, with more males (55 per cent) than females (45 per cent). Heads of households survey was carried out among five ethnic groups (Basarwa, Bambukushu, Bayeyi, Basubiya and Batawana), with the majority (37 per cent) of respondents being from the San (Basarwa) ethnic group. Table 17.1 shows the socioeconomic characteristics of these community trusts, including population size, gender make-up, ethnic composition, area covered by the community trust in square kilometres, and types of activities they engaged in.

BOTSWANA COMMUNITY BASED NATURAL RESOURCES MANAGEMENT (CBNRM) POLICY (2007)

In 2007 the Botswana government promulgated the CBNRM policy which represented a paradigm shift in rural development from state-led natural resource conservation to an all-inclusive people centred conservation approach that also fosters rural development through the sustainable use of natural resources. The CBNRM policy is described as: "a development approach that incorporates natural resources conservation, the ultimate aim of which is to manage and protect the natural resources base" (Government of Botswana, 2007, p. ii).

The policy recognises wildlife as the most valuable resource (Government of Botswana, 2007, p. 10; Centre for Applied Research, 2016). The CBNRM policy sought to establish a foundation for conservation-based development in which the need to protect biodiversity and ecosystems is balanced with the need to improve rural livelihoods and reduce poverty. This overall objective of the CBNRM policy was to be achieved through empowering communities to conserve and sustainably utilise the natural resources found in their areas. Through the same CBNRM policy, communities are required to apply for a Community Natural Resources Management Lease commonly known as 'head lease' which is acquired on a fifteen-year contract term from the land board (or Department of Lands). Through this lease, communities gain resource user rights but not land rights. Thus, communities would receive benefits from the use of natural resources in the area specified in the lease, referred to as a Controlled Hunting Area (CHA).

Communities gain such use rights as the management and sustainable use of natural resources (including wildlife) found in their concessions through promoting both consumptive and non-consumptive tourism, i.e. hunting and aesthetic experience respectively. It appears that because of the extension of usufruct rights, communities are expected to "warm up to the conservation efforts", and perhaps also required to agree with government policy changes (Lenao and Saarinen, 2016, p. 120).

CBNRM, ETHNICITY AND SOCIAL EXCLUSION OF HUNTER-GATHERERS

The San people are often described as the first people to inhabit sub-Saharan Africa, though paradoxically belonging to what is often defined as the fourth world – a group of the most underprivileged and oppressed populations of the world (Wagner, 2011). The statement below by a Government of Botswana litigation consultant shows how the San people's identity is often misconstrued even by government officials who enforce the laws and statutes – perhaps a sign that justice serves the interests of the strongest in the community.

> Masarwa (*sic*) have always been true nomads, owing no true allegiance to any chief or tribe … it appears to me that true nomad Masarwa can have no rights of any kind except rights to hunting. (Litigation consultant to the Attorney General cited in Thapelo, 2002, p. 140)

The statement above shows the deep rooted perception about and the actual plight of the San people in Botswana and other African countries where they live. Good (2002) notes that even the identity of the San people is socially conceived to suit their masters. For example, Le Roux

240 *Handbook on tourism and conservation*

(1999, p. 83) quotes a San woman in the Western Sandveld saying: "I know I am a Mosarwa, but I do not know why. It is probably because I am poor." This shows a deeply embedded sense of stigmatisation among this ethnic minority group.

Pradhan (2011) noted that there is a consensus among natural resources utilisation researchers that a focus on issues of social equity, inclusion and empowerment of the poor populations has dominated research over the years. The concept of social exclusion inherently involves the lack or denial of resources, rights, goods and services, and the inability to participate in the normal relationships and activities available to the majority of people in a society, whether in economic, social, cultural or political arenas. It affects both the quality of life of individuals and the equity and cohesion of society (Levitas et al., 2007). The social exclusion of the San in Botswana is largely attributable to the formulation of the land use policy by the colonial administration under the influence of indigenous agro-pastoral farmers in the last decade of the 1890s. The San were accused of cattle rustling and starting veldt fires especially by the alleged marauding bands of the San (Russell, 1976). From 1968 to about 1975 chunks of land which hitherto formed part of the hunting and gathering range for the San were institutionalised into land tenure, land use and water use through the introduction of the Tribal Land Act (TLA) which gave tribesmen over 71 per cent of the country for allocation at tribal level. Subsequently, the commercialisation of the rangeland through the amendment of the same TLA in 1970 (No. 6 of 1970) gave rise to the present-day Land Board entities which require as prerequisite individuals/applicants to belong to a tribe (Republic of Botswana, 1972, Regulation 8(1) (a)).

The San people were at the time already scattered across the country, at the periphery of settlements of white settlers and dominant Tswana groups. The British colonial administration which introduced tribal-based game reserves failed to allocate a reserve to the San. This was because of the interpretation of the TLA clause on tribesmen which excluded the San from such groupings due to their nomadic lifestyle even though they moved in a cyclical pattern (Thapelo, 2002). The predicament of the social exclusion of the San has not been helped by the fact that even though attempts were made to repeal some of the discriminatory laws in the 1990s, such exercise coincided with the increasing emphasis on tourism as the country's next economic engine of growth and diversification from a diamonds-based economy (Taylor, 2000, 2002; Thapelo, 2002; Chavallier and Harvey, 2016). Such a perception has worsened the plight of the minority especially the San everywhere in Botswana as land hitherto belonging to them has either been subsequently classified as Wildlife Management Areas (WMAs), or subjected to 'museumification', infrastructural development, mining and a host of tourism ventures forcing the San into a vicious cycle of landlessness and its associated poverty trap (Taylor, 2000; Molosi, 2015). The introduction of CBNRM meant that the Department of Wildlife and National Parks (DWNP) could then decentralise management responsibilities of wildlife in WMAs to local communities (DWNP, 2009). This effectively excluded them from access to and use of wildlife found in Protected Areas (PAs) and some communities either lost a chunk of land (e.g. Mababe) or were relocated to give way to the creation of PAs (Taylor, 2000). The CBNRM programme is practised in the WMAs and any such limited access to land inevitably excludes those subject to it. Nonetheless, the Government does not seem to have given the tourism sector the priority and resources it requires to successfully reach its full potential in terms of employment generation and economic diversification as the second engine of growth. The budget allocation for the year 2018 was BWP 617 million (for tourism and conservation) but the 2016 Tourism Satellite Account showed that internal tourism

Community-based natural resources management and poverty reduction 241

expenditure (i.e., international inbound and domestic tourism expenditure) totalled BWP 14.5 billion. Yet tourism direct gross value added (the value added generated by all industries in the provision of goods and services to visitors – known as TDGVA) was calculated as BWP 7.7 billion. Comparing this with Gross Value Added (GVA) in the National Accounts for 2016, which showed a total GVA of BWP 155.4 billion, the comparison gives a direct idea of the tourist expenditure of around 4.9 per cent to Botswana's GVA (like GDP), compared to 3.7 per cent in 2009 (Budget Speech, 2009, 2018).[1]

The FGDs with participants in Gudigwa and Khwai villages showed that the majority of villagers view the CBNRM programme as an extension of colonial policies, and a mechanism for their social exclusion from participating in activities that could improve their livelihood opportunities. This finding is like that of Power and Wilson (2000) who also established that grassroots programmes (such as CBNRM) can act to disconnect groups and individuals from social relations and actively participating in meaningful activities accessible to other groups though having the potential to improve their livelihoods. The concept of social exclusion is closely linked to cultural and political capitals used by key groups in a society to monopolise privileges, mark cultural distance and proximity and even exclude marginal groups from participating in high-value activities (Bourdieu 1984; Rozemeijer and van der Jagt, 2000; Bolaane, 2004). The theorists of social exclusion posit that even though ascribed status such as ethnicity and gender are acquired at birth, the state, though supposedly a benevolent guarantor of social service supply, can act in a way that results in a vicious cycle of discrimination and a life of struggle in accessing life opportunities for some groups such as the San (Kabeer, 1994; Fischer, 2008). The quotation[2] below suggests that the state through its Technical Assistance Committees (TACs) has exacerbated the marginalisation of certain ethnic tribes:

> The government has refused to allow us to form our own CBO [community-based organisation] because they want the Hambukushu to rob us of our God given treasures. They dominate decision-making since they form the majority in the OCT board. Government has excluded us, the San, by not allowing us to form our own CBO. (60-year-old man, former Okavango Community Trust board member, Gudigwa village)

Results from the heads of household survey also showed that multi-village CBOs were characterised by regular misappropriation of benefits and the emergence of local elites who are reported to have captured the CBOs' operations for self-interest. The CBO capture was expressed in the quotation below which advances some of the reasons why the CBNRM programme has failed despite having the potential to reduce poverty.

> How can poverty be reduced when the proceeds/benefits from CBNRM do not trickle down to the intended beneficiaries at individual household level? Only Trust board members, their families and those few employed community members benefit from CBNRM. (FGD[3] in Khwai village)

It appears that most community residents were excluded from accessing benefits from the CBNRM programme as expressed above by a drought relief worker who is a female head of household. Elite refers to "individuals who can exert disproportionate influence over a collective action process" (Beard and Phakphian, 2009, p. 11). In the CBNRM discourse, elite capture denotes a process where the elites manipulate the decision-making mechanism and public agenda to obtain more benefits than other locals contrary to the CBNRM principle of equitable benefits distribution to all members of the community. Elite power is perpetuated

242   *Handbook on tourism and conservation*

Table 17.2   Land tenure, devolution of authority and use of indigenous methods in conservation and sustainable utilisation of natural resources

CBNRM Objectives	Strongly agree (%)	Agree (%)	Neither agree nor disagree (%)	Disagree (%)	Strongly disagree (%)	Total (%)
Does your CBO have secure land tenure (land rights) to the delineated CHAs?	0.0	25.7	43.6	29.7	1.0	100
Do you agree or disagree that government has devolved authority to manage and utilise natural resources to your community?	0.0	22.8	42.6	33.7	1.0	100
It is easy for government to agree on utilisation of indigenous methods for conserving and managing the natural resources especially wildlife?	0.0	51.5	36.6	11.9	0.0	100

through "land holdings, family networks, employment status, wealth, political and religious affiliation, personal history and personality" (Dasgupta and Beard, 2007, p. 234). Taylor (2000) found that CBOs that are comprised of multi-villages tended to show characteristics of emergent elites (control and ownership of the economy along networks of patronage) at the expense of local people. This scenario projects the CBNRM programme as a socially designed mechanism that renders locals unwitting victims of other people's development rather than fulfilling its aim of reducing poverty in rural communities (Government of Botswana, 2007).

DEVOLUTION AND DECISION-MAKING MECHANISMS IN CBNRM IMPLEMENTATION

Even though the CBNRM policy (Government of Botswana, 2007, p. iii) sought to devolve natural resources user rights to local communities, the policy did not, in any way, grant *de jure* (legal) secure land tenure rights to communities, but communities got rights (with assistance from Technical Assistance Committees) to manage the wildlife in the trust areas. Communities do not own the wildlife which is instead owned by the state. This approach by government is not in keeping with international standards for effective implementation of the CBNRM programme which calls for devolution of power and control of natural resources within communities to be at the level of communities instead of central government (Steiner and Rihoy, 1995; Blaikie, 2006). Table 17.2 presents the local communities' opinions on the operation of their CBOs in their CHAs.

About 43.6 per cent of the households could not immediately tell whether they had security of land tenure in their CHAs or not followed by 30.7 per cent of those who disagreed or indicated that they did not have security of land tenure. This shows a lack of information sharing by CBOs to local communities. As if pre-empting what was to come, the Ngamiland CBNRM Forum (2012) had resolved in its bi-annual conference to lobby government on land tenure security so that CBNRM can thrive in the country. One of the core conditions which needed to be addressed was to make sure that there is "no elite capture of land that removes land from communal access" (Dikobe, 2012). At the time of interview (2016), CBOs were uncertain as their initial lease agreements had expired in 2013[4] and thus could not renew or engage private tourism operators in their delineated CHAs pending finalisation of the non-consumptive utilisation of renewable resource lease agreements by government. Most CBOs in the study areas did not know what the new lease agreement looked like, and how it will be implemented as they had only been asked to submit revised non-consumptive management plans to inform drafting of the new lease agreement strategy.

So far, only the Sankuyo Tshwaragano Management Trust (STMT) has been enjoined through joint venture agreement to a private safari operator. Previously, communities were allowed to choose their joint venture partners through the assistance of TACs as provided for in the CBNRM policy (Government of Botswana, 2007). Specifically, section 9.3 of the CBNRM policy states that, "the community in whose controlled hunting area (CHA) a concession is offered will be consulted to provide an input on their preferred joint venture partner". In the current situation, the government (Ministry) unilaterally decided and signed a 15-year lease agreement with Smart Stone investors without consulting the community for their high value tourism zone. The terms of the lease agreement were solely developed by government with the community helplessly compelled to sign so that at least their people could get employment. This sense of centralising natural resources by government is aptly captured by one STMT respondent[5] as follows:

> *Goromente wa gompieno ke ene fela a tsayang tshwetso ka tshomarelo ya ditsa tholego tsa rona, rona re fetogile balebeledi fela. Le fa re ka bona magodu a utswa diphologolo go ka nna thata gore re dire eng ka re sena thata.* [The current government has centralised natural resources management to a point where we have been turned into passive observers of our natural heritage. Even if we were to see poachers, it would be difficult to counter such illegal off-take as we are no longer empowered to do so.] (STMT male respondent, formally employed as a tracking guide during hunting)

Failure to recognise and involve local communities as key stakeholders in the planning and implementation of natural resources conservation and utilisation serves to marginalise and disempower communities. Clause 9.3 uses the words "will be consulted to provide an input …". The question is what is "consultation"? Is it simply a matter of being told and you nod your head? Is that not mock consultation? The above shows an institutionalised disempowerment of local communities by government through crafting of such legislations (Government of Botswana, 2007). The CBNRM policy (2007, Section 9.3) states that "The decision to award the tender to a particular joint venture partner will however rest with the TAC." Initially the award decision was vested upon communities being advised by the TAC. The CBNRM policy of 2007 changed this and gave such powers to the TAC, which is a disempowerment of CBOs by government (Mbaiwa and Thakadu, 2011). Also, there could be a possibility of communities revolting and consequently increased incidents of unsustainable utilisation of natural resources as per the preceding quotation from the STMT respondent. Failure by government to engage local communities in natural resources management goes against the goals of CBNRM which sought to devolve decision-making power to local people and thus restoring a "sense of ownership" (Mbaiwa and Stronza, 2010; Twyman, 2000). It should be noted that local communities' power to influence decision making at local institutions determines local communities' satisfaction with tourism (Diener, 1984; Grzeskowiak et al., 2003). Cassidy (2021, p. 2) expertly captures the operations of community trusts as follows:

244 *Handbook on tourism and conservation*

In Botswana, CBNRM was originally systematized through a community forming a trust to engage in joint venture partnerships (JVPs) on behalf of the community. The focus was initially on those communities located in wildlife management areas. Such communities are allocated a use-right head lease over a demarcated concession area, which only gives them the rights to commercial wildlife-based tourism in that area. The lease is known as a head lease because most communities sub-lease the tourism rights in the area to a JVP tour operator. The community collects lease fees in exchange for the JVP taking over the photographic or hunting (as determined by central government) rights for that area. This model precludes any true devolution of decision-making to the community level; resource use decisions are made at central government level, and then handed down to communities to implement. The only rights communities have had are to choose whether to use their designated hunting quota or not; to have final say over which tour operator they would sub-lease to and to decide how to spend the income received from their JVP.

Over the last fifteen years of trophy hunting in Botswana, joint ventures have proven to be neocolonial and/or not focused on community or national development, as noted by Cassidy (2021). The World Economic Forum (2018) and Blackie and Sowa (2019) argued that Botswana's favoured approach of 'high-value/low-volume tourism' as contained in the Botswana Tourism Master Plan of 2000 (Department of Tourism, 2000) is to blame for the disempowerment that has come to be associated with CBNRM practice. Similarly, Magole et al. (2008) and Mbaiwa (2004) have also previously argued that the Botswana style of tourism, which is focused on the promotion of high-end luxury tourism, created 'enclave tourism', the removal of profits from Botswana, the ownership of much tourism operators by foreigners and separating large parts of the rural population from natural resources.

CBNRM ACT AND REGULATIONS

Even though the CBNRM policy (Government of Botswana, 2007, p. 12) explicitly states that "government will provide regulations and management support to communities for the implementation of CBNRM", none exist as at August 2023. More than a decade of implementing the CBNRM policy government is yet to yield an aligned and specific CBNRM Act and subsequent CBNRM regulations to operationalise the CBNRM policy. If developed, the CBNRM Act could clarify the rights and responsibilities of the CBOs as well as stakeholders. A CBNRM Act would explicitly spell out actions to be taken in the event of breach of contracts and provide details of what needs to happen within the CBNRM environment. CBNRM regulations would address issues of natural resources products to facilitate the commercialisation of the rich natural resources that the country is endowed with. The foregoing assertion is in line with the premise that tourism is seen as one of the growing sectors of the economy. Tourism is responsible for biodiversity conservation as well as increased employment and revenue opportunities for local communities, promotion of cultural exchange and rural infrastructure development (Nunkoo and Gursoy, 2012).

There is currently no effective institutional framework to support the effective implementation of CBNRM save for the lone Ngamiland District CBNRM Forum in the country which is attended by CBOs in Ngamiland region only (Mbaiwa and Thakadu, 2011). However, it emerged during the interviews with key stakeholders that the idea of creating a CBNRM Support Association of Botswana (CSABO) with a broad or national mandate of supporting and coordinating all CBNRM functions in Botswana should be explored. It is anticipated that CSABO and the long awaited CBNRM Act would serve as a panacea to most problems

Community-based natural resources management and poverty reduction 245

bedevilling successful CBNRM implementation in Botswana. This is in line with arguments by Poteete and Ribot (2011) who maintained that local democratic empowerment relies on a strong institutional framework. CSABO is meant to improve access to decision-making authority backed by its advantage of having diverse strengths and resources of its members. In August 2011, the then Minister of Environment Wildlife and Tourism promised that government will support the CBNRM coordinating organisation by allocating it a concession area in one of the wildlife rich regions. These noble ideas have, to a large extent, remained abstract, even though the CBNRM policy of 2007 (9.5) allows for the creation of such an advocacy institutional framework to "coordinate and promote activities and aspirations of its members" (Government of Botswana, 2007, p. 13). Government together with other stakeholders particularly environmental NGOs should perhaps fast-track the establishment of the CSABO. The envisaged CSABO as an umbrella and semi-autonomous organisation should be charged with coordinating, managing and monitoring performance of CBNRM implementation in the country. The CSABO could be mandated with being the driving force towards product and geographical diversification of local communities' natural resources products and services. The organisation could also serve as watchdog and promoter of a transformative sustainable development agenda (Villamayor-Tomas and García-López, 2018, p. 115) as well as being the bargaining council for local communities and stakeholders with government instead of government having to deal with diverse CBOs as observed by the following key respondent:[6]

> *Lephata la Tshomarelo Tikologo le Bojanala le kgaogantse metse ya rona gore le re fenye ka re sa tlhole re bua ka lentswe le le lengwefela, malope ke one a fiwang dithuso.* [The Ministry of Environment, Natural Resources Conservation & Tourism is currently maximising on a 'divide and conquer' strategy since all CBOs are scattered and do not speak with a single voice. This has led to favouritism of some CBOs which connive with government for preferential treatment.] (60-year-old key respondent, Sankuyo village)

While the aims in establishing the CSABO may outweigh the disadvantages of having such an entity, caution should be exercised to prevent CSABO from being an agent of disempowering the already marginalised groups as well as the potential of becoming part of the top-down 'clique' as happened with the Namibian Association of CBNRM and support services (NACSO Strategic Plan, 2016–2020).

CONCLUSION

This chapter provided a context for discussing how local communities as intended beneficiaries of CBNRM implementation find themselves excluded from accessing and utilising communally held natural resources. This exclusion has occurred despite the fact that government policies and state apparatus such as the TACs and the Botswana Tourism Organisation are supposedly meant to empower these communities. The chapter examined how the natural resources management legislations which include CBNRM Policy of 2007, Wildlife Conservation and National Parks Act of 1992: Cap 38, and the associated Joint Venture Partnerships (JVPs) have deviated from being a source of local empowerment to agents of local exclusion over time. The study found that revenue from the implementation of CBNRM and other associated benefits have not significantly trickled down to the household level where poverty is mostly experienced. The study provides policy options in the form of recommenda-

246 *Handbook on tourism and conservation*

tions that are aimed at facilitating changes to structures that directly and/or indirectly perpetuate marginalisation and/or exclusion of certain sectors of local communities.

Ultimately, CBNRM has become a mechanism for managing relations between central government and rural communities. Government has inadvertently impoverished rural communities especially those living in close proximity to WMAs through recentralisation of natural resources management. These rural communities continue to bear the brunt of coexisting with wildlife which destroy their crops and property as well as cause injury and loss of life (Blackie, 2019). Even though various legislative instruments (such as the WCNP (Hunting and Licensing) Regulations of 1992, CBNRM policy (2007), and the WCNP (Prohibition of Hunting, Capturing or Removal of Animals) Orders, 2014 to 2018) were supposedly meant to promote conservation and sustainable utilisation of natural resources, they have achieved the opposite. The unintended consequences of the planned interventions (policies, programmes, plans, projects) governing natural resources conservation and usage are also recognised in some government quarters (MENT 2019, p. 46).

NOTES

1. Budget speech for 2009 presented by Minister of Finance and Development Planning Mr Baledzi Gaolathe; Mr Kenneth Mathambo as Minister of Finance and Economic Development presented the 2018 budget speech.
2. Interview held in June 2016.
3. FGD held in July 2016 with two board members and three other village leaders.
4. In 2019, Botswana reintroduced hunting of wild animals, and some communities such as Mababe and CECT have already received their 15-year lease agreements for hunting.
5. Interview held in June 2016.
6. Interview held in June 2016

REFERENCES

Beard, V., & Phakphian, S. (2009). Community-based planning in Chiang Mai, Thailand: Social capital, collective action and elite capture. Paper presented at Dialogical Conference 'Social Capital and Civic Engagement in Asia', University of Toronto, 7–10 May.

Blackie, I. R. (2019). The impact of wildlife hunting prohibition on the rural livelihoods of local communities in Ngamiland and Chobe district areas, Botswana. *Cogent Social Sciences*, 5(1), 1–20.

Blackie, I. R., & Sowa, J. (2019). Dynamics of social ecology of elephant conservation in Botswana and implications on environmental development. *Journal of African Interdisciplinary Studies*, 3(2), 4–25.

Blaikie, P. M. (2006). Is small really beautiful? Community based natural resource management in Malawi and Botswana. *World Development*, 34, 1942–1957.

Bolaane, M. (2004). The impact of game reserve policy on the River Basarwa/Bushmen of Botswana. *Social Policy and Administration*, 38(4), 399–417.

Bourdieu, P. (1984). *Distinction: A Social Critique of the Judgement of Taste*. Cambridge, MA: Harvard University Press.

Büscher, B. (2013). *Transforming the Frontier: Peace Parks and the Politics of Neoliberal Conservation in Southern Africa*. Durham, NC: Duke University Press.

Cassidy, L. (2021). Power dynamics and new directions in the recent evolution of CBNRM in Botswana. *Conservation Science and Practice*, 3(4), 1–8.

Centre for Applied Research (2016). *2016 Review of Community-Based Natural Resource Management in Botswana*. Report prepared for Southern African Environmental Programme (SAREP).

Chavallier, R., & Harvey, R. (2016). *Ensuring Elephant Survival Through Community Benefits*. Occasional Paper 243. South African Institute of International Affairs.

Creswell, J. W. (2014). *Research Design: Qualitative, Quantitative, and Mixed Methods Approaches*, 4th edition. London: Sage Publications.

Dasgupta, A., & Beard, V. A. (2007). Community driven development, collective action and elite capture in Indonesia. *Development and Change*, 38(2), 229–249.

Department of Tourism (2000). *Botswana Tourism Master Plan Final Report*. Gaborone.

Department of Wildlife and National Parks (2009). *Long Term Strategy: 2009–2016*. Gaborone.

Diener, E. (1984). Subjective well-being. *Psychological Bulletin*, 95(3), 542–575.

Dikobe, L. (ed.) (2012). *Consolidating Community-Based Natural Resources Management through Effective Stakeholder Engagement*. Proceedings of the 6th CBNRM Bi-Annual Conference: Botswana CBNRM National Forum, Gaborone.

Fischer, A. M. (2008). *Resolving the Theoretical Ambiguities of Social Exclusion with Reference to Polarisation and Conflict*. London School of Economics and Political Science. Working Paper Series, No. 08-90.

Glaser, B. G., & Strauss, A. L. (1967). *The Discovery of Grounded Theory: Strategies for Qualitative Research*. New York. Aldine de Gruyter.

Good, K. (2002). Routinized injustice: The situation of the San in Botswana. In K. Good (ed.), *The Liberal Model and Africa*. Basingstoke: Palgrave Macmillan, pp. 23–68.

Government of Botswana (2007). *Community Based Natural Resource Management Policy*. Government Paper No. 2 of 2007. Gaborone: Government Printer.

Grzeskowiak, S., Sirgy, J. M., & Widgery, R. (2003). Residents' satisfaction with community services: Predictors and outcomes. *Journal of Regional Analysis and Policy*, 33(2), 1–36.

Kabeer, N. (1994). *Reversed Realities: Gender Hierarchies in Development Thought*. London: Verso Publications.

Le Roux, W. (1999). *Torn Apart: San Children as Change Agents in a Process of Acculturation*. Windhoek: Kuru and WIMSA.

Lenao, M. & Saarinen, J. (2016). Political Ecology of Community Based Natural Resources Management: Principles and Practices of Power Sharing in Botswana. In Nepal, S., and Saarinen, J (eds), *Political Ecology and Tourism*. London: Routledge.

Levitas, R., Pantazis, C., Fahmy, E., et al. (2007). *The Multi-Dimensional Analysis of Social Exclusion*. London: Cabinet Office.

Magole, L. I., Magole, L., & Bapedi, T. (2008). The dynamics of benefits sharing in community based natural resource management (CBNRM) among remote communities in Botswana. In A. Ahmed (ed.), *Managing Knowledge, Technology and Development in the Era of Information Revolution*. London: World Association for Sustainable Development, pp. 195–206.

Mbaiwa, J. E. (2004). The success and sustainability of community based natural resources management in Okavango Delta, Botswana. *South African Geographical Journal*, 86(1), 44–53.

Mbaiwa, J. E. (2013). *Community Based Natural Resource Management (CBNRM) in Botswana*. CBNRM Status Report of 2011–2012.

Mbaiwa, J. E., & Stronza, A. L. (2010). The effects of tourism development on rural livelihoods in the Okavango Delta, Botswana. *Journal of Sustainable Tourism*, 18(5), 635–656.

Mbaiwa, J. E., & Thakadu, O. T. (2011). Community trusts and access to natural resources in the Okavango Delta, Botswana. In D. L. Kgathi, B. N. Ngwenya, & M. B. K. Darkoh (eds), *Rural Livelihoods, Risks and Political Economy of Access to Natural Resources in the Okavango Delta, Botswana*. Hauppauge, NY: Nova Science Publishers, pp. 275–304.

Ministry of Environment, Natural Resources Conservation and Tourism (MENT) (2019). MENT His Excellency the President's Briefing Report, February. Unpublished Performance Briefing Report.

Molosi, K. (2015). The world of development as experienced and perceived by the San through the RADP: The case of Khwee and Sehunong settlements. PhD thesis.

Nunkoo, R., & Gursoy, D. (2012). Residents' support for tourism: An identity perspective. *Annals of Tourism Research*, 39(1), 243–268.

Poteete, A. R., & Ribot, J. C. (2011). Repertoires of domination: Decentralization as process in Botswana and Senegal. *World Development*, 39(3), 439–449.

Power, A., & Wilson, W. J. (2000). *Social Exclusion and the Future of Cities*. CASE paper 35. Centre for Analysis of Social Exclusion, London School of Economics and Political Science.

Pradhan, M. S. (2011). Social exclusion and social change: Access to and influence of community-based collective action programs in Nepal. PhD dissertation, University of Michigan.

Republic of Botswana (1972). *Rural Development in Botswana*. Gaborone: Government Printer.

Republic of Botswana (2009). *Budget Speech 2009*. Gaborone: Government Printer. www.bankofbotswana.bw/sites/default/files/publications/budget_speech_2009.pdf.

Republic of Botswana (2018). *Budget Speech 2018*. Gaborone: Government Printer. www.botswanalmo.org.bw/sites/default/files/2018%20Budget%20Speech.pdf.

Rozemeijer, N., & van der Jagt, C. J. (2000). *Community Based Resources Management in Botswana: How Community-Based is Community Based Natural Resources Management in Botswana?* Occasional Papers, IUCN/SNV CBNRM Support Programme, Gaborone.

Russell, M. (1976). Slaves or workers? Relations between Bushman, Tswana and Bores in the Kalahari. *Journal of Southern African Studies*, 2(2), 179–197.

Steiner, A., & Rihoy, E. (1995). Comparative regional CBNRM. In *The Commons Without Tragedy? Strategies for Community Based Natural Resource Management Programme in Southern Africa*. Proceedings of the Natural Resources Management Programme Annual Conference: SADC Wildlife Technical Coordinating Unit, Kasane, pp. 1–36.

Stone, M. T., & Nyaupane, G. P. (2015). Protected areas, tourism and community livelihoods linkages: A comprehensive analysis approach. *Journal of Sustainable Tourism*, 24(5), 673–693.

Taylor, M. (2000). Life, land and power: Contesting development in northern Botswana. PhD dissertation, University of Edinburgh.

Taylor, M. (2002). As good as it gets? Botswana's 'Democratic development' from a critical perspective. Paper presented at the International Conference on Re-conceptualising Democracy and Liberation in Southern Africa, Windhoek/Namibia.

Thapelo, T. D. (2002). Markets and social exclusion: Post-colony and San deprivation in Botswana. *Pula: Botswana Journal of African Studies*, 16(2), 135–146.

Twyman, C. (2000). Participatory conservation? Community-based natural resources management in Botswana. *The Geographical Journal*, 166(4), 323–335.

Villamayor-Tomas, S., & García-López, G. (2018). Social movements as key in governing the commons: Evidence from community based resources management cases across the world. *Global Environmental Change*, 53, 114–126.

Wagner, P. (2011). Violence and justice in global modernity: Reflections on South Africa with world-sociological intent. *Social Science Information*, 50(3–4), 483–504.

World Economic Forum (2018). *Tourism Competitiveness Report*. Geneva.

18. Information communication technologies and community-based tourism organisations

Siamisang Sehuhula

INTRODUCTION

Information and communication technology (ICT) has revolutionised the world and has resulted in profound impacts on socio-cultural, economic and political lifestyles (Aramendia-Muneta & Ollo-Lopez, 2013; Gretzel et al., 2015; Safari & Spencer, 2016). ICTs empower consumers to identify, customise and purchase tourism products by providing tools for developing, managing and distributing offerings worldwide (Bethapudi, 2013; Guemide et al., 2019; Safari & Spencer, 2016). Increasingly, ICT plays a critical role for the competitiveness of tourism organisations and destinations as it enables "direct promotion and commercialisation of local tourism offerings in international markets, reducing dependence on big foreign intermediaries" (Petti & Passiante, 2009, p. 46). Globally, countries are competing to attract tourists through all means including ICT and this has become the focus for marketing tourism strategies (Buhalis & Amaranggana, 2015). To attract prospective tourists in this digitised world, ICTs have therefore become an essential partner as they increasingly determine the interface between consumers and suppliers (Wagaw & Mulugeta, 2018). The application of ICT in tourism also has the potential to boost rural community-based tourism organisations (CBTOs) by defining their global virtual presence thus turning their markets from local into global positions (Aramendia-Muneta & Ollo-Lopez, 2013; Spencer et al., 2012).

The recent advancements in ICT have also provided a means to develop CBTOs in Africa; however, the level of ICT in the continent remains low (Adeola & Evans, 2020; Karanasios, 2008; OECD, 2020; Shemi & Proctor, 2013). For instance, in 2019, the number of individuals using the internet in Africa was 299 million people, compared to 568 million in Europe and 1,901 million people in the Asia and Pacific region for the same period. Similarly, international bandwidth for Africa in 2020 stood at 11 Tbit/s**, compared to 153 Tbit/s** in Europe and 301 Tbit/s** for the Asia and Pacific region (ITU World Telecommunication/ICT Indicators Database, 2020). These challenges are making it difficult for enhanced ICT integration in African tourism (Lama et al., 2018; Sechele-Mosimanegape & Prinsloo, 2019).

In Botswana, the government remains a key player in the ICT market. On the ICT regulatory, institutional and legal front, the government has established the Botswana Communications Regulatory Authority (BOCRA) to regulate the communications sector, including the ICT sector. Botswana's government has also put in place the *Maitlamo* ICT policy of 2006, to draw up the ICT implementation road map for the country (Nkwe, 2012). Despite this, the adoption and use of ICT in CBTOs in Botswana has not yet been established. Research on ICT in Botswana is relatively in its infancy. The few studies which have been done on ICT in Botswana are in the fields of economic diversification (Bakwena & Kahaka, 2013), health (Ndlovu et al., 2014), labour issues (Kalusopa & Ngulube, 2012), small and medium-sized enterprises (Mutula & van Brakel, 2006) and youth (Lesitaokana, 2016). The extent of ICT

250 *Handbook on tourism and conservation*

use in tourism by CBTOs in Botswana remains uncertain and largely under researched. This chapter investigates the process through which ICTs empower tourism in Botswana through a case study of the Chobe Enclave Community Trust (CECT). It also explores the barriers preventing a full-fledged adoption of ICTs by CECT.

LITERATURE REVIEW

Community-based Tourism

The concept of community-based tourism (CBT) first appeared in the work of Murphy (1985) and it continues to be recognised as a viable rural development strategy which places an emphasis on active community participation in all tourism activities, including receiving a significant proportion of the benefits generated by such activities (Gan et al., 2016; Trejos & Chiang, 2009; Trejos et al., 2008). CBT has the potential to create jobs, develop skills and experiences and generate entrepreneurial opportunities mainly for rural communities (Harris & Vogel, 2014; López-Guzmán et al., 2010). It concentrates on community participation in all processes from idea formulation to planning, implementation, management, monitoring, evaluation and benefit sharing (Schott & Nhem, 2018).

Notwithstanding the benefits associated with CBT, literature also reveals the inherent challenges in the implementation and management of CBT projects (Schott & Nhem, 2018). For example, key processes in most CBT projects have been criticised for often being beyond community control (Tosun, 2000). Another noticeable criticism is that CBT products have been observed to have a very limited connection with the private sector or other mainstream tourism products, thus limiting CBT's ambition to relieve local poverty through tourism (Mitchell & Muckosy, 2008).

ICT in Tourism

There is now a substantial literature on the impact of ICT in tourism (Buhalis & Amaranggana, 2015; Karanasios, 2008; Lama et al., 2018). The application of ICT technologies in tourism dates back the early adoption of the Computer Reservation System (CRS) in airlines in the 1950s and to the transformation to Global Distribution Systems (GDSs) in the 1980s (Buhalis & Song, 2003). However, it is the proliferation of the internet that brought the revolutionary changes and accelerated the development and structure of the industry in a variety of ways including improved management/governance; enhanced tourist experience; and improved competitiveness of tourism firms and destinations (Buhalis & Song, 2003; Gretzel et al., 2015; OECD, 2020; Song, 2012). More than any other medium, ICTs empower consumers with easier access to travel products and the option of many new destinations (Adeola & Evans, 2020; Bethapudi, 2013; Gretzel et al., 2015; UNWTO, 2020).

Whilst a large body of literature has focused on the use and adoption of ICT in tourism, there are few studies to date that investigate the use of the technologies in CBTOs (Gan et al., 2016). Despite this limitation, the application of ICT initiatives in CBTOs has the potential to yield unprecedented opportunities such as (i) realising access to new markets, (ii) upgrading their position in global tourism value chains and integrating into digital ecosystems, and (iii) enabling the marketing and selling of tourism products on the internet (Gan et al., 2016;

Information communication technologies and community-based tourism organisations 251

Harris & Vogel, 2014; Mutula & van Brakel, 2006; OECD, 2020). Furthermore, as the world economy continues to move toward increased integration because of advances in ICT technology, and the increasing reduction in trade barriers, some of the greatest opportunities for CBTOs will derive from their ability to participate in the regional and international markets (Mutula & van Brakel, 2006).

CBTOs in Botswana

Compared to the developed countries, organised tourism in Botswana is a relatively new development, dating back to only around two decades (Lenao, 2015; Mbaiwa, 2017). In order to involve local communities in the tourism sector, Botswana officially adopted the Community-Based Natural Resource Management (CBNRM) programme in 1989 and started its implementation in 2007 (Government of Botswana, 2007). CBNRM is based on the premise that local communities are the custodians of the natural resources found within their immediate environs and, therefore, the management and use of such should be restored to them to ensure the conservation of natural resources while at the same time deriving economic benefits from them (Chevallier & Harvey, 2016; Gupta, 2014; Lenao, 2015; Mbaiwa, 2017). Botswana's CBNRM focuses on three domains: conservation, rural development, and democracy and/or good governance (Zuze, 2006). When focused on conservation, it is concerned with the wise and sustainable use of the resources. As a rural development strategy, CBNRM promotes income generation or improved livelihoods; and when focused on democracy and/ or good governance, CBNRM involves the devolution of authority from central government to communities (Zuze, 2006). CBNRM thus emphasises achieving benefits for the local communities derived from the management of resources in their localities, whilst at the same time developing a renewed sense of ownership and commitment towards conservation of such resources (Lenao, 2017; Mbaiwa & Stronza, 2011).

Under the CBNRM programme, villages and communities can form community trusts or CBTOs and apply for land to operate a tourism-based project (DWNP, 1999). CBTOs are registered legal entities created by communities with the aim of utilising natural resources in their local environment to generate jobs and revenue for the benefit of the community (Mbaiwa, 2017). In 2016, there were 147 CBNRM CBTOs, 94 of which were registered, 53 active, 16 not registered and 16 did not indicate their statuses (Centre for Applied Research (CAR), 2016).

In Botswana, CBTOs in Controlled Hunting Areas (CHAs) enter into Joint Venture Partnerships (JVPs) with private tourism operators through subcontracting rights and leases (DWNP, 1999). Such CBTOs are "allocated a use-right head lease over a demarcated concession area known as Controlled Hunting Area (CHA), which only gives them the rights to commercial wildlife-based tourism in that area. The lease is known as a head lease because most communities sub-lease the tourism rights in the area to a JVP tour operator" (Cassidy, 2021, p. 2). In return, the communities receive lease fees in exchange for the JVP taking over the photographic or hunting rights for that area. These JVPs not only pay the CBTOs for leasing the land but also generate local employment opportunities (Chevallier & Harvey, 2016; Zielinski et al., 2020). Some CBTOs such as Sankoyo Tshwaragano Management Trust, Mababe Zokaotsham Development Trust, Khwai Development Trust, Okavango Community Trust, Okavango Kopano Mokoro Community Trust and many others also operate tourism

252 *Handbook on tourism and conservation*

businesses such as campsites, boat (mokoro) safaris and cultural tourism activities managed by their board of trustees.

Although many studies have highlighted the importance of CBNRM, the practical challenges that thwart its successful implementation have seldom been articulated (Mbaiwa, 2017; Stone & Stone, 2020). For instance, some CBNRM projects failed largely because of a lack of financial viability as they failed to establish connections with the mainstream industry resulting in their businesses lacking in market and commercial orientation (Gupta, 2014; Mitchell & Muckosy, 2008; Stone & Stone, 2020). CBNRM projects in Botswana have also failed largely because of CBTOs transferring rights in their concession areas to private tourism companies who in return pay annual land rentals curtailing any prospect of communities being at the forefront of tourism business (Cassidy, 2021; Mbaiwa, 2015). Cassidy (2021) also argues that the CBNRM programme in Botswana reinforces state dependency as communities have limited rights in determining access and use of the allocated land concession rights.

Chobe Enclave Community Trust (CECT)

The first CBNRM project in Botswana began in 1993 with the registration of CECT (Chevallier & Harvey, 2016; Stone, 2015). CECT provides the Chobe Enclave communities an opportunity to participate in CBT projects in demarcated concession areas known as CHAs (Mbaiwa & Tshamekang, 2012). Over the years, the CECT has carried out tourism activities such as safari hunting tourism and photographic tourism in the two concession areas or CHAs in the Chobe Enclave (Mbaiwa & Tshamekang, 2012). CECT as a CBO is run by a board of trustees elected from each participating village (Stone, 2015).

Due to lack of human and financial capital, CECT has opted for JVP with private safari companies in the management of CH1, CH2 and Ngoma Lodge. In addition, the community also gets a portion of game meat from each kill by professional hunters, an arrangement meant to mitigate poaching. JVPs provide the community with revenue, employment and game meat. The CECT has been managing the annual wildlife quotas received from the DWNP since 1994, and it also has a 50-year head lease from the Land Board to conduct wildlife and tourism activities (Chevallier & Harvey, 2016). At the end of each financial year, the income generated through the various CBNRM mechanisms is distributed amongst the five villages in the enclave and each holds a forum to propose projects for funding (Chevallier & Harvey, 2016).

THEORETICAL FRAMEWORK

The Social Ecological Systems Framework (SESF) delineates tourism business as an integrated system with various levels and scales using different spatial and functional boundaries (Glaser et al., 2012). It can also be understood as a systemic and holistic approach, to deconstruct the structures, processes and outcomes of a complex tourism system (Ostrom, 2007; McGinnis & Ostrom, 2014). Tourism is composed of subsystems with different social and ecological interactions at different scales (Chia-Chi & Huei-Min, 2016). According to Glaser et al. (2012), a SESF comprises a complex, adaptive system consisting of a bio-geophysical unit and its associated social actors and institutions (see Figure 18.1). Within a SESF, a complex multitude of subsystems and internal variables, both social and ecological, continuously interact and transform the SESF as a whole (Ostrom, 2009). Analysing the characteristics of

Information communication technologies and community-based tourism organisations 253

its components and their interactions can help diagnose outcomes (Cole & Browne, 2015). It is primarily for this reason that the SESF framework is used for the analysis of ICT integration in the tourism business by the CECT.

The Chobe Enclave is a social ecological system with a defined geographical boundary, resource units, resource systems, actors and governance systems (see Figure 18.1). The focus of the SESF in this study is the integration and use of ICT by CECT in tourism businesses in the Chobe Enclave (see Figure 18.2). The integration of ICT in tourism by CECT reflects a continuous interaction in a complex SESF with a multitude of subsystems and internal variables, both socio-technical and ecological (Ostrom, 2009). The adoption process also adds a further dimension to SESF dynamics as ICT adoption in tourism represents both a significant outcome and driver of the socio-ecological changes associated with global interactions (Cole & Browne, 2015).

Tourism destinations such as the Chobe Enclave operate as complex systems, consisting of multiple interacting components that are non-linear, cross-scale, evolving and interdependent (Ostrom, 2009). The application of the SESF framework to this study is appropriate as it enables the simultaneous study of social, ecological and technical systems to facilitate a more detailed analysis of ICT adoption by the CECT (Ostrom & Cox, 2012). Ostrom and Cox (2012) point out that dynamics of structured SESFs are different and may be influenced by interventions differently. That is to say that change is inherent in all components of a SESF, including the social components which can be as unpredictable as their biophysical counterparts (Chia-Chi & Huei-Min, 2016). For instance, tourism is very seasonal, and arrivals can count in the millions. Furthermore, tourism has a diverse heterogeneity of users. New actors, such as tourism developers, legislators and tourists who become active users, will have widely disparate perspectives on ICT application in the industry. These ICT users can also have much greater social and economic power than most of their local counterparts, in this case CBTOs in rural peripheral areas in developing countries. The development of the tourism industry is also usually accompanied by a myriad of other transformations, including changes in traditional land use and tenure, climate change, ICT infrastructure and connectivity, and an escalation in human–wildlife interactions/conflicts. As it is difficult to deconstruct and isolate the tourism components within specific SESFs, the framework is applicable as it establishes the interdependence of relationships through their interaction with the resource, resource users and governance systems in the destination areas (Ostrom, 2009). This section shows how the SESF framework can be applied to understand how combinations of variables affect the actions of CBTOs under diverse ICT governance systems and infrastructure (Cole & Browne, 2015).

While the framework provides a useful perspective on local nature–human interactions, it has its limitations. The framework is designed to assess the diverging outcomes caused by interactions between four first-level core subsystems: resource systems, resource units, governance systems, and resource users (Ostrom, 2009). The framework ignores the global socio-economic and political interconnections with other places, and these are not often the subject of in-depth analysis (Partelow, 2018, p. 5). ICT adoption in tourism by the CBTOs builds upon the integration of SESFs into global production processes as it links the host communities with the global market. Therefore, we argue that a perspective is needed that highlights the integration of SESFs into global production circuits.

The focal SESF of the CECT (Figure 18.1) can be understood as a cluster of subsystems on functional and spatial scales of the four first-level core subsystems: resource systems, resource units, governance systems, and resource users (Ostrom, 2009). On the functional scale, which

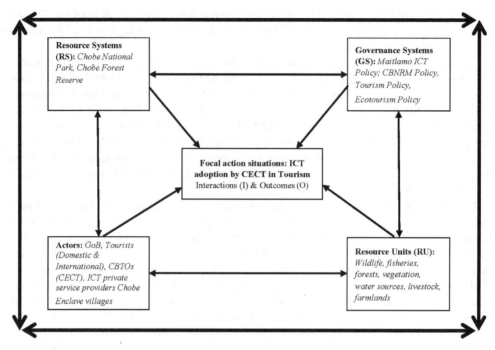

Source: Adapted from McGinnis & Ostrom (2014).

Figure 18.1 Social ecological system framework

refers to beneficial products of the ecosystem services, community members in the Chobe Enclave are actors with different levels of dependency on the resource units. For instance, pastoral farmers and fishermen are different actors on the functional scale and form two subsystems in the focal SESF because of their different interactions with the ecosystem (Figure 18.1). On the spatial scale, which is delimited by their spatial boundaries, each village within the focal SESF is a subsystem that may be affected differently by the adoption of ICT in tourism by CECT (Figure 18.1). The members of the five villages in the Chobe Enclave are actors of the subsystems on the spatial scale and their interactions with the focal action situations vary between and within themselves. There are no broad-brush solutions that can be applied to ICT adoption by CBTOs in the Chobe Enclave and it is therefore necessary to consider the unique conditions that apply to each (Gargallo & Kalvelage, 2021; Ostrom, 2007, 2009). For the purpose of this study, the resource system has been described along the second-tier variables to provide background, while interactions between resource units, governance systems and resource users are at the centre of the analysis. Assessing these interactions allows us to deepen the understanding of the ICT adoption by the CECT in tourism.

DESCRIPTION OF THE STUDY AREA

The Chobe Enclave (see Figure 18.2) lies in north of the country, forming the international boundary with Namibia to the north-west, Zambia to the north and Zimbabwe to the east (CDDP, 2003). The Enclave is surrounded by protected areas on three sides: (1) the Chobe National Park is on the western and eastern side; (2) the Chobe Forest Reserve on the southern boundary; and (3) the northern boundary is the Chobe River, which forms the border with Namibia (CDDP, 2003; Garekae & Shackleton, 2020). In 2011, the human population for the inhabitants of the Chobe Enclave stood at 4,128 people inhabiting the five main villages of Kachikau (1,356), Kavimba (549), Mabele (773), Parakarungu (845) and Satau (605) (Statistics Botswana, 2011). In terms of population distribution, the area is generally sparsely populated with an average density of 1.1 persons per square kilometre (Statistics Botswana, 2011).

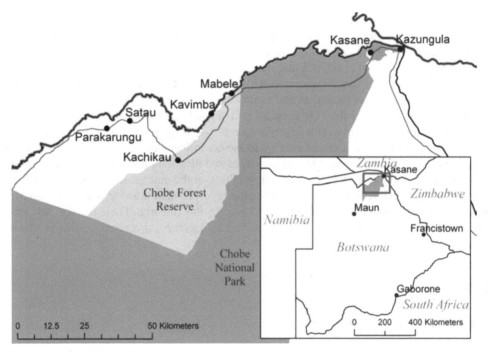

Source: Garekae et al., 2017.

Figure 18.2 Map showing location of Chobe Enclave and Kasane

METHODOLOGY

This chapter is based on a systematic literature review of secondary data sources comprising published articles, journals, books and reports. Google Scholar search was used to search for research articles dealing with "ICT use in tourism and CECT". The articles search was conducted between the months of December 2020 and February 2021 and focused on CECT since

256   *Handbook on tourism and conservation*

its establishment in 1993. This process produced 919 online articles. The search was further refined to focus on the "use of ICT by CECT tourism" and the search produced 43 articles. For easy data management, all articles that reflected some form of ICT use by the CECT were downloaded and saved as PDF copies in one folder. Abstracts were read and analysed to determine the relevance of articles to the study purpose. An additional 30 articles were excluded as they did not focus on "ICT use by the CECT in tourism". The final sample for the review analysed three selected papers that met the pre-specified criteria for inclusion in the review. A search for relevant government policies produced the following policy documents, namely: Botswana Tourism Policy of 2021, Tourism Act of 1992, Wildlife Conservation and National Parks Act of 1992, Botswana Tourism Master Plan of 2000, Ecotourism Strategy of 2002 and Community-Based Natural Resource Management Policy of 2007.

Data Analysis

Data management and analysis used in this chapter entailed the adoption of Miles and Huberman's (1994) model for the thematic analysis process. The analysis involved data reduction into themes and patterns that form a compressed picture on ICT adoption by CECT in tourism (Braun & Clarke, 2006). Data analysis was based on three stages: (a) data reduction, (b) data display and (c) data conclusion/verifying (Miles & Huberman, 1994). In the initial stage, the selected articles were read and evaluated in view of the factors that influence ICT use by the CECT in tourism. This exercise necessitated the process of code development. The process of coding involved the identification of words, phrases, sentences or paragraphs that conveyed a particular message relating to factors influencing ICT use in tourism by the CECT. In this way, processed data was displayed and classified according to its similarities and differences (Miles & Huberman 1994), resulting in the emergence of themes that have explanatory value (Ibrahim, 2012). The last step involved data drawing and conclusions to generate themes from the articles that were adopted by this research. These include: low level of internet penetration; lack of ICT and information management strategy and policy; and state dependency of the CECT.

RESULTS

Low Level of Internet Penetration

Relative to other countries in the region, Botswana boasts a very advanced legal and legislative system that is generally conducive to the proliferation of ICT and ICT-related industries (Chube, 2015). Botswana's highly developed legal system is an asset that can be used to assist with the orderly transition to an information-based economy. According to the SESF, this represents the existence of governance systems in a complex, multivariate and dynamic tourism industry (Ostrom, 2007). Despite this, Botswana government's commitment to promoting ICT as a priority is still at a very early stage and in places where it is accessible, cost is a barrier (Maitlamo, 2004). The country also suffers from considerable disparity between rural and urban access to information and services – a 'domestic information divide'. Most urban centres are relatively well supplied with radio, television, telephone and internet access. The picture is significantly different in remote and rural areas, where even access to basic

Information communication technologies and community-based tourism organisations 257

information such as radio, telephones and newspapers is problematic (CDDP, 2003). This is thought to be primarily due to prohibitive cost and limited access. In the Chobe Enclave, ICT service provision is provided by key stakeholders such as: Department of Information and Broadcasting, Botswana Post, Botswana Telecommunication Corporation (BTC), Botswana National Library Services, and private sector agencies such as Mascom, Orange, Be Mobile, private radio stations, internet service providers and private print media (CDDP, 2003). The coverage of BTC facilities is limited to Kasane, Kazungula and Pandamatenga leaving the CECT villages without coverage. However, through the CBNRM policy, CECT managed to connect V-sat public pay phones in villages of Chobe west. These phones are placed at the *kgotla* and are available for use by the public (CDDP, 2003).

Lack of ICT and Information Management Strategy and Policy

The findings show that CECT has neither an ICT strategy nor an information management policy (CDDP, 2003). This demonstrates the lack of ICT adoption by CECT in the management of tourism projects in their area. As an actor in the tourism industry, CECT lacks ICT based entrepreneurial and marketing skills and has consequently failed to market themselves to potential clients (CDDP, 2003; Gupta, 2014). The problem is partly attributed to the current CBNRM setup, where the enclave communities have formed a CECT and then engaged JVPs to manage their allocated concession areas of CH/1 and CH/2. In return, CECT have indirectly benefited from rental income, employment opportunities and game meat from trophy hunts (Cassidy, 2021; Mbaiwa, 2015). This demonstrates that CBTOs such as CECT lack the necessary entrepreneurial skills and experience in managing tourism enterprises (Mbaiwa, 2015). This alienates them from any meaningful participation in the management and marketing of the tourist destinations (Cassidy, 2021). For example, Ngoma Safari Lodge which is part of the CECT tourism projects is advertised through the various websites of JVPs and private entities such as Siyabona Africa and Chalo Africa Tours, and the findings do not show any instance where CECT directly advertised and marketed the lodge.

State Dependency of CECT

The CBNRM model emphasises the importance of devolved ownership over resources as a precondition for the local entrepreneurial development of products and markets for wild resources (Gupta, 2014). It is therefore important to understand the extent to which CECT have devolved authority and secure tenure over resources such as wildlife through the existing policy and legislation. Unfortunately, Botswana does not have a single policy or law that gives local communities secure rights over wildlife and tourism as resources, or rights over the land on which these resources are found. The current governance and legislative structures restrict the autonomy of the CECT in that decisions on the nature of their business model is imposed on them by higher-level authorities (Cassidy, 2021). The only rights communities are: (a) to choose whether to use their designated hunting quota or not, (b) have final say over which tour operator they would sub-lease to and (c), decide how to spend the income received from their JVP (Cassidy, 2021). There is a widespread sentiment among villagers that they are deliberately excluded from participating in the tourism industry and that the tourism sector is still monopolised by foreign-owned safari tourism (CDDP, 2003; Gupta, 2014). Furthermore, the tourism companies that operate on land leased to them through the CECT have not facilitated

258   *Handbook on tourism and conservation*

the development of spin-off small enterprises supporting their larger operations, as CBNRM advocates predicted would happen. Wildlife tourism therefore remains primarily the domain of outsiders with business expertise (Gupta, 2014, p. 125). The management of the tourism projects in the Chobe Enclave are carried out by the safari company to which CECT awarded the tender (Jones, 2002). While it might be argued that the company is carrying out the management on behalf of CECT, there is little real involvement or monitoring by CECT of the management activities carried out by the company (Jones, 2002). These institutional shortfalls represent a significant driver towards inequity in this SES.

DISCUSSION

The tremendous global growth of ICT, especially the internet, has fundamentally reshaped the way tourism related information is distributed and the way people search for, and consume, travel (Beldona, 2005; Xiang et al., 2008). Despite this global growth, ICT adoption in Botswana has been characterised by wide disparities as many citizens do not have access to ICT due to factors such as lack of appropriate products, high cost of computers, the lack of electricity in many rural locations, high charges for internet usage, lack of technical skills and lack of a robust regulatory framework for ICT (Statistics Botswana, 2011). The findings reveal that local support and capacity building prior to tourism development are essential for involving local entities such as the CECT. That is, the application of the SES framework has revealed that CECT, as an actor in a complex tourism system, lacks the ICT knowledge needed in tourism, nor does it have the tourism business entrepreneurship skills. The findings also reveal that CECT is a complex organisation as it represents the heterogeneous interests of the community, and their participation in ICT based tourism must be tailored to match their diverse needs. The results also show that despite investments in the governance systems such as policy and ICT infrastructure development, there is limited information on how small and medium enterprises such as CECT stand to benefit from such investment.

In the competitive tourism environment, hospitality enterprises strive to exploit various online distribution channels to increase their visibility and to support online purchasing, and this has not been the case with CECT. Despite the usefulness and growth of ICT based advertising of destination areas, there is still little understanding of how CECT has adopted the use of online platforms to market and advertise tourism projects in the Chobe Enclave. At the heart of the present study is the question on how CECT has used ICT platforms and services to *boost* tourism in the Chobe Enclave. Our results contrast the theoretical relevance of ICT application in CBT in the context of CECT which is to effectively market the tourism products and manage their business transactions in greater volumes (Angeloni & Rossi, 2020; Kiprutto et al., 2011).

The findings reveal that CECT is yet to integrate ICT adoption in their business. The current CBNRM setup has stifled any business potential for the CECT in that the dominance of JVPs has resulted in their alienation in the advertising, marketing and management of tourism businesses. The JVPs operate websites and "the value which they capture is a cause for concern as this replicates previous patterns of economic extraversion by offshoring or extracting social surplus, thereby potentially reducing domestic investment" (Anwar et al., 2014, p. 543). Citizen participation in tourism through CBTOs is necessary to reduce the repatriation of revenue from Botswana.

Information communication technologies and community-based tourism organisations 259

The findings of this study also identify important policy implications. ICT adoption should be seen as fundamental to successful tourism development in the Chobe Enclave, and all stakeholders need to pay close attention to ICT trends to ensure that the potential gains are fully maximised (Adeola & Evans, 2020). Stakeholders should be viewed as actors of the sub-systems on the spatial scale and their interactions with the focal action situation vary (Ostrom, 2007, 2009). As a result, it is therefore necessary to consider the unique conditions that apply to each about ICT adoption in tourism in the enclave. For instance, the tourism industry in Botswana needs a framework that allows for preferred treatment of domestic investors which will result in the economic development of the local communities thereby enhancing social sustainability (Gupta, 2014). The exclusion of CECT from the direct management of tourism projects in the enclave has diminished any effort of ICT adoption in their business to compete effectively in the international markets. This research demonstrates the utility of the SES framework to study coupled human–environment systems in tourist destinations.

CONCLUSION

As demonstrated in this study, the socio-ecological conceptual model can be practical for analysing the challenges faced by CBTOs such as CECT in the integration of ICT in their businesses. The case study analysis highlights the centrality of ICT in tourism for CBTOs in enhancing their marketing and distribution reach, with the potential of enabling them to attract large numbers of tourists. With the relevant ICT application in tourism, destination areas in rural areas need to achieve global presence to fully leverage the potential of the internet (Gan at al., 2016). Despite these opportunities, the findings also show that CECT are lagging in ICT adoption and this has alienated them from reaping the benefits associated with the innovations in the tourism industry. CBTOs that do not invest in their digitalisation will not survive, let alone thrive in the future. Destinations, businesses and the wider tourism sector need to fully embrace these new technologies to remain competitive, and to take advantage of the innovation, productivity and value creation potential. Policy makers have an important role to play to help tourism businesses of all sizes, including the more traditional and smallest firms, to engage with the digital revolution, and thrive in response to these paradigm-shifting technologies (OECD, 2020).

The study also suggests that sustainable governance of tourism resources warrants a holistic and multivariate approach that requires collaboration among public, private and community institutions acting across different levels and scales to provide a more inclusive, equitable and sustainable solution to the problem of ICT adoption by CBTOs (Gargallo & Kalvelage, 2021). Therefore, the SES framework can be used as a common framework to guide future studies as it does not assume an over-simplistic approach to ICT use in tourism. Rather, it takes a multifaceted non-linear approach that deals with complex interrelationships between governance systems, resource units, resource systems and the actors to analyse the integration of ICT by CBTOS.

260 *Handbook on tourism and conservation*

REFERENCES

Adeola O., & Evans, O. (2020). ICT, infrastructure, and tourism development in Africa. *Tourism Economics*, *26*(1), 97–114.

Angeloni, S., & Rossi, M.C. (2020). Online search engines and online travel agencies: A comparative approach. *Journal of Hospitality & Tourism Research*, *45*(4), 720–749.

Anwar, M. A., Carmody, P., Surborg, B., & Corcoran, A. (2014). The diffusion and impacts of information and communication technology on tourism in the Western Cape, South Africa. *Urban Forum*, *25*(4), 531–545.

Aramendia-Muneta, M. E., & Ollo-Lopez (2013). ICT impact on tourism industry. *International Journal of Management Cases*, *15*(2), 87–98.

Bakwena, M., & Kahaka, Z. (2013). The Botswana national information and communication technology policy and economic diversification: How have we fared thus far? *Botswana Notes and Records*, *45*, 206–213.

Beldona, S. (2005). Cohort analysis of online travel information search behavior: 1995–2000. *Journal of Travel Research*, *44*(2), 135–142.

Bethapudi, A. (2013). The role of ICT in tourism industry. *Journal of Applied Economics and Business*, *1*(4), 67–79.

Braun, V., & Clarke, V. (2006). Using thematic analysis in psychology. *Qualitative Research in Psychology*, *3*(2), 77–101.

Buhalis, D., & Amaranggana, A. (2015). Smart tourism destinations enhancing tourism experience through personalisation of services. In I. Tussyadiah & A. Inversini (eds), *Information and Communication Technologies in Tourism*. Cham: Springer, pp. 377–389.

Buhalis, D., & Song, H. (2003). ICTs and internet adoption in China's tourism. *International Journal of Information Management*, *23*(6), 451–467.

Cassidy, L. (2021). Power dynamics and new directions in the recent evolution of CBNRM in Botswana. *Conservation Science and Practice*, *3*(4), 1–8.

CDDP (2003). *Chobe District Development Plan 6: 2003–2009*. Gaborone: Botswana Government Printers.

Centre for Applied Research (2016). *2016 Review of Community-Based Natural Resource Management in Botswana*. Report prepared for Southern African Environmental Programme (SAREP).

Chevallier, R., & Harvey, R. (2016). Is community-based natural resource management in Botswana viable? South African Institute of International Affairs. *Policy Insights*, *31*, 1–12.

Chia-Chi, W., & Huei-Min, T. (2016). Capacity building for tourism development in a nested social ecological system: A case study of the South Penghu Archipelago Marine National Park, Taiwan. *Ocean & Coastal Management*, *123*, 66–73.

Chube, G. (2015). Effects of ICT Adoption by SMMEs Owners on Production in Borolong Area of Botswana. Master's dissertation, North-West University, Mafikeng, South Africa.

Cole, S., & Browne, M. (2015). Tourism and water inequity in Bali: A social-ecological systems analysis. *Human Ecology*, *43*(3), 439–450.

Department of Wildlife and National Parks (1999). *Joint Venture Guidelines*. Gaborone.

Gan, S., Inversini, A., & Rega, I. (2016). Community-based tourism and ICT: Insights from Malaysia. *Information and Communication Technologies in Tourism*. http://ertr.tamu.edu/enter-2016 - volume-7-research-notes/, accessed 2 July 2021.

Garekae, H., and Shackleton, C. M. (2020). Urban foraging of wild plants in two medium-sized South African towns: People, perceptions and practices. *Urban Forestry & Urban Greening*, *49*, 126581.

Garekae, H., Thakadu, O. T., & Lepetu, J. (2017). Socio-economic factors influencing household forest dependency in Chobe Enclave, Botswana. *Ecological Processes*, *6*(1), 40.

Gargallo, E., & Kalvelage, L. (2021). Integrating social-ecological systems and global production networks: Local effects of trophy hunting in Namibian conservancies. *Development Southern Africa*, *38*(1), 87–103.

Glaser, M., Christie, P., Diele, K., Dsikowitzky, L., Ferse, S., Nordhaus, I., Schlüter, A., Schwerdtner Mañez, K., & Wild, C. (2012). Measuring and understanding sustainability-enhancing processes in tropical coastal and marine social ecological systems. *Current Opinion in Environmental Sustainability*, *4*, 300e308.

Government of Botswana (2007). *Community Based Natural Resources Management Policy*. Government Paper no. 2. Gaborone, Ministry of Environment, Wildlife and Tourism/Government Printers.

Gretzel, U., Koo, C., Sigala, M., & Xiang, Z. (2015). Special issue on smart tourism: Convergence of information technologies, experiences, and theories. *Electronic Markets*, *25*, 175–177.

Guemide, B., Benachaiba, C., and Maouche, S. (2019). Integrating ICT-based applications for sustainable tourism development in Algeria. *Journal of Tourism Hospitality*, *8*(415), 1–11.

Gupta, C. (2014). Wildlife paying its way? A critical analysis of community-based natural resource management in the Chobe Enclave, Botswana. In M. Sowman & R. Wynberg (eds), *Governance for Justice and Environmental Sustainability: Lessons across Natural Resource Sectors in Sub-Saharan Africa*. New York: Routledge.

Harris, R., & Vogel, D. (2014). E-commerce for community-based tourism in developing countries. www.researchgate.net/publication/228843439, accessed 2 July 2021.

Ibrahim, A. M. (2012). Thematic analysis: A critical review of its process and evaluation. *West East Journal of Social Sciences*, *1*(1), 39–47.

International Telecommunication Union (ITU) (2020). *World Telecommunication/ICT Indicators Database*. www.itu.int/en/ITU-D/Statistics/Pages/publications/wtid.aspx, accessed 20 February 2022.

Jones, B. T. B. (2002). *Chobe Enclave, Botswana: Lessons Learnt from a CBNRM Project 1993–2002*. Gaborone: IUCN/SNV CBNRM Support Programme.

Kalusopa, T., & Ngulube, P. (2012). Developing an e-records readiness framework for labour organizations in Botswana. *Information Development*, *28*(3), 199–215.

Karanasios, S. S. (2008). An e-commerce framework for small tourism enterprises in developing countries. DPhil thesis, Victoria University, Melbourne, Australia.

Kiprutto, N., Kigio, F. W., & Riungu, G. K. (2011). Evidence on the adoption of eco-tourism technologies in Nairobi. *Global Journal of Business Research*, *5*(3), 56–66.

Lama, S., Pradhan, S., Shrestha, A., & Beirman D. (2018). Barriers of e-tourism adoption in developing countries: A case study of Nepal. Australasian Conference on Information Systems, Sydney, Australia.

Lenao, M. (2015). Challenges facing community-based cultural tourism development at Lekhubu Island, Botswana: A comparative analysis. *Current Issues in Tourism*, *18*(6), 579–594.

Lenao, M. (2017). Community, state and power-relations in community-based tourism on Lekhubu Island, Botswana. *Tourism Geographies*, *19*(3), 483–501.

Lesitaokana, W. (2016). Young people and mobile phone technology in Botswana. *Journal of Media and Communication Studies*, *8*(1), 8–14.

López-Guzmán, T., Borges, O., & Cerezo, J. M. (2010). Community-based tourism and local socio-economic development: A case study in Cape Verde. *African Journal of Business Management*, *5*(5), 1608–1617.

Maitlamo (2004). Botswana's National ICT Policy: Legislative Framework & Change Report (December). https://ictpolicyafrica.org/fr/document/khdaorfc689, accessed 12 May 2022.

Mbaiwa, J. E. (2015). Community-based natural resource management in Botswana. In R. van der Duim, M. Lamers, & J. van Wijk (eds), *Institutional Arrangements for Conservation, Development and Tourism in Eastern and Southern Africa*. Dordrecht: Springer, pp. 59–80.

Mbaiwa, J. E. (2017). Poverty or riches: Who benefits from the booming tourism industry in Botswana? *Journal of Contemporary African Studies*, *35*(1), 93–112.

Mbaiwa, J. E., & Stronza, A. L. (2011). Changes in resident attitudes towards tourism development and conservation in the Okavango Delta, Botswana. *Journal of Environmental Management*, *92*(8), 1950–1959.

Mbaiwa, J. E., & Tshamekang, T. E. (2012). Developing a viable community-based tourism project in Botswana: The case of the Chobe Enclave Conservation Trust. *World Sustainable Development Outlook*. www.worldsustainable.org.

McGinnis, M. D., & Ostrom, E. (2014). Social-ecological system framework: initial changes and continuing challenges. *Ecology and Society*, *19*(2), 1–13.

Miles, M. B., & Huberman, A. M. (1994). *Qualitative Data Analysis: An Expanded Sourcebook*, 2nd edition. Thousand Oaks, CA: Sage.

Mitchell, J., & Muckosy, P. (2008). *A Misguided Quest: Community-Based Tourism in Latin America*. London: Overseas Development Institute.

Murphy, P. E. (1985). *Tourism: A Community Approach*. London: Methuen.

262 *Handbook on tourism and conservation*

Mutula, S. M., & van Brakel, P. (2006). E-readiness of SMEs in the ICT sector in Botswana. *The Electronic Library*, *24*(3), 402–417.

Ndlovu, R., Park, E., Dikai Z., & Kovarik, C. L. (2014). Scaling up a mobile telemedicine solution in Botswana: Keys to sustainability. *Frontiers in Public Health*, *2*(275), 1–6.

Nkwe, N. (2012). E-government: Challenges and opportunities in Botswana. *International Journal of Humanities and Social Science*, *2*(17), 39–48.

Organisation for Economic Co-operation and Development (2020). *OECD Tourism Trends and Policies 2020*. Paris: OECD Publishing.

Ostrom, E. (2007). A diagnostic approach for going beyond panaceas. *Proceedings of the National Academy of Sciences*, *104*(39), 15181–15187.

Ostrom, E. (2009). A general framework for analyzing sustainability of social-ecological systems. *Science*, *325*(5939), 419–422.

Ostrom, E., and Cox, M. (2012). Moving beyond panaceas: A multitiered diagnostic approach for social-ecological analysis. *Environmental Conservation*, *37*(4), 451–463.

Partelow, S. (2018). A review of the social-ecological systems framework: Applications, methods, modifications, and challenges. *Ecology and Society*, *23*(4), 36.

Petti, C., & Passiante, G. (2009). Getting the benefits of ICTs in tourism destinations: Models, strategies and tools. *International Arab Journal of e-Technology*, *1*(1), 46–57.

Safari, E., & Spencer, J. P. (2016). A model for the contribution of ICT to the tourism value chain for pro poor benefits in Rwanda. *The Business and Management Review*, *7*(5).

Schott, C., & Nhem, S. (2018). Paths to the market: Analysing tourism distribution channels for community-based tourism. *Tourism Recreation Research*, *43*(3), 356–371.

Sechele-Mosimanegape, P., & Prinsloo, J. J. (2019). Competitiveness of Botswana as a tourist destination. Paper presented at the International Conference on the Future of Tourism (ICFT), Arusha, Tanzania, 16–17 April. https://core.ac.uk/download/pdf/245883048.pdf, accessed 12 May 2022.

Shemi, A., & Proctor, C. (2013). Challenges of e-commerce adoption in SMEs: An interpretive case study of Botswana. *Botswana Journal of Business*, *6*(1), 17–30.

Song, H. (2012). *Tourism Supply Chain Management*. New York: Routledge.

Spencer, A. J., Buhalis, D., & Moital, M. (2012). A hierarchical model of technology adoption for small owner-managed travel firms: An organizational decision-making and leadership perspective. *Tourism Management*, *33*(5), 1195–1208.

Statistics Botswana (2011). *Chobe Sub-District Population and Housing Census 2011: Selected Indicators for Villages and Localities*. Gaborone: Statistics Botswana.

Stone, M. T. (2015). Community empowerment through community based tourism: The case of Chobe Enclave Conservation Trust in Botswana. In R. van der Duim, M. Lamers, & J. van Wijk (eds), *Institutional Arrangements for Conservation, Development and Tourism in Eastern and Southern Africa*. Dordrecht: Springer, pp. 81–100.

Stone, M. T., & Stone, S. L. (2020). Challenges of community-based tourism in Botswana: A review of literature. *Transactions of the Royal Society of South Africa*, *75*(2), 181–193.

Tosun, C. (2000). Limits to community participation in the tourism development process in developing countries. *Tourism Management*, *21*(6), 613–633.

Trejos, B., & Chiang, L. N. (2009). Local economic linkages to community-based tourism in rural Costa Rica. *SinGore Journal of Tropical Geography*, *30*(3), 373–387.

Trejos, B., Chiang, L. N., & Wen-Chi Huang, W. (2008). Support networks for community-based tourism in rural Costa Rica. *The Open Area Studies Journal*, *1*(1), 16–25.

United Nations World Tourism Organization (UNWTO) (2020). *International Tourism Highlights*. Madrid: United Nations World Tourism Organisation. www.e-unwto.org/doi/book/10.18111/9789284422456, accessed 9 September 2022.

Wagaw, M., & Mulugeta, F. (2018). Integration of ICT and tourism for improved promotion of tourist attractions in Ethiopia, *Applied Informatics*, *5*(6), 1–12.

Xiang, Z., Wöber, K., & Fesenmaie, D. R. (2008). Representation of the online tourism domain in search engines. *Journal of Travel Research*, *47*(2), 137–150.

Zielinski, S., Jeong, Y., Kim, S., & Milanés, C. B. (2020). Why community-based tourism and rural tourism in developing and developed nations are treated differently? A review. *Sustainability*, *12*(5938), 1–20.

Zuze, C. (2006). Conservation education vs community based natural resources management. DWNP Community Extension and Outreach Division Workshop, Gaborone.

19. Assessing the role of the central government and communities in alleviating poverty through ecotourism
Agnes Tshepo Nkone and Thekiso Molokwane

INTRODUCTION

The proposal for fusing conservation and tourism was first put forward by Budowski (1976) and the term ecotourism emerged shortly after in the 1980s (Orams, 1995). The negative impact of mass tourism on the environment was just being realised, and in order to sustain their livelihoods, tourist companies were beginning to explore tourism management options and the concept of 'eco-friendly' tourism (Mestanza-Ramón and Jiménez-Caballero, 2021; Jamal et al., 2006). The idea was supported by the public who expressed much interest in experiencing pristine natural habitats that were continually being protected from exploitation and extinction as well as degradation. Several factors led to the birth of ecotourism, and the desire to forge relationships between several major ideas and concepts resulted in the conception of ecotourism. Sustainable funding for conservation was a primary motive for ecotourism; however, ecotourism planners and researchers quickly realised the diversity and numbers of stakeholders affected by ecotourism.

The World Ecotourism Summit in 2002 narrowed ecotourism down to eight postulates that are currently used today. Among the outcomes of the summit were views that: ecotourism should respect cultures existing in the respective areas; provide ongoing contributions to the local communities; and contribute to the conservation of natural areas and cultural heritage at the same time elevating the livelihoods of the local communities (Crabtree et al., 2002). Ecotourism is conceived to identify and brand tourism ventures that provide benefits to conservation, environmental education and local communities (Diamantis, 1999). Furthermore, in Africa, ecotourism is witnessed in eco-adventure tours based on the hunting of wildlife resources. The rapid extinction of some animal species in the African continent such as elephants and leopards marked a huge realisation on the part of ecotourism, environmentalist and animal activists that it was necessary to protect the animal species to avoid extinction (Ivanova et al., 2022). Efforts to rebuild animal populations were made through government laws and policies.

In Botswana, ecotourism was realised in 2002 with the adoption of the country's Ecotourism Strategy aimed at conserving Botswana's natural resources and wildlife. Many years ago, conservationist host communities, academics and tourism practitioners perceived ecotourism as a panacea for conservation and poverty problems in the tourism designated areas (Mbaiwa, 2015). In the Okavango Delta, ecotourism has mixed results because it has succeeded in alleviating poverty in some areas while failing in others. Where ecotourism succeeds, it generated economic benefits such as income and employment opportunities, leading to positive attitudes of residents towards ecotourism and conservation. Where ecotourism fails, the lack of entrepreneurship and managerial and marketing skills of local communities are cited as some of the key factors contributing to the failure of projects. Through this, ecotourism has proved to be a tool that can be used to achieve improved livelihoods and conservation; however, this depends on the socio-economic and political dynamics of the host communities in a specific ecotourism designation area (Mbaiwa, 2015).

THEORETICAL FRAMEWORK

Ecotourism is an industry that seeks to take advantage of market trends which concern external benefits and costs (Svensson and Wood, 2002). The framework by Das and Chatterjee (2015) in Figure 19.1 emphasises the expected outcomes of ecotourism.

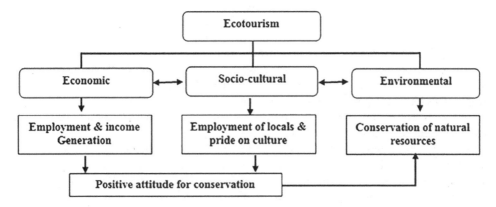

Source: Das and Chatterjee (2015).

Figure 19.1   Framework of the study

The conceptual framework (Figure 19.1) was developed considering the objectives of ecotourism. Ecotourism aims at conservation of nature through tangible improvement in the local economy, and increased respect for local culture and traditions. The first section of the literature review starts with the evolution of ecotourism. Ecotourism broadly aims at conservation of natural resources through providing economic benefit, social empowerment and cultural pride. It aims at improving livelihood activities and income which in turn will help in the conservation of natural resources (Das and Chatterjee, 2015).

IMPROVING LIVELIHOODS THROUGH ECOTOURISM

Promotion of livelihoods through ecotourism has been widely considered as an important tool and policy instrument for biodiversity conservation (Stronza et al., 2019, p. 241), provision of financial benefits and empowerment for local people. The Quebec Declaration on Ecotourism of 2002 recognises the principle of sustainable tourism concerning the economic, social and environmental impacts of tourism. The declaration goes on to argue that ecotourism contrib-

266 *Handbook on tourism and conservation*

utes actively to local and indigenous communities in its planning, development and operation and contributes to their well-being. Ecotourism is a strategy for supporting conservation and providing income for communities in and around protected areas (United Nations, 2002). It contributes to economic development by providing local employment and inculcating a sense of ownership (Jalani, 2012). In 2019 for instance, international tourist arrivals globally reached 1.5 billion; this marked a decade of uninterrupted growth, and accounted for 10.3 per cent of global GDP, generating 330 million jobs (Mestanza-Ramón and Jiménez-Caballero, 2021). Wildlife areas and heritages constitute a significant market for ecotourism based on natural resources and local culture. Ecotourism is an integral tool of conservation of natural resources and development of indigenous communities (Stronza, 2007). It supports livelihood diversification which is important in rural areas.

In many parts of the world, ecotourism has contributed to the dual goal of poverty eradication and conservation of natural resources. Promotion of ecotourism creates a larger amount of employment opportunities for local people who remain engaged in a variety of activities related to tourism. Ecotourism creates significant opportunities for conservation of natural resources (Das and Chatterjee, 2015). However, Banerjee (2010) posits that the present policies of ecotourism benefit neither conservation nor local communities. The concept remains poorly understood and much abused. Ecotourism simply neglects local communities and people. This has led to policy failures hence communities remain marginalised in the name of conservation. The communities face problems of inappropriate government interventions and declining local participation in the projects (Lee, 2001).

Ecotourism has external benefits to help protect biodiversity and a series of positive impacts on the local community. Establishing a market for ecotourism is nonetheless limited. When the consumption of market goods and the accompanying environmental deterioration is excessive, this can result in negative impacts to local communities in terms of environment, socio-culture and economy. In countries like Botswana, ecotourism is promoted to protect the country's cultural and natural heritage, as well as to enhance the active involvement of all Batswana and the host communities in all facets of the industry's development and management. The challenges to realising this have included a lack of relevant marketing and promotional tools for communities involved in tourism and the misappropriation and mismanagement of funds (Stone et al., 2017). Similarly, Das and Chatterjee (2015) argue that the communities concerned might be involved in ecotourism activities and supposedly being drivers of ecotourism projects but lack involvement and participation in the industry, as well as lacking access to the protected areas. The few successes achieved are in areas like the Okavango in Botswana where income generated strategies through ecotourism were developed. These strategies contribute to local economic development as is the case with the establishment of Santawani Lodge, Cultural Village (Mbaiwa, 2005).

Furthermore, Botswana has resorted to community-based tourism and makes use of traditional knowledge systems and activities. This approach utilises the available natural resources and enables the community to have access to the land and natural resources (Madzwamuse, 2007). Ecotourism management benefits have facilitated rural development through employees' acquisition of modern facilities such as the internet, television, modern houses, informal sector businesses and the opportunity to engage indirectly with tourism and in turn sustain local livelihoods and eradicate poverty. Local communities believe that solely depending on tourism is too risky, however, hence revenue from ecotourism is used to improve agricultural production (Stone and Nyaupane, 2016).

Belsky (1999) and Butler and Hinch (1996) are of the view that the major concern is the issue of local benefits, and significant social and political issues such as the issue of resource distribution in which inequalities in representation and power are often ignored. Local people do not necessarily benefit from ecotourism, instead, residents tend to bear the problems associated with ecotourism development while at the same time most benefits are siphoned away to outside stakeholders, and it is very uncommon for ecotourism activities to be situated in the homelands of indigenous people. The development and maintenance of protected areas is very likely to supersede local benefits as it may involve enforced changes in the local population's pattern of subsistence because of new regulations to preserve the environment (Weaver, 2007).

COSTS AND BENEFITS OF ECOTOURISM

In Botswana, forest reserves are located within an area that has shortages of arable land which leaves local communities with 31 per cent of the land available for communal use. This has resulted in dire need for land since more land (39 per cent) has been devoted to protected areas, with 17 per cent being national parks and game reserves, where there is zero utilisation by communities (Othusitse, n.d.). The remaining percentage is wildlife management areas and forest reserves, which allow some form of utilisation by the communities. It should be noted that the 40 per cent that is mentioned here only accounts for national parks and game reserves, wildlife management areas and forest reserves. There are, however, privately owned game reserves, and educational parks whose areas have not been included here (Othusitse, n.d.).

Local communities have limited access to the forest reserves (van der Sluis et al., 2017, pp. 92–95, 167). They are only allowed to enter forest reserves with a permit. The permit is purchased from the capital city, Gaborone. Botswana's local communities acknowledge the importance of private sector as well as small and medium-sized enterprise involvement in ecotourism activities especially if it focuses on packaging and marketing of their products (Blackie, 2006, pp. 1942–1947; Manwa et al., 2017, p. 3). The joint venture between local communities and private sector in the running of ecotourism activities is seen as a better way of improving living standards of the locals because employment is created as well (Andries et al., 2021; Scheyvens, 2009). In addition, ecotourism has educational benefits as the public learns about certain locations and the importance of conserving the natural resources. Again, ecotourism helps build cultural awareness by fostering respect for the place travelled to and the communities visited.

CAUSES OF POVERTY

Poverty obviously constitutes a danger to prosperity. For purposes of this study, poverty relates specifically to rural poverty as the study is conducted in villages situated in rural parts of the Kweneng district. The phenomenon is seen in all its manifestations as a denial of the opportunities and the choices that underpin improved standards of living and human development. On the same note, development requires the removal of major sources of unfreedom. These include poverty, tyranny, poor economic opportunities, systematic social deprivation, neglect of public facilities, as well as intolerance or over activity of repressive states despite

268 *Handbook on tourism and conservation*

unprecedented increases in overall opulence. The contemporary world denies elementary freedoms to vast numbers which directly relates to poverty (Sen, 1999).

Klitgaard (1997) hypothesised that corruption is more likely to occur in an environment where officials have monopolistic control over state resources such as who can gain access to the resources. Due to administrative excesses and arbitrary behaviour, it would add to the suffering of the people. On a similar note, Gupta et al. (2002) argue that corruption increases income inequality through several channels which decrease economic growth. Secondly, poverty is caused by low access to opportunities for developing human capital and education; education serves both a constitutive and instrumental role in development, hence reducing poverty. Mauro (1998, p. 263) laments that poor countries increase their level of poverty due to lack of training skills, while productive knowledge and education creates valuable human capital. Education can be a life empowering experience for all and what the poor need most is empowerment.

Furthermore, the features of poverty linked to education include non-participation or low rates of participation of children in schooling, high rates of dropouts and failures, low rates of continuation in schooling, low rates of achievement and finally exclusion of the poor from education (Shaul, 2002). Therefore, due to income poverty, very few achieve a high level of learning or proper education. Other factors contributing to poverty are political instability, wars and civil wars; countries which are rich in natural resources may be more prone to civil wars due to rent-seeking activities. Countries with both dependences upon primary commodity exports and a large diaspora significantly increase the risk of conflict (Collier and Hoeffler, 2004; Elbadawi, 1999a). Civil wars lead to poverty mainly due to destruction of capital, displacement of people and increased insecurity, and households whose houses are destroyed or lose land are at a greater risk of falling into poverty (Elbadawi, 1999b).

Natural and geographical characteristics also cause poverty. The African continent is tropical with slow economic growth; furthermore being landlocked reduces a nation's annual growth rate by around half of 1 per cent (Sachs et al., 2004). Africa's adverse climate also causes poor health. The life expectancy has been historically low, with infant mortality often caused by tropical diseases such as malaria. Given these geographically unfavourable situations, some African countries are already at a disadvantage when trying to escape the poverty trap (Bloom and Sachs, 1998).

Lastly, poverty is caused by ineffective governance and government policies. Some of the African governments have been undemocratic for much of the time. Notwithstanding, the ineffective local governance and government policies seem to prevent the chronically poor from escaping the poverty trap (Lwanga-Ntale and McClean, 2004). A typical pattern is that the governments are controlled by the ruling elites, and educated, urban resident populations are resistant or indifferent to pro-poor policies (Hillman, 2002). These political elites use the poor as hostages to personally benefit from aid resources and debt relief (Crook, 2003).

Poverty can also be caused by general exclusion of people from social life. Exclusion reflects discrimination, which is a process that prevents individuals fully participating in material exchange or interactions (Sindzingre, 2000). In Botswana, more than half of the rural populations have incomes that are inadequate to meet their basic needs. Since most of the population live in the rural areas, the poverty headcount ratio stands at 24.2 per cent (Statistics Botswana, 2022). Among other causes of poverty and the widening gap between the rich and the poor in Botswana are droughts which have led to the concentration of livestock ownership in a few hands, alienation of communal land and the curtailment of facilities for hunting and gathering as part of environmental management (Peke, 1994).

ECOTOURISM IN KWENENG WEST

Kweneng West is one of the two sub-districts comprising Kweneng District. This study's research locale included several villages, namely: Letlhakeng; Botlhapatlou; Ditshegwane; Dutlwe; Moshaweng; Khudumelapye; Serinane; Mantshwabisi; Motokwe; Ngware; Salajwe; Takatokwane; Tsetseng; Tswaane; Monwane; Malwelwe; Maboane; Sesung; Sorilatholo; Kotolaname; Kaudwane; and Diphuduhudu. According to Botswana's 2011 Population and Housing Census, the sub-district has a population of 47,797 inhabitants. During the census, the highest proportion (20.7 per cent) of Kweneng West's population was in Letlhakeng village. This was followed by Takatokwane and Salajwe with 7.8 per cent and 6.8 per cent respectively. Also, the villages of Diphuduhudu and Monwane each contributed less than 2 per cent to the total district population. Other smaller localities with less than 500 people contributed a combined proportion of 7.9 per cent (Republic of Botswana, 2011).

In Kweneng West, the local citizens comprise a combination of the hunting and gathering community, namely the San and the Bakgalagadi. When the government of Botswana introduced ecotourism activities to Kweneng West residents were meant to earn a living from ecotourism. The community-based tourism was created to offer an opportunity for communities in remote areas to generate income and employment from the use of the few resources they have within their areas (Sebele, 2010). In some areas of Kweneng West, the government identified community-based projects as possible avenues for diversification from agricultural production. Ecotourism is a strategy for supporting conservation and providing income for communities in and around protected areas. It contributes to economic development and conservation of protected areas by generating revenues that can be used to manage protected areas and it provide locals with employment and inculcate senses of community ownership (Moeljadi, 2015). There have been some successes but the list of failures is very high as rural dwellers continue to live below the poverty line while their natural resources are exploited to generate income which benefits a minority.

On the same note, operating ecotourism as a proxy market means that external costs are imposed on others. It ignores the operation of market exchange which in turn leads to negative impacts on the environment, socio-culture and economy of local communities. This study mainly focused on scrutinising the role of ecotourism in poverty alleviation in areas of Kweneng West. The locations were chosen based on the unpleasant situations prevailing amongst the rural area dwellers who live below the poverty line. Their livelihood influences, beliefs and behaviours do not correlate with the environment and actual utilisation of natural resources insofar as conservation is concerned. Communities are detached from mainstream socio-lingual, socio-economic and socio-cultural activities due to inadequate infrastructural development, low level of education of parents, minimal opportunities for economic empowerment and lack of information (Pansiri, 2008). The living standards of people in Kweneng West have been declining for years despite the ecotourism activities continuing to prevail in the area. Many factors have contributed to this decline, the most common factor being marginalisation of locals by their natural resource exploiters.

Handbook on tourism and conservation

Natural resources found in rural areas benefit the urban minority elites while the hosts of common pool resources are economically marginalised. The locals in Kweneng West are unable to utilise the local natural and cultural heritage resources because of the environmental policy put in place to conserve the natural resources. The lack of capital investment and advertisement of natural resources for the ecotourism market in Kweneng West creates another stumbling block for ecotourism in alleviating poverty in the sub-district. The government's failure to empower the Kweneng West dwellers exacerbates the poverty and dependency syndrome among inhabitants. Dependency in Kweneng West is growing because residents are unable to utilise natural resources available in their area. This puts pressure on government budgets and in the process delays other developmental projects that could provide Kweneng West communities with food baskets. The natural resources available in Kweneng West can enable the communities to be self-reliant only if appropriate policy reforms are made to stimulate self-empowerment.

METHODOLOGY

This study adopted a case study strategy and constructivist research paradigm. The study thus analysed and explored the research environment to acquire appropriate and relevant data. This study focused on the research problem from specific settings to explore the true reality of the issues through the feelings and thoughts of the study subjects. A qualitative research approach was adopted in this descriptive type of study. Qualitative descriptive studies are the least 'theoretical' of all the qualitative approaches to research. In addition, qualitative descriptive studies are the least encumbered by a pre-existing theoretical or philosophical commitment compared to other qualitative approaches. By comparison, qualitative descriptive studies tend to draw from naturalistic inquiry, which attempts to study something in its natural state to the extent that is possible within the context of the research arena (Lambert and Lambert, 2012, pp. 255, 256). A comprehensive summarisation of everyday encounters and specific events experienced by individuals or groups was documented. Adopting a qualitative descriptive study assisted a lot in conducting this research by revealing the reality pertaining to ecotourism and poverty alleviation strategies in Kweneng West. Individual respondents shared their experiences and opinions creating a balance between issues presented by other respondents. Data in this study was analysed qualitatively. There are various factors that influence the analysis of qualitative data. These factors include nature and type of study as well as the study population. In qualitative data analysis, numerous approaches apply and these include: data interpretation; description and descriptive method; explanation as well as organisation of data.

FINDINGS

Findings of this study are based on the information gathered using in-depth interviews with the village leadership, villagers (residents) and Village Development Committee (VDC) members in Ditshegwane, Metsibotlhoko and Letlhakeng villages. These villages were purposively selected. Below is the presentation of the findings of the study.

The State of Poverty in Kweneng West

Findings from Letlhakeng village revealed that poverty is prevalent in the village. Interviewed residents argued that some people are still without shelter in their area. The respondents indicated that they fully rely on government for social support in the form of food baskets. The youth mentioned that they are unemployed and some pass time through indulging in home-brewed beverages. Findings reveal that some of the youth have never been to school because they felt the school environment neglected them due to lack of multicultural education in the curriculum. The perception of the students was that teachers from mainstream Setswana ethnic groups discriminate against them. These factors contribute to the prevalence and sustenance of poverty in Kweneng West.

Residents decried that their village leaders, mainly Dikgosi (village chiefs), have been deprived of their powers, and that government policies are sometimes irrelevant to their situations. To illustrate, at Ditshegwane village, the VDC Chairperson lamented that poverty is experienced by toddlers mainly because the residents are ignoring the importance of being literate. It was also mentioned that parents neglect children who then drop out from school. Girl children were said to be victims of teenage pregnancy and exposed to diseases such as HIV/AIDS and other STIs because they are vulnerable to sexual abuse by older males in their effort to earn a living. Boarding schools have increased poverty hence children often run away from school and return to their homes and families. Those who remain in school pay little attention to learning due to poverty challenges.

Respondents further indicated that most of the residents were poor, except for the elderly and those with special needs. Further to this, respondents noted that there were many home-based liquor spots which aggravated the already high poverty rates in the sub-district. They said that some locals resorted to selling the food received from the Social Welfare Office to acquire liquor. Residents of Metsibotlhoko said that poverty in their area was high because they are considered a very small village with small population that cannot be given particular socio-economic benefits. Residents mentioned that this alone demoralised their efforts at emancipation and this leaves them poor. They also stated that they were discouraged to travel distances to other developed areas such as Molepolole or Letlhakeng in search of opportunities. In his language one resident lamented that: "*he ikamoogeego*" which translates as, "We have accepted our situation."

Indicators of Poverty in Kweneng West

Residents of Metsibotlhoko village felt that poor infrastructure, illiteracy, unemployment especially among the youth and dependency on government food baskets are signs that there is poverty in their area. They explained that most people cannot afford to have a house built out of corrugated iron and proper bricks from cement. Instead they use mud bricks and thatch to build their huts. The few individuals who afford such houses are those who belong to the working class. The observation made was that such individuals usually belong to different ethnic groups than those of resident communities of the area. They further stated that nutritious meals for their children are only the supplements given at the clinic to children under 5 years. A 78-year-old man who is a village elder lamented that:

272 *Handbook on tourism and conservation*

the fact that our children are academically underperforming is that they lacked proper diets to develop their brains, this alone indicate poverty in our area.

On a similar note the 68-year-old leader of Metsibotlhoko village complained that:

people cannot afford to get their own stand pipes at their homes, bills and installation of water pipes are too high for residents to afford, some elderly people use their small government pension funds to purchase public stand pipe water vouchers for their families, and when the amount purchased finishes they cannot afford travelling fees to offices in Letlhakeng to buy again. Villagers flog to Ipelegeng registration mainly because it is the only means of getting money, youth for that matter.

In another note, the VDC member of Letlhakeng posits that:

in our area poverty is indicated by lack of basic needs by some of our residents, the youth unemployment is very high; the youth are hopeless and have resorted to alcoholism. Parents also try hard as subsistence farmers to feed their families because they do not have any other means, this at the end leave children to be left alone in the village, these children end up dropping out of school making poverty not to be reduced.

A 50-year-old VDC member at Letlhakeng argued that:

In our schools, especially the secondary schools, almost all our children are exempted from paying school fees, this is a sign that parents cannot afford even small amount of school fees payment, this is poverty indication.

Role of the Government in Alleviating Poverty in Kweneng West

The residents of Ditshegwane village stated that the government has put in place the poverty alleviation programmes and provides them with food baskets. Government also gives the poor and the orphans free school uniforms and has exempted them from paying school fees. Residents also said that their children commute to school from the settlement surrounding Letlhakeng. A 30-year-old female resident added that:

our children are transported by a bus provided by the Ministry of Basic Education and Skills Development. Children have to wake up very early in the morning not to miss the bus. But it is a better arrangement even though we don't want our children to attend school at a boarding school.

Residents further indicated that the government has given them cards for monthly food baskets and in a few instances built houses for some of the residents but failed to provide them with a permanent solution to their poverty. One of the residents even indicated that the food baskets provided might only be the mandate of the government of the day; in the event the government changes they could face starvation. Furthermore, residents indicated that poverty eradication programmes put in place by the government did not bear fruits. This is because it is not possible to access the offices which are located in Letlhakeng which is several kilometres from their settlement. It was mentioned that even application for businesses was very difficult as the government prints forms in English language making language a barrier for those who are illiterate. On a similar note, residents of Letlhakeng village argued that the government policies meant to assist in alleviating poverty such as backyard gardens were a failure. Residents argued that the lack of water to maintain the gardens and high bills by the water company made this particular programme fail.

The Community Efforts to Raise Themselves Out of Poverty in Kweneng West

In terms of community efforts to lift themselves out of poverty, residents of Metsibotlhoko village indicated that the community could only get themselves out of the situation if they were given capital investment. Residents further noted that they had brilliant and relevant business plans but lacked funding to start the businesses. Residents noted that encouraging others, grouping themselves together and contributing money for a period of time in order to try and start their businesses could improve their living standards. However, they argued that they were unable to do this because of the long process involved in applying for licences. This left them with the only option of relying on government anti-poverty schemes such as Ipelegeng.

Residents of Letlhakeng village opined that if residents were cooperative enough, a proposal could be made to have households that are experiencing better economic status contribute some money towards starting projects that would benefit the less privileged. The residents are of the view that access to natural resources in Letlhakeng would enable revenue raising activities that could make a difference in their living standards. Examples of the revenue raising activities mentioned included building lodges and experiential tourism activities among others. Further to these, the village leadership of Metsibotlhoko village said that their villagers could utilise the nearby lake to generate income. They also mentioned that the activity would bring about benefits such as: improved infrastructure, transport and communications employment as well as self-employment.

The Expectations of Kweneng West Residents of the Government Efforts to Improve Living Standards

The residents made the recommendations that they needed to be enlightened on how to make use of the natural resources to earn a living. The village leadership at Metsibotlhoko lamented in their language that "*batcho habadze bilo jha tholego dzheri mo kgalagadi, he thokgha heri gore he anamiswe batcho bedhe be idzhe bilo tjhecho ha chona heri thokhomela*", meaning that, people do not know about the natural resources and attractions found in Kgalagadi. In addition, residents must be given the responsibility of taking care of such resources so that they remain attractive. They also urged the government to make policy reforms regarding use of land. They needed policies that could easily allow their communities to have access to use of the lake in their area. The Letlhakeng VDC member noted that Kweneng West residents needed to change their attitudes especially the dependency syndrome. She revealed that most residents were not doing anything for a living simply because the government does not provide employment opportunities. So, in her view the government should encourage self-reliance. One of the residents raised concern that residents need to be self-reliant and reduce pressure on government budget. The respondent went on to indicate that only if the government can give residents poverty eradication programmes that are sustainable and suitable for their conditions would they be able to strive to improve their living standards. Furthermore, the argument was that declining literacy levels should be taken care of by the government, and students should be encouraged to love school so that they can emancipate themselves and work to eradicate the poverty in their areas.

274   *Handbook on tourism and conservation*

How Ecotourism Can Help in Alleviating Poverty in Kweneng West

Residents of Metsibotlhoko village revealed that ecotourism could be of more help if only they had access to natural resources. Residents lamented that the government controls all the natural resources. The stated that individuals could not be given licences to operate in some areas and earn a living. It was said that Pans and Valleys were for viewing by tourists yet locals in no way benefited from these natural resources. Residents raised a concern that when they cut a few trees for wood carving, they were met with complainants from the environmental group called Somarelang Tikologo, the Department of Forestry and Rangeland Resources' Green Scorpion officials who charged them for deforestation. In addition, when they dug soil specifically for pottery, the government officials through by-laws charged them for degrading the land. Residents also revealed that the only ecotourism activity left was music and dance. They argued that because the media does not help them publicise their festivals, the events often fail. As such, music and dance did not alleviate poverty in their area as expected.

On a similar note, residents of Letlhakeng village revealed that:

> land remains the property of the government hence the village has a lot of mud and wood huts. If the small lake in the areas was owned by the community, they would be using it to generate income through tourism for the benefit of the community. So many conservational laws are limiting the community in the use of natural resources.

Ecotourism Resources Ownership and Usage in Kweneng West

The residents of Kweneng West revealed that the resources belong to the government. Residents argued that they did not have as much access as they expected to have because the government controls utilisation of the heritage and natural resources in their area. Government policies on environmental conservation deny them access to natural resources. Residents argue that they are told to apply for the licences in order to utilise some of the resources available, but the application forms are difficult to complete because the language used is discriminatory and it caters only for the literate individuals.

The village VDC member of Metsibotlhoko lamented that:

> the government does not allow the residents to show how capable they are in earning their living. Government policies meant for conservation are misinterpreted by those who are supposed to imple-ment the policies. They always exaggerate what the policies are supposed to mean. Residents are not allowed even to dig mud for making clay as the by-law says that is land degradation, no cutting of few trees for wood carving as that is taken as deforestation, a lot of vocabulary that is in place is making the residents to have fear in making use of the natural resources. As a result, the community is not benefiting from use of natural resources.

DISCUSSION

The study's findings reveal multiple prevailing factors that hinder residents in access and utilisation of ecotourism resources in Kweneng West. This has contributed to idleness by some of the residents who evidently choose not to set up businesses. Village leaders mentioned that residents mostly indulge in non-life progressing activities and that they choose to rely on food hampers from government and thus do not develop much interest in seeking gainful employment. The VDCs also believed that the government through the social welfare office nurtured a dependency syndrome.

Regarding making gainful income from natural resources, one of the village leaders in Letlhakeng village indicated that government should bestow powers on them to control and allocate land for their people. They should be able to utilise ecotourism resources such as landscapes, rocks and vegetation available in their area to alleviate poverty. Residents explained that the policies put in place for conservation disadvantage them using the ecotourism resources available in their area. Findings further illustrate a link between high rates of unemployed youth as well as school dropouts and absentee parents who often leave to seek temporary employment to provide for their families. Some of the youth attributed their situations to the absence of their parents. Lack of inclusion by government in formulating policies and programmes meant that residents in Kweneng West are seen as passive targets and recipients of government policy. This potentially significantly reduces the effectiveness of policies meant to alleviate poverty in the area.

When applying for poverty alleviation programmes, residents indicated that the language on the application forms was English and that places them at a disadvantage and more so because most are illiterate. Findings also revealed music and dance as an ecotourism resource but that still doesn't help the residents to get out of poverty because there is lack of publicity and sponsorship to nurture their performing arts activities. When responding to a question about who the ecotourism belongs to, some residents indicated that the ecotourism resources solely belong to government and that such resources are protected by the conservation laws hence barring residents from accessing them.

Findings further revealed that at the time of data collection, residents did not have gainful employment and as such could not raise themselves out of poverty. Residents further noted that there was a lack of cooperation by some community members in forming groups to identify and begin ecotourism business activities that could provide a livelihood. Such residents were said to refuse to make monetary contributions to set up small businesses which could help them reduce dependency. Moreover, the findings indicate that though minor, some members of the community could contribute funds to start small scale ecotourism businesses to get themselves out of poverty. If the community was given access to natural resources, they would invest in ecotourism by inviting the private entities to establish partnerships that would directly benefit the residents. Potential functions and activities would include: getting direct employment; and improvement of infrastructure which would elevate the livelihoods of community members. The findings also revealed that there are many ecotourism resources in the study locale such as the valleys, lakes, rocks and mud for making pottery, wood for carving and music and dance. The study further revealed that apart from direct poverty alleviation policies, the government is not doing much in using ecotourism resources to alleviate poverty of locals.

CONCLUSION

While ecotourism is a widely known source of stable income elsewhere, the situation in Kweneng West is different as there is an absence of active engagement by residents in activities and businesses that can give them gainful employment through ecotourism. Rather than being allowed access to resources that could potentially provide Kweneng West residents with employment and income, the emphasis is placed on conservation. Allowing the residents to

276 *Handbook on tourism and conservation*

make use of natural resources while maintaining the conservation objective would help them to escape from poverty. Failure to organise by locals does not help their situation as they stand a better chance of setting up businesses as groups rather than as individuals. Citizen engagement is key in public policy formulation as when engaged, there would be ownership on the part of citizens and thus help implementation of policy. Lastly, support in publicising cultural heritage would accord the residents the much-needed exposure in converting the talents of local people into gainful employment.

REFERENCES

Andries, D. M.. Arnaiz-Schmitz, C., Díaz-Rodríguez, P., Herrero-Jáuregui, C., and Schmitz, M. F. (2021). Sustainable tourism and natural protected areas: Exploring local population perceptions in a post-conflict scenario. *Land*, 10(331), 1–19.

Banerjee, G. (2010). Tourism and environmental conservation: Conflict coexistence or symbiosis? *Environmental Conservation*, 3(1), 27–31.

Belsky, J. M. (1999). Misrepresenting communities: The politics of community based rural eco-tourism in Gales Point Manatee, Belize. *Rural Sociology*, 64(4), 641–666.

Blackie, P. (2006). Is small really beautiful? Community-based natural resource management in Malawi and Botswana. *World Development*, 34(11), 1942–1957.

Bloom, D. E. and Sachs, J. (1998). Geography, demography and economic growth in Africa. *Brookings Papers on Economic Activity*, 2, 207–273.

Budowski, G. (1976). Tourism and the environmental conservation: Conflict coexistence, or symbiosis? *Environmental Conservation*, 13(1), 27–31.

Butler, R. and Hinch, T. (eds) (1996). *Tourism and Indigenous People*. London: International Thomson Business Press.

Collier, P. and Hoeffler, A. (2004). Greed and grievance in civil war. *Oxford Economic Papers*, 56(4), 563–595.

Crabtree, A., O'Reilly, P., and Worboys, G. (2002). Sharing expertise in ecotourism certification: Developing an International Ecotourism Standard. World Ecotourism Summit – Quebec.

Crook, P. C. (2003). Decentralization and poverty reduction in Africa: The politics of local-central relations. *Public Administration and Development*, 23(1), 77–88.

Das, M. and Chatterjee, B. (2015). Ecotourism: A panacea or a predicament? *Tourism Management Perspectives*, 14, 3–16.

Diamantis, D. (1999). The concept of ecotourism: Evolution and trends. *Current Issues in Tourism*, 2(2).

Elbadawi, I. A. (1999a). External aid: Help or hindrance to export orientation in Africa? *Journal of Africa Economies*, 8(4), 578–616.

Elbadawi, I. A. (1999b). Civil wars and poverty: The role of external interventions, political rights and economic growth. Presentation at the World Bank's Development Economic Research Group (DERG) launch conference on "Civil Conflicts, Crime and Violence", World Bank, Washington, DC, 22–23 February.

Gupta, S., Davoodi, H., and Alonso-Terme, R. (2002). Does corruption affect income inequality and poverty? *Economics of Governance*, 3(1), 23–45.

Hillman, A. L. (2002). The World Bank and the persistence of poverty in poor countries. *European Journal of Political Economy*, 18(4), 783–795.

Ivanova, S., Prosekov, A., and Kaledin, A. (2022). Is ecotourism an opportunity for large wild animals to thrive? *Sustainability*, 14(5), 1–15.

Jalani, J. O. (2012). Local people's perception on the impacts and importance of ecotourism in Sabang, Palawan, Philippines. *Procedia – Social and Behavioural Sciences*, 57(9), 247–254.

Jamal, T., Borges, M., and Stronza, A. (2006). The institutionalization of eco-tourism: Certification, cultural equity and praxis. *Journal of Eco-Tourism*, 5(3), 145–175.

Klitgaard, R. (1997). Cleaning up and invigorating the civil service. *Public Administration and Development*, 17(5), 487–509.

Lambert, V. A. and Lambert, C. E. (2012). Editorial. Qualitative descriptive research: An acceptable design. *Pacific Rim International Journal of Nursing Research*, 16(4), 255–256.

Lee, J. P. (2001). Ecotourism, community development, and local autonomy: The experience of Shan-Mei aboriginal community in Taiwan. MA Thesis, Oregon State University.

Lwanga-Ntale, C. and McClean, K. (2004). The face of chronic poverty in Uganda from the poor's perspective: Constraints and opportunities, *Journal of Human Development*, 5(2), 177–194.

Madzwamuse, M. (2007). Natural resources management, land tenure and land rights: The case of Basarwa communities. In B. Schuster and O. T. Thakadu (eds), *Natural Resources Management and People*. CBNRM Support Programme Occasional Paper No. 15, pp. 46–52. Gaborone, Botswana.

Manwa, H., Saarinen, J., Atlhopheng, J. R., and Hambira, W. L. (2017). Sustainability management and tourism impacts on communities: Residents' attitudes in Maun and Tshabong, Botswana. *African Journal of Hospitality, Tourism and Leisure*, 6(3), 1–15.

Mauro, P. (1998). Corruption and the composition of government expenditure. *Journal of Public Economics*, 69(2), 263–279.

Mbaiwa, J. E. (2005). Enclave tourism and its socio-economic impacts in the Okavango Delta, Botswana. *Tourism Management*, 26(2), 155–172.

Mbaiwa, J. E. (2015). Eco-tourism in Botswana: 30 years later. *Journal of Eco-Tourism*, 14(2–3), 204–222.

Mestanza-Ramón, C. and Jiménez-Caballero, J. L. (2021). Nature tourism on the Colombian–Ecuadorian Amazonian border: History, current situation, and challenges. *Sustainability*, 13(8), 4432. https://doi.org/10.3390/ su13084432.

Moeljadi, A. S. (2015). Eco-tourism development strategy Baluran National Park in the regency of Situbondo, East Java, Indonesia. *International Journal of Evaluation and Research in Education*, 4(4), 185–186.

Orams, M. (1995). Towards a more desirable form of eco-tourism. *Tourism Management*, 16(1), 3–8.

Othusitse, B. (n.d.). Action plan for implementing the conservation on Biological Diversity's programme of work on protected areas. Department of Wildlife and National Parks Ministry of Environment, Wildlife & Tourism Gaborone, Botswana. Submitted to the Secretariat of the Convention on Biological Diversity.

Pansiri, O. (2008). Improving commitment to basic education for the minorities in Botswana: A challenge for policy and practice. *International Education Development*, 28(4), 446–459.

Peke, D. (1994). Inequality prevalent in Botswana. *The Botswana Guardian*, 22 September.

Republic of Botswana (2011). *Kweneng West Sub District, Population and Housing Census Selected Indicators 2011*. Vol. 4(1). Statistics Botswana, Gaborone.

Sachs, J. D., McArthur, J. W., Schmidt-Traub, G., Kruk, M., Bahadur, C., Faye, M., and McCord, M. (2004). Ending Africa's poverty trap. *Brookings Papers on Economic Activity*, 35(1), 117–240.

Scheyvens, R. (2009). Pro-poor tourism: Is there value beyond the rhetoric? *Tourism Recreation Research*, 34(2), 191–196.

Sebele, L. S. (2010). Community-based tourism ventures, benefits and challenges: Khama Rhino Sanctuary Trust, Central District, Botswana. *Tourism Management*, 31(1), 135–144.

Sen, A. (1999). *Development as Freedom*. Oxford: Oxford University Press.

Shaul, M. S. (2002). *School Dropouts: Education Could Play a Stronger Role in Identifying and Disseminating Promising Prevention Strategies*. Report to the Honourable Jim Gibbson's General Accounting Officer, Washington, DC.

Sindzingre, K. (2000). *Exclusion and Poverty in Developing Countries*. Koherdonrfer Bosig Workshop Series, German Foundations of International Development, Berlin.

Statistics Botswana (2022). *Poverty, 2016–2022*. www.statsbots.org.bw/poverty, accessed 30 January 2019.

Stone, M. T. and Nyaupane, G. P. (2016). Protected areas, tourism and community livelihoods linkages: A comprehensive analysis approach. *Journal of Sustainable Tourism*, 24(5), 673–693.

Stone, L. S., Stone, M. T., and Mbaiwa, J. E. (2017). Tourism in Botswana in the last 50 years: A review. *Botswana Notes and Records*, Special Issue on Environment, Tourism and Contemporary Socio-Economic Issues in the Okavango Delta and other Ecosystems, 49(SI), 57–66.

Stronza, A. L. (2007). The economic promotion of eco-tourism for conservation. *Journal of Eco-Tourism*, 6(3), 210–230.

Stronza, A. L., Hunt, C. A., and Fitzgerald, L. A. (2019). Eco-tourism for conservation. *Annual Review of Environment and Resources*, 44, 229–253.

Svensson, G. and Wood, G. (2002). A conceptual framework of corporate and business ethics cross organizations: Structures, processes and performance. *The Learning Organization*, 18(1), 21–35.

United Nations (2002). Québec Declaration on Ecotourism defines basis for its international development. www.un.org/press/en/2002/unep113.doc.htm.

van der Sluis, T., Cassidy, L., Brooks, C., Wolski, P., VanderPost, C., Wit, P., Henkens, R., van Eupen, M., Mosepele, K., Maruapula, O., and Veenedaal, E. (2017). *Chobe District Integrated Land Use Plan*. Wageningen Environmental Research, Wageningen, Report 2813.

Weaver, D. B. (2007). Twenty years on: The state of contemporary ecotourism research. *Tourism Management*, 28(5), 1168–1179.

20. Sense of place and tourism in cultural landscapes

Joseph E. Mbaiwa and Gladys B. Siphambe

INTRODUCTION

Globally, the tourism industry has experienced sustained growth and development over the last decades. International tourist arrivals increased by 3.8 per cent annually between 2010 and 2020 (WTTC, 2020). Similarly, cultural tourism has experienced significant growth, leading to commodification and globalisation of many cultural landscapes. Cultural tourism accounts for 37 per cent of global tourism and it grows at a rate of 15 per cent per year (UNWTO, 2014). Thus, in many places cultural tourism can present an ideal vehicle for local and regional development where local people can experience the direct economic benefits such as employment (Smith, 2003). Cultural tourism is thus a major global industry that brings income and promotes national identity and preservation of cultural heritage (Richards, 2007).

Africa is known for nature-based tourism; however, in recent years, it has started to receive its own share of cultural tourists in cultural landscapes. Shackley (2000, p. 80) notes that: "the concept of a cultural landscape closely involves the spiritual and contemporary uses of a landscape with the management of its archaeological heritage". Though underdeveloped, the history, heritage attractions and lifestyles of residents in Botswana's landscapes have become tourist attractions and products in the last decade. The challenge is that the relationship between tourism development in cultural landscapes and the lives of residents has not been given adequate attention. Cultural landscapes and local communities have a bond and a relationship with each other which needs to be understood through research. This bond and relationship define a sense of place, place attachment and identity of resident communities.

Previous studies on socio-cultural effects of tourism in destination areas focused on subjects such as the consumption of cultural heritage products (Chhabra, 2010) and related issues such as commodification and authenticity of such products (MacCannell, 1973; Cohen, 1988, 1989; Ateljevic & Doorne, 2003; Steiner & Reisinger, 2006), with little attention to residents' sense of place, and place attachment and identity. Research has shown that globalised, standardised tourism is characterised by the replacement of real authenticity with a 'staged' authenticity in which local cultures and traditions become manufactured or simulated for tourist consumption (MacCannell, 1973, 1992; Cohen, 1988; Reisinger, 2012). The commodification of local traditions, and a sanitisation of culture that is delocalised from the actual social, political and cultural context, leads to the disappearance of genuine local tourism and is the direct result of cultural homogenisation (Sorkin, 1992; Kearns and Philo, 1993; Reisinger, 2012). Globalisation and commodification of culture affects residents' sense of place, place attachment and identity (Nijman, 1999). Globalisation and commodification of cultural landscapes impinges on the realness or genuineness of culture and its practices (MacCannell, 1973, 1992; Cohen, 1988; Mbaiwa, 2011). As a result, cultural commodification could impact on the structure and composition of cultural heritage tourism at sites such as the Goo-Moremi

280 *Handbook on tourism and conservation*

Landscape. Commodification of cultural heritage tourism affect the residents' sense of place, identity and attachment in Goo-Moremi Cultural Landscape. This chapter uses the concept of sense of place to analyse the socio-cultural effects of globalisation and the commodification of culture on residents' sense of place, place attachment and identity at Goo-Moremi Cultural Landscapes in Eastern Botswana. This chapter examines whether the process of commodification and globalisation of cultural landscapes for the tourism market automatically result in the loss of sense of place, place attachment and identity or promote their preservation and sustainability.

THE CONCEPT OF SENSE OF PLACE

The concept of 'place' has emerged as important in realms of cultural scholarship and professional practice (Smith, 2014). Research about place, experiences around place and the impact of place on people and their quality of life has been ongoing for almost two decades (Hashem et al., 2013a, 2013b). The concept received attention first in architecture and urban design studies, environmental psychology and geography. In tourism studies, the concept has not received adequate attention even though it is in tourism that the concept should play a significant role (Smith, 2014).

The concept of place is critical in tourism studies since tourism development often has pressures and impacts on community interests and their surrounding environment and cultural landscapes. The concept of sense of place has been defined differently by scholars. For example, Feld and Basso (1996) in their edited volume of a collection of essays entitled *Senses of Place* addressed various anthropological interpretations of sense of place. The volume presented several perspectives about sense of place and concluded that sense of place is not a static concept and has more than one meaning (Cighi, 2008). Casey (1996, p. 23) defined sense of place as a "plenary presence permeated with culturally constituted institutions and practices". Similarly, Feld (1996) described sense of place in terms of reflections on the sensuousness of place as soundscape. Kahn (1996) defined sense of place in terms of the emotions people have about their landscape. Because of these many definitions, Stedman (2003) notes that the concept of sense of place is ambiguous and difficult to define or measure.

Despite the debate around the concept of sense of place, scholars agree that place refers to the "dimension formed by people's relationship with physical settings, individual and group activities, and meanings" (Najafi and Shariff, 2011, p. 1054). Stedman et al. (2004) argue that human experiences, relationships, emotions and thoughts bring life into space, transforming it into place. Therefore, sense of place is said to be associated with sentiments of attachment that people develop and have about a physical place or environment.

The different definitions of sense of place by scholars may involve an understanding of the meaning and emotions residents have about a physical, geographical location. Sense of place is thus the relationship between human beings, their sense of self and environmental characteristics (Hashem et al., 2013b). Hashem et al. (2013b) further argue that the concept of sense of place is rooted in subjective experience of people especially their memories, traditions, history, culture and society. Sense of place thus implies the interaction people/residents have with their physical environment and place. The concept of sense of place was found suitable for this study because it helps unpack the interaction between cultural tourism development at Goo-Moremi Cultural Landscape and Goo-Moremi residents' sense of place, place attachment and identity. The use of the concept in this study will also provide insight into how tourism development and traditional norms and values of Goo-Moremi residents are intertwined.

Factors that Define Sense of Place

Sense of place refers to both descriptive and emotional aspects of the environment experiences (Hashem et al., 2013a, 2013b). Steele (1981) argues that factors which create a sense of place are divided into two categories: cognitive and perceptual factors; and physical characteristics (Figure 20.1). Cognitive factors include the meanings which people attach to a place (Steele, 1981). Sense of place is, therefore, an emotional connection between people and place (Hashem et al., 2013a, 2013b). Steele (1981, p. 56) states that "The physical parameters of sense of place include those that explain Identity, History, Fun, Mysterious, Pleasant, Wonderful, Security, Vitality and memories on the way people communicate with places." In this regard, people's satisfaction with environmental elements is assumed to be an influential factor that creates conditions for measuring place attachment.

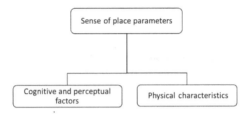

Source: Steele (1981).

Figure 20.1   *Sense of place factors*

The concept of sense of place is considered suitable to analyse the relationship between tourism development and the emotional attachment the people of Goo-Moremi Village have with their physical environment. Culture is considered sacred and ancestor worship controls the daily lives of people of Goo-Moremi Village (Mbaiwa, 2011). There are, therefore, ambivalence, tensions and local resistance by sections of the community especially the elderly in the process of commodifying and globalisation of culture for the tourism market at Goo-Moremi (Mbaiwa, 2011; Siphambe et al., 2017). All these factors make it critical to analyse the relationship between sense of place, place attachment and identity on the one hand and commodification and globalisation of cultural landscapes for the tourism market on the other.

Related to the concept of sense of place is the concept of nature bonding (Figure 20.2). Nature bonding is defined as an implicit or explicit connection people have to some part of the non-human natural environment, based on history, emotional response or cognitive representation such as knowledge generation (Kals et al., 1999; Clayton, 2003; Schultz, 2001; Schultz et al., 2004: Raymond et al., 2010). This is because nature or the physical setting provides the container for social experiences and the bonds which form through these experiences (Raymond et al., 2010). The historical interaction between the people of Goo-Moremi Village and Goo-Moremi Gorge can thus guide an analysis required in this research about sense of place, neighbourhood attachment, sense of belonging and sense of familiarity.

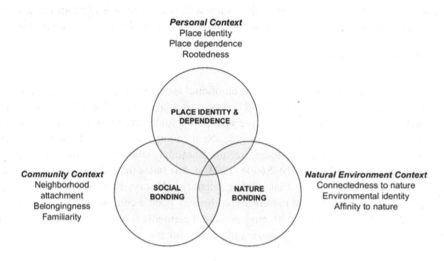

Source: Raymond et al. (2010).

Figure 20.2   The concept of nature bonding

GOO-MOREMI CULTURAL LANDSCAPE

This research was carried out at Goo-Moremi Cultural Landscape in eastern Botswana (Figure 20.3). Goo-Moremi Cultural Landscape includes Goo-Moremi Village and Goo-Moremi Gorge.

Goo-Moremi Village is located 76 km east of Palapye Township in eastern Botswana. The Goo-Moremi community originates from South Africa and migrated to Botswana in the 1800s (White, 2001). Kgosi (Chief) Mapulane led his people to settle in present day Goo-Moremi Village. A total of 597 people live in the village (Statistics Botswana, 2012). Agriculture is the dominant source of livelihoods in the village (Mbaiwa, 2011). Since 1999, Goo-Moremi residents became involved in tourism development through the Community-Based Natural Resource Management programme. The people of Goo-Moremi believe in ancestors whose dwelling place is Goo-Moremi Gorge. The ancestors are believed to watch over the community's socio-economic, socio-cultural and political development. Goo-Moremi Village was thus found to be a suitable site for the analysis of the interaction between cultural tourism development and residents' culture, with a focus on residents' sense of place, place attachment and identity.

Goo-Moremi Gorge occupies a fenced area of about 1797.3880 hectares. It is located about 342 km from Botswana's capital city of Gaborone, 4 km from Goo-Moremi Village and 76 km east of Palapye (Maruping, 2017). The Gorge is considered by Goo-Moremi residents as the spiritual site of their ancestors (White, 2001; Dichaba, 2009). The Gorge area has historical and archaeological attractions, rich biological diversity, spectacular streams and water pools. Historical remains at Goo-Moremi Gorge include the graves of Kgosi Mapulane and his sons. Because of Goo-Moremi Cultural Landscape's rich cultural history, it has become one of the key cultural tourism sites in Botswana. The landscape is therefore a suitable site for analysing

Sense of place and tourism in cultural landscapes 283

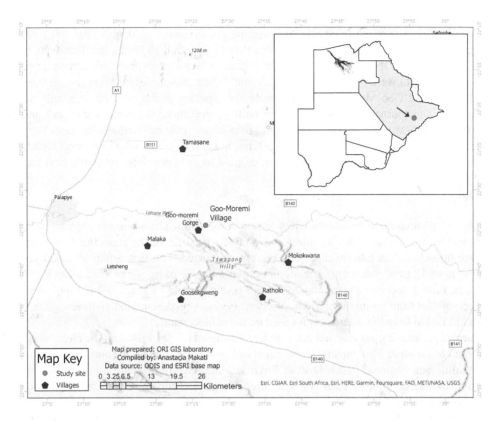

Source: Anastacia Makati, ORI GIS Laboratory.

Figure 20.3 Map of Botswana showing the location of Goo-Moremi Cultural Landscape

the relationship between sense of place, place attachment and identity in an era of commodification and globalisation of culture for the tourism market.

METHODS

Data Collection

The chapter utilises qualitative research methods to collect data. Primary data was collected from face-to-face interviews with household representatives, focus group discussions (FGDs) and key informants such as village leaders and decision-makers. Household interviews were conducted using a structured questionnaire that comprised both closed and open-ended questions. Open-ended questions were used to solicit the views of all the household representatives on the historical and cultural changes at Goo-Moremi Village. Data collected included that on the interaction of cultural tourism development and how it impacts on local traditional values of ancestor worship and daily lives. Data on how residents interact with Goo-Moremi Gorge,

284   *Handbook on tourism and conservation*

their sense of place, identity and resident attachment to their culture and Goo-Moremi Gorge were collected. The open-ended questions required respondents to identify, list, explain and describe the historical and cultural attachment they have with their landscape. Residents were also asked to narrate their experiences and interaction with ancestors and tourism development.

Secondary data were obtained from both unpublished and published literature on the cultural heritage of Goo-Moremi Cultural Landscape. Specific literature used included: policy documents and journal articles on cultural heritage and tourism development, and annual reports of Goo-Moremi culture. Historical data about cultural heritage at Goo-Moremi Cultural Landscape was also used. Historical data made it easy to track the cultural history of Goo-Moremi residents and the emergence of cultural tourism development in the landscape.

Sampling

A total of 100 household representatives was randomly sampled from a total of 165 households at Goo-Moremi Village. This sample represented 61 per cent of the total 165 Goo-Moremi village households. A total of three FGDs were conducted. The first group included young people aged 18 to 35 years old. The second group included older adults aged 36 to 92 years old. The third group was a combination of both young and old people. FGD is a tool used for collecting data from group discussions (Nielsen, 1997). A predetermined interview guide was used in FGD to keep the discussion focused on the subject matter.

Purposive sampling was carried out with key informants. They included: the Department of National Museum & Monuments, Department of Lands, Department of Tourism, Department of Wildlife and National Parks, District Environmental Health, Ngwato Land Board, Tribal Administration at Goo-Moremi Village, the Botswana Tourism Organization, the chief of Goo-Moremi Village and the spiritual mediators in the village. The sampling of key informants was done taking into consideration their positions of influence; knowledge of the ancestral culture; and their experience and long-term knowledge of the cultural and historical characteristics of their landscape and the tourism development process in Goo-Moremi Village. As a result, key informants were representatives "who are in a privileged position to provide detailed information on local area processes" (Pauwels & Hardyns, 2009, p. 404).

Data Analysis

This study was qualitative hence content analysis was used. Content analysis involved the summarisation of data into themes and patterns. This approach made it easier to interpret data and inductive conclusions were drawn. Leininger (1985, p. 60) argues that in thematic analysis, themes are identified by "bringing together components or fragments of ideas or experiences which often are meaningless when viewed alone". As a result, in this research, themes that emerged from informants' stories about the interaction of culture and tourism in their village were put together into themes and patterns. Aronson (1995) argues that in thematic or content analysis, themes are pieced together to form a comprehensive picture of collective experience. Qualitative data from households, key informants' interviews and FGDs were summarised into specific themes and patterns on the historical and cultural commodification and globalisation at Goo-Moremi Cultural Landscape. Themes were also identified from interaction between tourism development at Goo-Moremi Landscape and the residents' sense of place, place attachment and identity.

Sense of place and tourism in cultural landscapes 285

Table 20.1   Goo-Moremi Gorge used by ancestors as home

Responses from household respondents	Frequency	Percentage
Gorge is used by our gods as their home	99	99.0
Not sure	1	1.0
Total	100	100.0

Table 20.2   Goo-Moremi Gorge used as burial grounds for ancestors

Responses from household respondents	Frequency	Percentage
Gorge is burial site for our chiefs who transform to ancestors at death	97	97.0
I do not know	2	2.0
I am not sure	1	1.0
Totals	100	100.0

RESULTS

Community Views on the Use of Goo-Moremi Gorge

Goo-Moremi Gorge and its immediate environment have sentimental value and are held with high respect by residents. Residents of Goo-Moremi Village regard the Gorge as a "seat of the Gods" (Mbaiwa, 2011, p. 294). During FGDs on 25 April 2018, a village elder noted that: "the gods have told us that they drink water at the Gorge and they should not be disturbed". The significance of Goo-Moremi Gorge to residents further prompted this research to ask households what they consider to be the main use of the Gorge. Responses were in two forms: first, as shown in in Table 20.1, 99 per cent of the households stated that the main use of Goo-Moremi Gorge is as a dwelling place for ancestors and gods; only 1.0 per cent was unsure.

Table 20.2 shows that a total of 97.0 per cent said that the Gorge is used as a burial ground for their chiefs who transform to become ancestors at death. Only 1.0 per cent and 2.0 per cent did not want to commit themselves in fear of punishment from gods and answered in the negative.

These results are interesting in that none of the household representatives stated tourism as a use for the Gorge even though tourism has been developed around the site, including chalets, campsites, gate houses, walking trails, interpretation signs, road networks and ablution block as tourist amenities. These results therefore illustrate the attachment the people of Goo-Moremi have with their ancestors and Moremi Gorge and their disregard of the growing tourism industry in their local environment.

Because of the historical attachment the people of Goo-Moremi Village have with the Goo-Moremi Gorge, the Gorge is considered sacred and a place of their ancestors. Goo-Moremi residents have a traditional belief system where kings and chiefs do not die but get transformed into spiritual beings that continue to oversee the welfare of their people. Kgosi Mapulane and his sons who ruled Goo-Moremi Village over a century are therefore assumed to have been transformed into these spiritual beings. Such beings have become their ancestors or gods and live in the Gorge to watch over the community's welfare (personal communication with Komana leader, 25 April 2018). This ancestral worship and the cultural belief system has created a sense of identity and sense of place for Goo-Moremi people. There is a bond between Goo-Moremi residents and their physical environment that constitutes place attachment.

286   *Handbook on tourism and conservation*

Ancestor Worship at Goo-Moremi Cultural Landscape

The ancestor worship at Goo-Moremi Village was found to be very strong and religiously observed by everyone in the community. At the centre of ancestral worship at Goo-Moremi Village is a male group of Bazina (spiritual mediators) and their ancestral spirit council known as the Komana where they make decisions. The Bazina hold regular spiritual council meetings to direct the affairs of the village. The Komana is headed by a spiritual leader who directly communicates with ancestors or gods. The Bazina interpret the wishes and concerns of the ancestral spirits to the people of Goo-Moremi Village and vice versa (Mbaiwa, 2011). The Bazina are responsible for organising ceremonies and sacrifices to ancestors. They also serve intermediary roles between the community and the ancestral spirits. As a result, prior to undertaking any major development project at Goo-Moremi Village for example, village electrification, water reticulation, school development and road construction, consultations with the ancestors must be initiated through Bazina. These consultations are done through a traditional ceremony known as Mophaso. During the Mophaso a day is set aside for libations, singing and dancing at the village public gathering. The Bazina approach the ancestors and consult them about proposed development projects. The ancestors, in turn, communicate with Bazina and offer their responses. During interviews, the chief spiritual mediator of Bazina noted, "we need developments in our village, but all development projects should be requested from ancestors" (25 April 2018). The chief spiritual mediator acknowledged that he is a messenger of the ancestors to the people of Goo-Moremi Village. This shows the power which ancestors have over the lives of the local people. The spiritual mediators are a group composed of people believed to be descendants of Kgosi Mapulane. Maruping (2017, p. 1) notes that:

> Not only is the gorge akin to some places found in the Bible but it is also a highly sensitive area. Traditionally and even today, communication with the Badimo has always taken place with the use of the Komana (community members living in the village similar to soothsayers) who are responsible for interpreting the wishes and concerns of the Badimo.

These results indicate that the Goo-Moremi community believes that their ancestors (Badimo) are omnipresent, and this belief remains powerful in the daily lives of the community as noted by Kiyaga-Mulindwa (1980), Werbner (2004) and White (2001). Ancestral worship and a strong spiritual belief system play an important role in the community's daily activities and how they organise their lives. Key informants revealed that there is usually communication between the ancestors (Badimo) and the community on several issues through the Mophaso ceremony. This is performed to appease the ancestors and rituals are conducted to ask for rain, advise on infrastructural developments, and seek direction for ailments and diseases. Key informants said that communication with ancestors is also done during funerals and during announcements of the death of a community member. These activities – except for announcement of death – happen at night or during the Mophaso ritual ceremony held at the main Kgotla and coordinated by the members of the Komana group (Dichaba, 2009).

Key informants also indicated that during Mophaso, the ancestors would communicate about ailments and/or diseases that the concerned people are seeking help for. Thus, they will direct the patient to where they can get help – from medical doctors and/or traditional healers. Badimo are said to be able to direct the person seeking help to the place where such help will come from and are even able to tell if there is no hope that the person may be healed by saying that they have long received the spirit of such a person and it would be known that there is no

Sense of place and tourism in cultural landscapes 287

Table 20.3   Role of ancestors for Goo-Moremi community

Role of ancestors	Household responses in percentage, n=100			
	Yes	Not sure	No	Total
Protect our community from disasters, e.g. diseases and drought	95	3	2	100
Are consulted in everything the community does, e.g. ploughing and harvesting	98	1	1	100
They are part of everyday lives of our community	96	2	2	100
Socio-economic status of the community depends on them	88	4	8	100
Guide and watch over community wellbeing	96	2	1	100

hope of healing and the person would eventually die. The ancestors can also predict tragedies that will befall the community. Thus, if a member of the community gets injured or dies away from home, for example working in the mines, then the ancestors would convey a message through the Komana, even before the concerned family is notified through telephones or other means (Dichaba, 2009). The community believe that the ancestors protect them from disasters such as diseases and droughts.

Role of ancestors before and after commodification
Households were asked whether the role of ancestors before and after the globalisation and commodification of their landscape has changed. The results indicate that all the 100 households (100.0 per cent) interviewed said the roles have not changed in the two eras (i.e., before and after 1992). Households were further asked to state what they consider to be the role of ancestors in their community. The results in Table 20.3 indicate community responses describing the role of ancestors in their community. They protect the community from disasters such as diseases and drought; they are consulted in everything the community does such as ploughing and harvesting; they guide the community's everyday lives; the socio-economic status of the community depends on them; and they watch over the community's wellbeing. These responses can be summed in one statement, which is, "ancestors guide the socio-cultural, economic and political lives of Goo-Moremi residents".

In FGDs, respondents were asked to confirm household responses on the role of ancestors in their community; similar responses were obtained, and they include the following:

- Ancestors guide the village on socio-economic activities.
- Ancestors convey death messages to the community.
- When ancestors are requested, they provide rain and can tell when the rain will come.
- Ancestors direct the community on where to seek medical attention, i.e., either from traditional doctors or from modern doctors.
- Ancestors tell the community when to start ploughing and crop production each year.

The role of ancestors at Goo-Moremi were also confirmed through informal interviews with the Chief of Goo-Moremi Village, Kgosi Major Tshito Days, when he noted that:

> the ancestor culture is the backbone of Goo-Moremi community's economy, it also brings dignity to the community at the same time attracting a lot of tourists to Goo-Moremi Cultural Landscape.

These results indicate that the Goo-Moremi community have unique traditions and customs which define them differently from the rest of the people of Botswana. These traditions and customs have also become a tourist product for both national and international tourists.

288   *Handbook on tourism and conservation*

Table 20.4   Any rules governing community from ancestors

Household response	Frequency	Percentage
Yes, there are rules (i.e., taboos and norms)	96	96.0
Not sure	2	2.0
I do not know	1	2.0
Total	100	100.0

Table 20.5   Household knowledge of taboos and norms from ancestors

Household response	Frequency	Percentage
I have knowledge on taboos and norms	95	95.0
Not sure	1	1.0
I do not know	4	4.0
Total	100	100.0

Taboos and Social Norms at Goo-Moremi

Results in this study indicate that the traditions of the people at Goo-Moremi and the ancestors' role in governing the community are rooted in social taboos and norms which guide and dictate how people should live. These taboos and norms form a strong cultural and spiritual belief system which every community member should observe without question and they are passed from generation to generation. Households were asked whether there are any rules and regulations in the form of taboos and norms for their community from ancestors. The results in Table 20.4 indicate that a total of 96.0 per cent of the households acknowledged that ancestors in their village govern them through rules and regulations in the form of taboos and norms. Only 4.0 per cent of the respondents responded in the negative mainly because they did not want to commit themselves in fear of discussing secrets of the village with outsiders.

These results indicate that ancestors play a significant role in governing the lives of the people of Goo-Moremi Village. This is achieved through taboos and norms which residents of Goo-Moremi are expected to respect. Research has shown that social taboos exist in most cultures, both Western and non-Western (Colding & Folke, 2001). Taboos serve as informal institutions, hence norms, rather than government laws and rules determining human behaviour and living (Colding & Folke, 2001). Colding and Folke (2001) argue that in many traditional societies throughout the world, taboos guide human conduct in the use of the natural environment. Taboos and norms are informal institutions that govern a society that are not dependent on the state for enforcement (Colding & Folke, 2001; Jones et al., 2008). In the case of Goo-Moremi Village, taboos and norms are said to be instructions from the gods who live in Goo-Moremi Gorge and the Hills. Households were further asked to state whether they are familiar with specific taboos and norms designed for various aspects of social life such as the behaviour of women and men, human movement, sexual activity, farming and tourism. Results in Table 20.5 indicate that a total of 95.0 per cent of the households noted that they know of all the taboos and norms from the ancestors governing aspects of their social life. Five per cent of the households were afraid to divulge information on this subject for fear of punishment from the ancestors hence they preferred to give "not sure" and "no" responses. These results further demonstrate that the people of Goo-Moremi Village know the expectations of their ancestors as to how they want them to live and conduct their daily lives.

The taboos and norms identified in our research have been categorised into those governing behaviour of women, projects, travelling, etc. They are discussed below.

Taboos and norms governing everyday life
Results indicate that there are several taboos and norms meant specifically for the day-to-day conduct and behaviour of the people of Goo-Moremi Village, which include the following:

- All development projects in Goo-Moremi area should be submitted to ancestor spirits for approval. The spiritual mediator should organise a spiritual ceremony known as Mophaso and request permission from ancestors for the implementation of projects. Without this permission, no project should be carried out in the Goo-Moremi area.
- Ancestors' secrets must not be divulged to unauthorised people and outsiders. These secrets include those related to ancestral culture. During interviews, some respondents could not answer questions about ancestral culture for fear of reprisal by the spirits.
- A day reserved for the Mophaso should be observed by every community member. No one is allowed to do any work except participation in the ceremony.
- Noise making at Goo-Moremi Village at night is prohibited.
- Whistling while at the Gorge is said to be a serious misdemeanour since ancestors communicate through whistling. Key informants note that if one whistles in the Gorge, the ancestors may appear, and the person will not know how to handle them and this can result in the person disappearing for ever, getting injured or becoming mute forever.

Respondents acknowledged that their ancestors are able to see whatever members of the community do or say in their everyday lives. White (2001, p. 16) notes that violations of the "rules are not subject to sanctions by human authority but by the ancestors" who may bring bad luck to people who contravene the rules. The ancestors must then be propitiated by sacrificing either a black ox or goat whose portions will then be cooked and placed at the Kgotla as an offering to the ancestors for someone to escape bad luck. The elders revealed that signs of disobedience to the rules guiding behaviours are relayed through the Komana to alert the whole community of such disobedience. The person or people who disobeyed will then be summoned to the Kgotla for flogging or punished according to the dictates of the ancestors. The ancestors may decide to forgive a person if the person shows remorse for what they did, or they can punish the individual or a group of people.

Taboos and norms for women
There are taboos and norms that govern women, and these include the following:

- Women on their monthly menstruation periods and those who are pregnant are not allowed to go to the Kgotla during the Mophaso ceremony.
- Women should not visit Goo-Moremi Gorge during menstruation. This is because of the belief that a woman in her menstruation period is unclean. The gods or ancestors do not want to be defiled.
- Babies who have not yet teethed are not allowed at the Kgotla. The Komana men as representatives of the ancestors are not allowed to touch such babies, as they are deemed unclean and would defile them.

290 *Handbook on tourism and conservation*

Traditional beer meant for the Mophaso ceremony is only prepared by elderly women who have reached menopause as they are considered clean by the ancestors.

Taboos and norms governing sexual activity

Results also indicate that there are taboos and norms that govern sexual activity for couples either married or unmarried; these include:

- The whole community is supposed to abstain from sexual activities on the night of the Mophaso ceremony.
- All couples, whether married or unmarried, should avoid sexual activity the night before visiting Goo-Moremi Gorge or on the day the Mophaso ceremony is held. Sexual activities should also be avoided at the Gorge.

Respondents indicated that offences for engaging in sexual activities during Mophaso and before visiting the Gorge are punished. Offenders receive punishment such as that of giving birth to an albino or a disabled baby. Women who ignore the requirement that they should not visit the Gorge during their menstruation period may slip and fall into the pools and even drown or get injured. The Head of Komana during interviews noted that recently, a young woman in her monthly period ignored the requirement and visited the Gorge. She was punished by the ancestors as she fell into one of the pools at the Gorge and was injured.

Taboos and norms governing movement in the village

The movement of people at Goo-Moremi Village especially at night is governed by taboos and norms. As a result, all people in the village are expected to observe the following:

- Ancestor spirits that guard the village are considered not to approve of lights from vehicles and fires in the village after 18:00 hours.
- No lights and noise are allowed from either people talking, radios or cars driving at night. Ancestors are said to be against vehicles' lights and no movement is allowed.

Respondents noted that cars driving into Goo-Moremi Village after 18:00 hrs have been reported to have had engine failure and stop running. However, the same vehicle would have its engine start running in the event the driver starts in a different direction away from the village. Key informant interviews revealed that moving around the village at night during Mophaso is prohibited because it is believed that one may meet the ancestors when they are on their way to and from the Kgotla and this may result in injury. However, key informants said that if a person meets the ancestors at night, they should stand still until the ancestors have passed. The punishment from the ancestors for people who move about at night during Mophaso is the twisting of the neck or making the person disappear forever.

Taboos and norms governing tourists

There are taboos and norms that govern behaviour of visitors at Goo-Moremi Gorge which include:

- Swimming in the pools is not allowed. There are six deep pools receiving water from a permanent waterfall and a stream running across the gorge. The temptation for tourists who visit the Gorge is to swim in these cool pools. The pools are the bathing places for the ancestors' spirits not humans.

- All tourists, whether married or unmarried, should avoid sexual activity the night before visiting Goo-Moremi Gorge. Sexual activities should also be avoided during a stay at the Gorge.
- Female tourists who might be in their monthly menstruation periods should not visit the Gorge. Such a woman is considered unclean and the ancestors do not want to be defiled.

No killing of snakes or any animal in the gorge is allowed, nor picking of anything people see at the gorge. The requirements of visitors at Goo-Moremi are further illustrated by journalist M. Maruping after a tour of Goo-Moremi Gorge who noted that:

> One important rule interpreted by the Komana that visitors should be aware of is that the pools in the gorge are reserved for the ancestors (badimo) and swimming is not allowed in them. Furthermore, let me stress that smoking cigarettes or drinking alcohol is not allowed in the gorge. You are also not allowed to pick or leave anything foreign in the gorge. These are the rules which must be followed. (Maruping, 2017, p. 2).

Goo-Moremi elders noted that it is taboo to swim in the Gorge because it is a drinking place for the ancestors hence the water should not be defiled. The Gorge has always been protected by the strength of the spiritual and cultural beliefs that are associated with it. As a result most residents are reluctant to visit the Gorge either for firewood collection, cutting poles for house construction or even to hunt in it (White, 2001).

These results indicate that cultural values, taboos and beliefs of Goo-Moremi Village have come to define the identity of residents. This belief system forms the sacred culture, dignity and history of the people of Goo-Moremi Village. It creates an identity not found anywhere else in Botswana. In-depth interviews with the spiritual mediator (head of Komana), village chief, village elders and other community members indicated that no one within the Goo-Moremi community questions these taboos and beliefs because they are interpreted as instructions from ancestors who are the guardian spirits in the village. The people of Goo-Moremi Village were found to be proud of their traditional customs of ancestor worship. During FGD at the Kgotla, young people who commented demonstrated pride about their history and were knowledgeable about their traditions. This indicates that oral history has been passed from one generation to another within the village. This was particularly so even though to an outsider social norms and taboos may appear to be repressive to the people. This defines the emotional attachment people have with their environment and their daily lives that are governed by the ancestors.

Stories of Development Projects With or Without Ancestor Permission

Data gathered from FGDs indicate that there is a history of development projects, which were requested from ancestors and were allowed. Conversely, permission was not requested from ancestors for other projects because implementers ignored the role of ancestors in project development at Goo-Moremi Village. None of these development projects succeeded. Only those projects that observed the taboos and norms of the village succeed. Previously, government ignored taboos and norms of Goo-Moremi residents when implementing development projects. The result was that such projects either failed or stalled until permission was sought from the ancestors. Examples of such projects are given below.

292 *Handbook on tourism and conservation*

Kgosi Tshekedi Khama Road and car saga

The legendary Tshekedi Road linking Moeng College to Palapye was initiated by Kgosi Tshekedi Khama but later abandoned because the ancestors were not happy with it (Lesotlho, 1983; Dichaba, 2009). Moeng College was the first secondary school in pre-independence Botswana and most of the country's first leaders attended it. They include key personalities such as the former President Festus Mogae, Minister of Finance, Mr Baledzi Gaolatlhe and Foreign Affairs Minister, Dr Gaositwe Chiepe. Kgosi Tshekedi was a regent to one of the main ethnic groups in Botswana, the Bangwato. He was also uncle of Botswana's founding President Sir Seretse Khama (1966–1980). He became regent and ruler of the Bangwato tribe as the rightful leader, Sir Seretse Khama, was still studying in Britain.

The elders during their FGD revealed that Kgosi Tshekedi Khama was instrumental in establishing Moeng College, and therefore needed a road that would connect Palapye to the college, hence his choice of a passage on top of the Gorge to construct the road. Palapye by then was the nearest major township in the area with connections to other centres such as Gaborone and Francistown through a railway line. Palapye also linked these centres with a road network together with Serowe the capital of the area where Kgosi Tshekedi resided. Key informants noted that road workers would work hard during the day clearing the bushes and cutting down trees to prepare the way for the road. However, in the morning when they came back to continue with the road construction, they found the trees growing as if they had never been chopped. Kgosi Tshekedi used mephato (a regiment) which is a group of men in a community of the same age group and often hundreds in number. Key informants alleged that Tshekedi Khama did not ask for permission from the ancestors before he started the road to Moeng College. Kgosi Tshekedi Khama finally abandoned the road construction project and sought a different route that goes around the hill and that is the present road which is used to travel to the College.

The second incident that involved Kgosi Tshekedi at Goo-Moremi Gorge was when his car could not drive forward but only backwards at the Gorge. FGD respondents revealed that Kgosi Tshekedi Khama's car got stuck at the Gorge and could not drive forward into the Gorge where the ancestors live. The car only made it backwards in the direction of Palapye where it came from. This is interpreted to mean that the ancestors were not amused by his visit to their dwelling place but spared his life by allowing him to go back where he came from. The failure of Kgosi Tshekedi's car was noted by key informants to have been a result of the fact that ancestors do not like car fumes, hence they stopped his car from proceeding through the Gorge. Kgosi Tshekedi had not followed proper procedure of communicating with ancestors through spiritual leaders and Mophaso. Respondents noted that Kgosi Tshekedi violated all the taboos and norms at Goo-Moremi because he claimed to have jurisdiction over the area including all chiefs and ancestor worship. However, his road plans failed.

Fencing of the Motlhodi (water spring) fiasco

Fencing of the Motlhodi (a water spring coming from the gorge) was another project that the gods objected to and was later abandoned. In the 1990s, the Department of Water Affairs (DWA) wanted to fence and secure a water spring at the Goo-Moremi Gorge known as Motlhodi. The elders revealed that the fence team would put up the fence during the day and the following morning when they got there they would find the fence pulled out and packed nicely. Respondents noted that the fence team did not give up and persisted in building the fence. This angered the ancestors who responded by releasing mosquitoes that have never been

Sense of place and tourism in cultural landscapes 293

experienced in the area before to bite the team. The team was camped near the Motlhodi spring and it had to run away at night abandoning the project.

Drilling for water by Botswana government
The second story that involved the DWA was that of 1994 when the department wanted to drill and dig trenches collecting water from Motlhodi water spring at the Gorge. The project was meant for water reticulation for Goo-Moremi Village. The story is that government or the DWA did not seek permission from the ancestors to start the project. During the day, DWA workers would drill and equip the borehole, however, the following morning, these workers would find the borehole had been dismantled and all the equipment parked neatly for them. DWA workers persisted and continued drilling and assembling the borehole as well as digging trenches for water pipes. Their persistence angered the ancestors who attacked all the DWA workers with a terrible and continuous diarrhoea which made them abandon the project. However, after a Mophaso was held, the ancestors gave permission to the DWA for the project and the borehole is still working to this day.

Construction of Goo-Moremi Road delay
The 10-kilometre tarred road from the Lesenepole junction to Goo-Moremi Village is another example of a failed project. The road construction began in 2001. However, the Department of Roads and the construction company building the road ignored advice to seek permission from the ancestors by conducting a Mophaso. Key informants noted that during the day, the construction company would work on putting tarmac on the road, but the tar would peel off almost immediately after. In some instances, the machinery used by the construction company if left on the site, the following morning would be dismantled with each piece packed nicely. It took seven years for the Department of Roads and the construction company to acknowledge the need to seek permission from the ancestors. However, after permission was sought and a Mophaso ceremony was undertaken to appease the ancestors (Dichaba, 2009), the road construction was completed. The road is now in good condition and it is the main road that connects Goo-Moremi Village with the outside world particularly with Palapye.

Goo-Moremi tourism projects and culture
In addition to ancestral worship, Goo-Moremi Gorge has historical and archaeological attractions, rich biological diversity such as flora and fauna, spectacular streams and pools, and wildlife which have become tourism products in the area. As a result, natural and cultural products at Goo-Moremi Landscape have been commodified and globalised for the tourism market since 1992. This process of commodifying culture and nature at Goo-Moremi was led by the Kalahari Conservation Society, Botswana Government and Botswana Tourism Organization (BTO) (Siphambe et al., 2017). Geoflux (Pty) Ltd was the consultancy company commissioned by BTO for tourism infrastructure development at Goo-Moremi Cultural Landscape. The company recommended that permission and a Mophaso ceremony should be held for the project to be undertaken (Geoflux, 2009). The BTO observed the recommendation, and the ceremony was conducted. Results from key informants' interviews and FGDs also acknowledge that infrastructural developments for tourism development at Goo-Moremi Village sought permission through Mophaso. The ceremony was sponsored by the BTO and a black cow was killed. The permission and cooperation sought to develop tourism at Goo-Moremi Gorge was acknowledged by the Minister of Environment, Natural Resource Conservation

294 *Handbook on tourism and conservation*

and Tourism, Mr Tshekedi Khama, who noted at the official opening of Goo-Moremi Resort on 24 February 2017 that:

> I can confidently state that this is what we did when setting up this facility, led and guided by the community elders. The spirituality of the place and its respective sacred nature which we took on board when developing this place is embedded in the story that is shared with visitors to the place as it is what makes Goo-Moremi Resort unique.

Recognising the cultural sensitivity at Goo-Moremi Gorge, Mr Khama expressed Botswana Government's appreciation of the norms and heritage of the people around which developments take place. He noted that when setting up tourism operations, the cultures and norms of the people adjacent to the developments should be observed and respected. Minister Khama said:

> The development of this Resort was preceded by an Environment Impact Assessment (EIA), and an Archaeological Assessment Report, to ensure that all the valuable resources such as archaeological, cultural, environmental, and social aspects of the place were observed and preserved. And that mitigation measures are put in place. (Maruping, 2017, p. 2)

Permission was sought for the following projects: Goo-Moremi Resort which offers two presidential suites; six standard rooms with en suite facilities with double vanities, showers and baths; and six developed campsites, each with a holding capacity of 35 people. These campsites come with hot shower and fully equipped gas braai facilities. There is also an à la carte restaurant which can hold up to 120 people in a cocktail set-up and 60 seated; a swimming pool; a bridge across the Lotsane River; a fence around the Goo-Moremi Gorge; and chalets in the Gorge area with electricity connection. In confirming that the construction of tourism facilities and daily operations of the tourist activities adhere to the cultural taboos and norms of the area, the BTO Acting Chief Executive Officer, Mr Zibanani Hubona, noted that:

> Goo-Moremi Resort opened to diversify the tourism product from the wildlife and wilderness experience, to a more community based cultural and heritage tourism product, which aims to benefit both the local community and the various tourist markets, either local, regional, or international. (Maruping, 2017, p. 2)

Previously, ancestors were very much against these projects; however, after the Mophaso ceremony, ancestors granted permission to projects which were all completed in 2014 and a cultural tourism project is currently operational at Goo-Moremi Gorge. This shows how seriously the community and other stakeholders such as government and other organisations respect the taboos and the traditional beliefs that reign supreme in this conservation area. It is in this context that Mbaiwa (2011, p. 296) noted that "ancestor worship appears to exert a strong influence upon the lives of the people of Goo-Moremi Village".

Even in modern Botswana, occurrences at Goo-Moremi Gorge which modern science could interpret as being caused by physical natural conditions such as exfoliation and erosion are interpreted as acts of ancestors and gods. One example is the Sir Seretse Khama Alarm Stone, a big boulder which apparently fell from the Gorge above on the morning the founding President of Botswana Sir Seretse Khama died in the small hours of 13 July 1980 (Maruping, 2017). A framed poster placed on the monumental rock reads:

On the morning of the 13th of July 1980 between 2am and 5am, a loud rumbling was heard by the villagers of Goo-Moremi as this stone fell from the gorge above. The stone falling is said to be a sign that someone of great importance had passed away. Later that morning, it was formally reported to the village that Sir Seretse Khama had passed away.

What is emerging from these results is that even though a 5-star tourism resort is promoted at Goo-Moremi Gorge, tourism planners and developers are cautious not to disturb the traditions and belief system of the place. This indicates that Goo-Moremi culture remains respected by all newcomers to the areas and is unchanged. As Glasson et al. (1995, p. 59) argue, "tourism is, by its very nature, an agent of change", however, it has been unable to change ancestor (Sedimo) culture at Goo-Moremi Cultural Landscape.

DISCUSSION

Globalisation and commodification of the tourism products at Goo-Moremi respects the culture of residents. Goo-Moremi residents have called for a tourism industry in their village which respects the taboos and norms of their culture. Tourism development at Goo-Moremi thus adheres to local customs and thus enhances a sense of place, place attachment and identity for the people of Goo-Moremi Village. Bender (1993) places 'people' at the centre of landscapes. The community at Goo-Moremi see themselves as a part of the Goo-Moremi Landscape (Dichaba, 2009). This is the landscape given to them by their ancestors. This confirms Bender's (1993) argument that landscapes are created by people through their experiences and engagement with the world around them, even as they move from place to place. Bender further argues (1993, p. 3) that: "The landscape is inert, people engage with it, re-work it, appropriate it and contest it. It is part of the way in which identities are created and disputed, whether as an individual, group or nation state".

People and their culture are intertwined (Crumley and Marquardt, 1990) hence provide a sense of place to residents. Goo-Moremi Gorge and its cultural history provides the people of Goo-Moremi with place attachment. Place attachment is the emotional bond between person and place (Florek, 2011). The people of Goo-Moremi do not only have an emotional attachment with Goo-Moremi Gorge and its culture, they also have experiences to share that form part of their identity. Thus, place attachment develops from positive experiences and the satisfactory relationship between a person and a place (Casakin et al., 2013; Proshansky et al., 1983). The people of Goo-Moremi have a cultural bond with their ancestors who live in Goo-Moremi Gorge. Residents believe that the gorge and their ancestors should not be disturbed. This confirms the view that place attachment has an emotional impact on individuals who are attracted to a physical environment through emotional and cultural bonds (Abbas et al., 2013a, 2013b). Place attachment is a symbolic relationship between a physical environment and people who live in and around it. This therefore defines people's emotional meanings and sense of place, their perceptions of the place and how they relate to this physical environment (Low & Altman, 1992; Abbas et al., 2013a, 2013b). The relationship between Goo-Moremi residents and their environment and culture can be described as very strong because it is sustained by taboos and norms that are not significantly affected by the growing cultural tourism industry in the area.

Studies have concluded that the nature and strength of attachment to community, and to surrounding landscapes, may influence how residents perceive potential impacts of a growing

296 *Handbook on tourism and conservation*

tourism industry. And this may be an important determinant of successful coexistence between residents and the tourism industry (Williams et al., 1995; McCool & Martin, 1994; Sheldon & Var, 1984; Um & Crompton, 1987). It is from this perspective that Goo-Moremi residents are ambivalent about tourism development in the area. When comparing tourism and ancestor worship at Goo-Moremi Gorge, almost all the households noted that the Gorge is used by their gods and disregarded tourism development. This indicates the strong relationship people have with their environment as it gives them identity and meaning. Place identity comes from beliefs, meanings, emotions, ideas and attitudes assigned to a place (Casakin et al., 2013; Proshansky et al., 1983). Therefore, it is the ancestor worship in their culture that gives them identity and defines them as a people.

Sense of place has been used as an indicator of community sustainability (Stedman, 1999). This suggests that there is a crucial relationship between residents' sense of identity and behaviour towards sustainability (Uzzell et al., 2002). If the residents' sense of place is strong, the more stable their community is likely to be (Sullivan et al., 2009). The preservation of sense of place enables destinations to retain their unique character and allows the local community to maintain their sense of belonging and identity. There is no doubt that Goo-Moremi residents have kept their unique culture for generations even though it has been commodified for the tourist market.

Because of the sense of place, place attachment and identity which the people of Goo-Moremi have with their physical environment and their ancestors who live at Goo-Moremi Gorge, developments by government are commissioned after permission has been sought from the ancestors – a traditional requirement by residents. This is illustrated by projects such as the water reticulation project by the DWA; the road construction project by the Department of Roads; and the tourism project by the Botswana Tourism Organization which involves stakeholder departments such as: Department of Tourism, Department Museum and National Monuments, Department of Wildlife and National Parks, Ngwato Land Board, and the Central District Council. Results of this study therefore suggest that commodification and globalisation of cultural landscapes for the tourism market does not always and automatically result in the modification or the destruction of cultural authenticity as noted in cultural tourism literature. Instead, at Goo-Moremi, globalisation and commodification promotes cultural preservation and sustainability.

CONCLUSION

The sense of place, place attachment and identity of resident communities to their cultural landscape which also happens to be a tourism destination cannot be overemphasised. The people of Goo-Moremi have taboos, beliefs and traditional practices which have kept their community intact for generations. These taboos, beliefs and traditional practices are built on a culture known as the Sedimo or ancestor worship. The ancestors of Goo-Moremi Village reside at Goo-Moremi Gorge. The ancestors watch over the village and determine the welfare of the people. All the generations of the Goo-Moremi community have respected and hailed this ancestor culture without questioning it. This tradition provides the people of Goo-Moremi with a sense of place, sense of attachment and identity.

Tourism development at Goo-Moremi thus has insignificant negative effects on residents' culture. As result, residents' sense of place, place attachment and identity is only made more

Sense of place and tourism in cultural landscapes 297

authentic by tourism development. The commodification and globalisation of cultural land-scapes for the tourism market does not always result in the destruction of cultural authenticity. In the case of Goo-Moremi residents, globalisation and commodification of their cultural landscape has resulted in the promotion and preservation of their culture sustained through taboos and norms built on ancestor worship. In conclusion, the cultural tourism development in cultural landscapes needs to recognise residents' identity, attachment and sense of place. That is, if tourism is to be sustainable, it needs to observe the socio-cultural, economic, political and environmental conditions of resident communities.

REFERENCES

Abbas, Y. S., Akbar, H. A., & Nazgol, B. (2013a). Comparison the concepts of sense of place and attachment to place in architectural studies. *Malaysia Journal of Society and Space*, 9(1), 107–117.

Abbas, Y. S., Akbar, H. A., Nazgol, B., & Mohammad, E. (2013b). Effect of place attachment in creating sense of place case study: Tajrish old Bazaar and new commercial center. *International Research Journal of Applied and Basic Sciences*, 4(4), 855–862.

Aronson, J. (1995). A pragmatic view of thematic analysis. *The Qualitative Report*, 2(1), 1–3. http://nsuworks.nova.edu/tqr/vol2/iss1/3, accessed 7 October 2018.

Ateljevic, I., & Doorne, S. (2003). Culture, economy and tourism commodities: Social relations of production and consumption. *Tourism Studies*, 3(2), 123–141.

Bender, B. (ed.) (1993). *Landscape: Politics and Perspectives*. Oxford: Berg Publishers.

Casakin, H., Ruiz, C., & Hernandez, B. (2013). Differences in place attachment and place identity in non-natives of cities of Israel and cities of Tenerife. *Estudios de Psicologia*, 34(3), 287–297.

Casey, E. S. (1996). How to get from space to place in a fairly short stretch of time: Phenomenological prolegomena. In S. Feld & K. H. Basso (eds), *Senses of Place*. Santa Fe: School of American Research Press, pp. 13–52.

Chhabra, D. (2010). *Sustainable Marketing of Cultural and Heritage Tourism*. London: Routledge.

Cighi, C. I. (2008). Senses of place. Master's thesis, University of Massachusetts Amherst.

Clayton, S. (2003). Environmental identity: A conceptual and an operational definition. In S. Clayton & S. Opotow (eds), *Identity and the Natural Environment: The Psychological Significance of Nature*. Cambridge, MA: MIT Press, pp. 45–65.

Cohen, E. (1988). Authenticity and commoditization in tourism. *Annals of Tourism Research*, 15, 371–386.

Cohen, E. (1989). 'Primitive and remote': Hill tribe trekking in Thailand. *Annals of Tourism Research*, 16(1), 30–61.

Colding, J., & Folke, C. (2001). Social taboos: "Invisible" systems of local resource management and biological conservation. *Ecological Applications*, 11, 584–600.

Crumley, C., & Marquardt, W. H. (1990). Landscape: A unifying concept in regional analysis. In K. M. S. Allen, S. W. Green, & E. B. W. Zubrow (eds), *GIS and Archaeology*. London: Routledge, pp. 73–79.

Dichaba, T. S. (2009). From monuments to cultural landscapes: Rethinking heritage management in Botswana. MA thesis, Rice University, USA.

Feld, S. (1996). Waterfalls of song: An ecoustemology of place resounding in Bosavi, Papua New Guinea. In S. Feld & K. H. Basso (eds), *Senses of Place*. Santa Fe, NM: School of American Research Press, pp. 91–136.

Feld, S., & Basso, K. H. (eds) (1996). *Senses of Place*. Santa Fe: School of American Research Press.

Florek, M. (2011). No place like home: Perspectives on place attachment and impacts on city management. *Journal of Town & City Management*, 1(4), 346–354.

Geoflux (2009). *Consultancy Services for Environmental Impact Assessment for the proposed Tourism Development of Moremi Manonnye Conservation Area*. Final EIA Report. Gaborone: Botswana Tourism Board.

298   *Handbook on tourism and conservation*

Glasson, J., Godfrey, K., & Goodey, B. (1995). *Towards Visitor Impact Management: Visitor Impacts, Carrying Capacity and Management Responses in Europe's Historic Towns and Cities*. Aldershot: Avebury.

Hashem, H., Abbas, Y. S., Akbar, H. A., & Nazgol, B. (2013a). Comparison the concepts of sense of place and attachment to place in architectural studies. *Malaysia Journal of Society and Space*, 9(1), 107–117.

Hashem, H., Heidari, A. A., & Hoseini, P. M. (2013b). Sense of place and place attachment. *International Journal of Architecture and Urban Development*, 3(1), 5–12.

Jones, J. P. G., Andriamarovololona, M. M., & Hockley, N. (2008). The importance of taboos and social norms to conservation in Madagascar. *Conservation Biology*, 22(4), 976–986.

Kahn, M. (1996). Your place and mine: Sharing emotional landscapes in Wamira, Papua New Guinea. In S. Feld & K. H. Basso (eds), *Senses of Place*. Santa Fe, NM: School of American Research Press, pp. 167–196.

Kals, E., Schumacher, D., & Montada, L. (1999). Emotional affinity toward nature as a Motivational basis to protect nature. *Environment and Behavior*, 31(2), 178–202.

Kearns, G., & Philo, C. (1993). Culture, history, capital: A critical introduction to the selling of places. In G. Kearns & C. Philo (eds), *Selling Places: The City as Cultural Capital, Past and Present*. Oxford: Pergamon, pp. 1–32.

Kiyaga-Mulindwa, D. (1980). *Origins of the Batswapong*. Gaborone: Tswapong Historical Research Project.

Leininger, M. M. (1985). Ethnography and ethno-nursing: Models and modes of qualitative data analysis. In M. M. Leininger (ed.), *Qualitative Research in Nursing*. Orlando, FL: Grunne & Stratton, pp. 33–72.

Lesotlho, J. (1983). Badimo in the Tswapong Hills. *Botswana Notes and Records*, 15(1), 7–8.

Low, S., & Altman, I. (1992). *Human Behaviour and Environments: Advances in Theory and Research*. Volume 12: *Place Attachment*. New York: Plenum Press.

MacCannell, D. (1973). Staged authenticity, arrangements of social space in tourist settings. *American Journal of Sociology*, 79(3), 589–603.

MacCannell, D. (1992). *Empty Meeting Grounds: The Tourist Papers*. London: Routledge.

Mbaiwa, J. E. (2011). Cultural commodification and tourism: The Goo-Moremi community, Central Botswana. *Tijdschrift voor Economische en Sociale Geografie*, 102(3), 290–301.

Maruping, M. (2017). The adventurous and inspiring Goo-Moremi Gorge memoir. *Botswana Unplugged Magazine*. Gaborone, Botswana.

McCool, S. F., & Martin, S. R. (1994). Community attachment and attitudes toward tourism development. *Journal of Travel Research*, 32(3), 29–34.

Najafi, M., & Shariff, M. K. B. M. (2011). The concept of place and sense of place in architectural studies. *International Journal of Humanities and Social Sciences*, 5(8), 1054–1060.

Nielsen, J. (1997). The use and misuse of focus groups. *Software, IEEE*, 14(1), 94–95.

Nijman, J. (1999). Cultural globalization and the identity of place: The reconstruction of Amsterdam. *Ecumene*, 6(2), 146–164.

Pauwels, L., & Hardyns, W. (2009). Measuring community (dis)organizational processes through key informants analysis. *European Journal of Criminology*, 6(5), 401–417.

Proshansky, H. M., Fabian, A. K., & Kaminoff, R. (1983). Place-identity: Physical world socialization of the self. *Journal of Environmental Psychology*, 3(1), 57–83.

Raymond, C. M., Brown, G., & Weber, D. (2010). The measurement of place attachment: Personal, community, and environmental connections. *Journal of Environmental Psychology*, 30(4), 422–434.

Reisinger, M. (2012). Platform competition for advertisers and users in media markets. *International Journal of Industrial Organization*, 30(2), 243–252.

Richards, G. (2007). *Cultural Tourism: Global and Local Perspectives*. New York: Haworth Press.

Schultz, P. W. (2001). Assessing the structure of environmental concern: Concern for the self, other people, and the biosphere. *Journal of Environmental Psychology*, 21, 327–339.

Schultz, P. W., Shriver, C., Tabanico, J. J., & Khazian, A. M. (2004). Implicit connections with nature. *Journal of Environmental Psychology*, 24(1), 31–42.

Shackley, M. (2000). The cultural landscape of Rapa Nui (Easter Island, Chile). In M. Shackley (ed.), *Visitor Management: Case Studies from World Heritage Sites*. Oxford: Butterworth-Heinemann, pp. 66–81.

Sheldon, P. J., & Var, T. (1984). Resident attitudes to tourism in North Wales. *Tourism Management*, 5(1), 224–233.

Siphambe, G., Mbaiwa, J. E., & Pansiri, J. (2017). Cultural landscapes and tourism development in Botswana: The case of Moremi Gorge in Eastern Botswana. *Botswana Journal of Business*, 10(1), 117–137.

Smith, K. A. (2003). Literary enthusiasts as visitors and volunteers. *International Journal of Tourism Research*, 5(2), 83–95.

Smith, S. L. (2014). *Tourism Analysis: A Handbook*. Abingdon: Routledge.

Sorkin, M. (1992). *Variations on a Theme Park: The New American City and the End of Public Space*. Basingstoke: Macmillan.

Statistics Botswana (2012). *National Population and Housing Census 2011*. Gaborone: Ministry of Finance and Development Planning.

Stedman, R. C. (1999). Sense of place as an indicator of community sustainability. *The Forestry Chronicle*, 75(5), 765–770.

Stedman, R. C. (2003). Is it really just a social construction: The contribution of the physical environment to sense of place. *Society and Natural Resources*, 16, 671–685.

Stedman, R., Beckley, T., Wallace, S., & Ambard, M. (2004). A picture and 1000 words: Using resident employed photography to understand attachment to high amenity places. *Journal of Leisure Research*, 36(4), 580–606.

Steele, F. (1981). *The Sense of Place*. Boston: CBI Publishing Company.

Steiner, C. J., & Reisinger, Y. (2006). Reconceptualising object authenticity. *Annals of Tourism Research*, 33(1), 65–86.

Sullivan, L. E., Schuster, R. M., Kuehn, D. M., Doble, C. S., & Morais, D. (2009). Building sustainable communities using sense of place indicators in three Hudson river valley, tourism destinations: An application of the limits of acceptable change process. *Proceedings of the 2009 Northeastern Recreation Research Symposium*, pp. 173–179.

Um, S., & Crompton, J. L. (1987). Measuring residents' attachment levels in a host community. *Journal of Travel Research*, 26(1), 27–29.

UNWTO (2014). *Tourism Barometer: 2013 International Tourism Results and Prospects for 2014*. Madrid: Tourism Market Trends Programme.

Uzzell, D., Pol, E., & Badenas, D. (2002). Place identification, social cohesion, and environmental sustainability. *Environment and Behavior*, 34(1), 26–53.

Werbner, R. (2004). Scared centrality and flows across town and country: Sedimo in Botswana's time of AIDS. In R. Probst and G. Spittler (eds), *Between Resistance and Expansion: Exploration of Local Vitality in Africa*. London: Transaction Publishers, pp. 389–413.

White, R. (2001). *Integrated Development and Management Plan for Moremi Gorge*. Gaborone: Moremi Manonnye Conservation Trust.

Williams, D. R., McDonald, C. D., Riden, C. M., & Uysal, M. (1995). Community attachment, regional identity and resident attitudes towards tourism. In *Proceedings of the 26th Annual Travel and Tourism Research Association Conference*. Wheat Ridge, CO: Travel and Tourism Research Association, pp. 424–428.

WTTC (2020). *Tourism Expenditure*. https://tool.wttc.org/.

21. Co-management of world heritage sites for community benefit

Olekae T. Thakadu, Wame L. Hambira, Gaseitsiwe Smollie Masunga, Barbara N. Ngwenya, Abigail Lillian Engleton, Dandy Badimo and Ineelo Mosie

INTRODUCTION

The need for active multi-stakeholder involvement and participation in the conservation of biodiversity resources and protected areas management is no longer debatable. This is necessary for local communities that are living within the landscapes where the protected areas and resources are found. International organisations, for example, the International Union for Conservation of Nature (IUCN), World Wildlife Fund (WWF), the United Nations Educational, Scientific and Cultural Organisation (UNESCO), the Integrated Conservation and Development Projects (ICDP), and the Man and Biosphere Programs (MABPs), have emphasised the need to integrate conservation and sustainable development (Baird et al., 2018; Larsen, 2012; Murphy, 2010). Implementation of community-based initiatives worldwide has shown that when local communities are empowered to co-manage environmental resources and derive benefits from the management of the resource, the conservation of biological resources and protected areas will be assured. The World Heritage Convention (WHC) has also evolved over time to embrace community participation and engagement in the process of World Heritage Site (WHS) designation. The Convention advocates for the participation of local and indigenous communities in the designation of World Heritage Sites starting from the pre-nomination stage. It also emphasises "the right-based approaches that link conservation, sustainable development and protection of human rights" (Brown & Hay-Edie, 2014, p. 11).

In 2007, the World Heritage Committee advanced these initiatives by adopting a strategic objective on community engagement, thereby underscoring the key role played by local people in the conservation of WHS. This shift necessitated mechanisms to facilitate meaningful participation of local people in the process of nomination, their active involvement in the conservation of the WHS, capacity strengthening and ensuing benefits accruing to the local people for the improvement of sustainable livelihoods (Brown & Hay-Edie, 2014). The Community Management of Protected Areas Conservation (COMPACT) was subsequently established as an innovative approach towards community engagement and co-management of World Heritage Sites and protected areas (Brown & Hay-Edie, 2014).

COMPACT Programme

The goal of the COMPACT programme is to add significant value to existing conservation programmes through community-based approaches to conserve globally significant biodiversity. The programme has been piloted since its inception in 2000 and has demonstrated that "community-based initiatives can significantly increase the effectiveness of biodiversity conservation in WHS while helping to improve the livelihoods of the local people" (Brown & Hay-Edie, 2014, p. 18), the very essence of sustainable development.

The COMPACT model, having been piloted globally, including six WHS in Africa, is now being replicated within the Okavango Delta WHS, situated in the Ngamiland district, Botswana and a tourist hub. The preparatory process to develop the Okavango WHS COMPACT site strategy was undertaken simultaneously with the review of the Okavango Delta Management Plan (ODMP) of 2008 (Department of Environmental Affairs, 2008). By design, the COMPACT model and framework address the recommendations made during the ODMP midterm review which underscored the importance of embracing all stakeholders in the development of plans and processes earmarked for the management of the Okavango Delta (USAID SAREP, 2013).

When the Okavango Delta was inscribed as a WHS in June 2014, the need to "enhance governance mechanisms to empower stakeholders in the management of the property" (World Heritage Committee, 2014, p. 159) was highlighted. In line with this, the State Party was specifically requested to:

1. expand and strengthen programmes which accommodate traditional resource use for livelihoods, user access rights, cultural rights and access to opportunities to participate in the tourism sector, in keeping with the property's Outstanding Universal Value, and
2. continue efforts to address a range of other protection and management issues including governance, stakeholder empowerment, management planning, management capacity and control of alien invasive species (World Heritage Committee, 2014, p. 159).

Replication of the COMPACT model in Botswana is therefore one of the attempts to address the concerns raised by the World Heritage Committee. COMPACT implementation elsewhere has shown that the model promotes community engagement, co-management and governance through broad-based participation (Brown & Hay-Edie, 2014). It has also demonstrated that when local communities are empowered and engaged in the co-management of protected areas, they can significantly contribute to biodiversity conservation as well as the socio-economic development of the communities.

CBNRM and COMPACT

Community participation in conservation is not a new approach in Botswana. In 1989, the community-based natural resources management (CBNRM) programme, "a development approach that incorporates natural resources conservation" (Government of Botswana, 2007, p. ii) was initiated in northern Botswana. The programme later spread throughout the entire country, with about 145 community-based organisations (CBOs) recorded in 2016 (Centre for Applied Research, 2016). Northern Botswana, especially the Ngamiland district, is a hive of active CBNRM CBOs, some of which are operating within the Okavango Delta WHS. Implementation of COMPACT replication projects in the Okavango Delta WHS will thereby be done mostly within local communities either already engaged in the CBNRM programme or who are aware of the programme.

Both the conceptual origins of the CBNRM and COMPACT are based on the assumption that natural resources and protected areas management and conservation will benefit much from local communities' active participation and meaningful benefit. They postulate that when

302 *Handbook on tourism and conservation*

communities are empowered to manage resources within their areas, and they derive benefits that will contribute to livelihood improvements from such management and conservation, resource sustainability will be ensured. The CBNRM programme in Botswana has been implemented in the wildlife management areas (WMAs), mostly buffer zones around national parks and game reserves. The COMPACT project will be implemented within a WHS or property. The modalities of operations are similar, only that COMPACT can be viewed as a form of CBNRM in a WHS or property.

The COMPACT project will be initiated in the Panhandle area of the Okavango Delta WHS in Botswana. The overall long-term project is envisaged to accomplish the following:

- Establishment of the COMPACT community conservation programme constituting a permanent structure to support the involvement of local communities in the conservation of the WHS.
- Provision of small grants for community-based conservation and livelihood projects.
- Provision of targeted capacity-building, networking and exchange activities.

The COMPACT project replication at the Panhandle of the Okavango Delta WHS was initiated in 2019, to expand and strengthen programmes which accommodate traditional resource use for livelihoods, user access rights, cultural rights and access to opportunities to participate in the tourism sector, while keeping up with the WHS's Outstanding Universal Value (World Heritage Committee, 2014). It is anticipated that the project will "continue efforts to address a range of other protection and management issues including governance, stakeholder empowerment, management planning, management capacity and control of alien invasive species" (World Heritage Committee, 2014, p. 7). The Government of Botswana (through the Department of National Museum and Monuments and the Department of Environmental Affairs), collaborating with the GEF Small Grants Programme, and the UNESCO World Heritage Centre, initiated a preparatory process to develop planning frameworks for the implementation of the COMPACT replication project at the Panhandle of the Okavango Delta WHS.

This chapter therefore examines the potential for COMPACT programme implementation within the Panhandle area of the Okavango Delta WHS. This will be achieved by exploring COMPACT programme stakeholders, institutional and policy framework to facilitate its implementation and propose governance structures. The specific objectives are to (i) explore the individuals and groups with interest and influence over the Panhandle area of the Okavango Delta WHS, (ii) examine available institutional and policy framework to facilitate COMPACT programme implementation and (iii) recommend governance structure to support COMPACT implementation.

STUDY AREA

The Panhandle area of the Okavango Delta WHS in Botswana is part of the Okavango River Basin that has the bulk of its catchment area in Angola (Figure 21.1). The Okavango Delta is situated in the Ngamiland District, northwestern Botswana. The Panhandle area is situated between the Popa Falls on the Kavango River in Namibia, being the northern terminal end of the Panhandle, and extends downstream to Seronga in Botswana.

Co-management of world heritage sites for community benefit 303

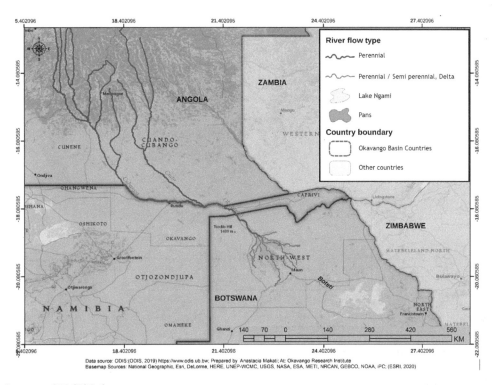

Source: ORI GIS Lab.

Figure 21.1   Okavango River Basin

The Gumare fault divides the Panhandle from the alluvial fan (Figure 21.2). The Panhandle, being the Okavango River, is about 120 km in length and 15–20 km wide. It consists of a single meandering river surrounded by vast floodplains. In the upper Panhandle reach, the channel is about 3–5 m deep and decreases to about 4 m deep at the lower end. The total area of the Panhandle is 820 km².

The Okavango Delta is a Ramsar site, designated in 1996 under the Ramsar Convention on Wetlands (Department of Environmental Affairs, 2008). Under the Convention, the Delta is considered a wetland of international importance and significant value locally and globally. Part of the Okavango Delta Ramsar Site (ODRS) was inscribed as the 1000th WHS in June 2014. The inscribed World Heritage property encompasses an area of 2,023,590 ha with a buffer zone of 2,286,630 ha (UNESCO, 2019a). As a WHS, it is a protected cultural and natural heritage site of international importance under the UNESCO World Heritage Convention (UNESCO, 2019b). The double designation status underscores the significance of the Delta on a global map and demonstrates the integration of conservation and preservation with sustainable use of the wetland.

304 *Handbook on tourism and conservation*

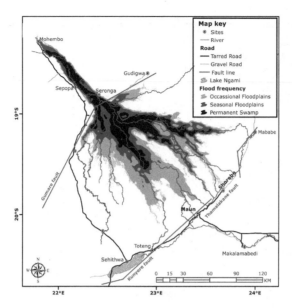

Source: ORI GIS Lab.

Figure 21.2   Panhandle and Okavango Delta

Study Sites

The Panhandle, situated in the Ngamiland West, comprises two sections separated by the Okavango River, the eastern and the western side. Ngamiland West occupies an area of 22,730 km^2 (MLH, 2009). There are about 14,075 and 29,727 people living along the eastern and western sides of the Panhandle respectively (Statistics Botswana, 2022), with the former having experienced a slight population decrease from 2011 and the latter an increase (Table 21.1). In the east, Seronga has the highest population while in the west Shakawe has the largest population, even being the highest among the Panhandle communities.

The majority of the settlements are distributed and concentrated along the Okavango River channel, presenting a linear settlement pattern (Figure 21.2). The settlement pattern signifies the reliance of the local populace upon the Okavango River for water. The river system is a source of water for both domestic and livestock use. The river is also a source of the Panhandle community's livelihood, as they depend upon aquatic resources. The major villages making the eastern and western parts of the Panhandle are shown in Table 21.1. All of the main Panhandle communities are situated in either NG7,[1] NG11 or NG12. The northernmost village in the Ngamiland west is Gudigwa.

Ethnic groups living in the Panhandle include HamBukushu, Wayei, the San (Bugakhwe and Xanekwe), Dxeriku and BaTawana (VanderPost et al., 2003). The Etsha communities are predominantly HamBukushu while Seronga is Wayei, with the Gudigwa being mostly Bugakhwe San. These ethnic groups have coexisted and acculturated over time within the Ngamiland and Okavango Delta with varied resource use strategies. The HamBukushu, Dxeriku and the Wayei, typical of the Bantu, practised mixed economy, while the San commu-

Co-management of world heritage sites for community benefit 305

Table 21.1   Main Panhandle communities

East	Population		West	Population	
	2011	2021		2011	2021
Mohembo East	550	785	Mohembo West	1,988	2,500
Kauxwi	2,233	1,888	Shakawe	7,420	10,589
Xakao	1,799	1,594	Samochema	1,155	1,667
Sekondomboro	629	978	Xhaoga	879	526
Ngarange	1,447	1,336	Nxamasere	1,584	1,500
Seronga	3,716	2,793	Kajaja	247	580
Gonutsuga	953	928	Sepopa	2,283	2,090
Eretsha	912	774	Ikoga	1,222	755
Beetsha	1,585	1,502	Etsha 1	1,297	1,554
Gudigwa	725	909	Etsha 6	5,234	4,565
Tobere	475	588	Etsha 13	2,694	3,401
Total	15,024	14,075		26,003	29,727

Source: Statistics Botswana, 2011, 2022.

nities were hunter-gatherer societies (Bock, 1998). The Bugakhwe traditionally utilised both forest and riverine resources while the Xanekwe (the swamp or river Bushmen) relied mainly on riverine resources (VanderPost et al., 2003).

To demonstrate the historical ethnic groups' attachments to the Panhandle area, several historical and cultural landscapes of significance have been identified in the area. These include amongst others Goxa Island, //uakao, N/oaxom, Tcoyi, Gombo Island, Mahaya, Khwaxa, Mokgatsha, and Kyauo (Matswiri, 2017).

Ngamiland West, owing to its rich ethnic diversity, also shows the same pattern in language as in ethnic groups. In terms of the actual numbers of language used in the district by 2011, Setswana was spoken by 19,934 people in the Ngamiland West compared to 65,208 in the Ngamiland East. SemBukushu was spoken by 25,685 people in the Ngamiland West while SeHerero was spoken by 7,342 in the Ngamiland East (Statistics Botswana, 2015). These are the languages used by a minimum of 5,000 people only. The remaining other languages' usage was lower.

Livelihoods

The mainstay livelihood activities within the Panhandle communities include tourism, arable and pastoral farming, subsistence fishing, collection of veldt products, and crafts (Kgathi et al., 2006).

Tourism is considered the major economic activity in the District, with the Okavango Delta being the beacon of tourist attraction. The double listing of the Okavango Delta as a Ramsar Site and WHS places the tourism appeal of the District very high in the globe. The Panhandle area was identified among the seven Tourism Development Areas (TDAs) as a distinct area through the Ngamiland District Tourism Development Plan (2007), which is now integrated into the Ngamiland Integrated Land Use Plan.

The current tourism facilities along the Panhandle area include several lodges and guesthouses, mostly situated directly along the Okavango River and Shakawe. There are also houseboats operating within the Okavango River. These facilities currently serve self-drive tourists en route either to the lower end of the Panhandle or Popa Falls in Namibia. The

306 *Handbook on tourism and conservation*

Table 21.2   CBOs in the Panhandle

Name of Trust	Villages Covered	Target Area	Operational	
			Yes	No
Okavango Jakotsha Community Trust	Jao, Ikoga, Etsha 3, Etsha 6, Etsha 13	NG24	x	
Okavango Community Trust	Seronga, Gunutsoga, Eretsha, Beetsha, Gudigwa	NG22/23	x	
Bukhakhwe Cultural Conservation Trust	Gudigwa	NG12		x
Tcheku Community Trust	Kaputura, Tovere, Kyeica	NG13		x
Okavango Panhandle Community Trust	Shakawe, Mohembo West, Samochima, Nxamasere, Xhauga	NG7		x
Itekeng Community Trust	Ngarange, Mogotho, Sekondomboro, Xakao, Kauxwi, Mohembo East	NG11		x
Teemashane Community Trust	Mohembo East, Mohembo West, Kauxwi, Xakao, Sekondomboro, Tobere, Ngarange, Mogotho, Shaikarawe	NG11		x
SETHAMOKA Community Trust	Sepopa, Tamachaa, Mowana, Kajaja	NG7		x
Okavango Polers Trust	Seronga, Gunutsoga	NG10	x	
Makgobokgobo Youth Trust	Seronga, Gunutsoga, Eretsha, Beetsha, Gudigwa	NG23A	x	
Okavango Panhandle Nature Conservation Okavango Knowledge Trust	Samochima, , Xhauga, Nxamasere, Kajaja Gudigwa	NG7		x

Source: Fieldwork.

tourism potential of the Panhandle area and the adjoining Controlled Hunting Areas (CHAs) has not yet been fully exploited. The area's tourism attraction includes the Panhandle area now part of the larger Okavango Delta WHS, the abundance and diversity of fish stocks, the wildlife associated with the river and its wetlands, abundant birdlife, crocodile breeding areas and relatively large numbers of elephants (NWDC, 2007).

The area offers prospects for community-based tourism initiatives, more so because there are several Community Trusts formed by local communities within the Panhandle (Table 21.2). There are several CHAs that sandwiched the Okavango River – NG10. The Panhandle falls within CHA NG10, and it is bordered on the western side by NG7, with the eastern side being NG 11, NG12, and NG13 and the southern tail end being NG24 (Figure 21.2).

NG13, situated above NG 11, has been recommended for rezoning into a community-based Wildlife Management Area through the Ngamiland Integrated Land Use Plan (MLH, 2009). It is envisaged that in so doing, community benefits from tourism will be maximised through wildlife movement in the area. Currently, NG13 is earmarked for use by Tcheku Community Trust (TCT), pending the development of the management plan by the Trust (DWNP, 2019). The Trust is made up of settlements of Kaputura, Tovere and Kyeica. The Tawana Land Board has given approval for the utilisation of NG13 by the Trust on the condition that the Trust develops a management plan for the area. In NG11, a Community Trust, Itekeng Community Trust (ICT), made up of Ngarange, Mogotlho, Sekondomboro, Xakao, Kauxwi and Mohembo East, intends to benefit from the area through CBNRM.

Arable agriculture comprises both *molapo* (flood-recession) and dryland farming, though the area is also a hot spot for human–wildlife conflicts, mainly elephants and has poor soils

(MLH, 2009). Dryland farming in the Okavango has less yield compared to *molapo* farming. However, it continues to be core for more arable farmers in the area. Common crops grown for subsistence include millet, maize, sorghum, melons and watermelons, beans, sweet reeds, pumpkins and groundnuts (Songhurst, 2011). The Panhandle is a permanent water source for livestock and the Panhandle communities. The floodplains serve as a good source of dry-season grazing, and the sandveld is used for rainy-season grazing (MLHE, 2001).

Pastoral farming, mainly cattle, is common in the Panhandle area as a mainstay subsistence livelihood activity. The Panhandle area is a livestock area, mainly cattle posts, while sections within the northern veterinary fence are livestock-free areas. Most people are engaged in pastoral farming even though it is declining because of cattle diseases and mortalities due to droughts and climate change effects. Cattle farming in the eastern Panhandle takes place mainly along the periphery of the Okavango River due to the presence of surface water (MLH, 2009).

The sector is not performing well in the district due to human–wildlife conflicts, crop depredation, endemic animal diseases (FMD) and recurring droughts. The Panhandle, like the rest of the District, is prone to Foot and Mouth Disease (FMD) and other diseases such as Contagious Bovine Pleuro-Pneumonia (CBPP). Other livestock kept in the Panhandle area includes sheep, goats, donkeys and horses being used for transport.

METHODOLOGY

The study used a mixed-method design for data collection, using primary and secondary data sources. The utility of the mixed methods is that it allows for use of qualitative and quantitative approaches, hence leading to results that can be triangulated to enhance validity, breadth and depth. Mixed methods further contribute to complementarity, development, initiation and expanding the breadth and range of the inquiry by using different methods for different components of the study (Ary et al., 2018).

The design was coupled with the COMPACT approach which embraces participation, adaptive management and landscape approach as guiding pillars in the COMPACT implementation. Participation in this project was achieved through engaging local communities in *Kgotla*[2] meetings, focus group discussions (FGDs), one-on-one interviews and stakeholder workshops.

The concurrent exercise of the ODMP Review and COMPACT Strategy development helped to give the project a landscape approach. The landscape approach acknowledges connectivity among ecosystems and the increasingly widespread and complex environmental, economic, social and political challenges that transcend traditional and man-made boundaries (Arts et al., 2017). Adaptive management in this context was achieved through learning during the consultations and adapting the consultation processes, platforms and approaches to ensure that all stakeholders were involved. For example, FGDs were deliberately selected to represent indigenous communities, as this stakeholder group does not often attend conventional public meetings, mainly the *Kgotla*. The COMPACT programme is a product of the adaptive management approach as it evolved, being tested for at least 13 years in diverse settings (Brown & Hay-Edie, 2014). Adaptive management is described as a rigorous approach to learning through deliberately designing and applying management actions as experiments to better understand the ecosystem to achieve results in uncertainty (Murray & Marmorek, 2004).

Primary data were collected using the following methods:

308 *Handbook on tourism and conservation*

Stakeholder Consultations

Stakeholder consultations were undertaken using different approaches, methods and fora such as *Kgotla* meetings, stakeholder consultative workshops, FGDs, and key informants' interviews. The different methods used during stakeholder consultations helped in ensuring a broad-based participation approach and reach espoused by the COMPACT model.

Kgotla meetings
A series of *Kgotla* meetings were conducted in select Panhandle communities. The villages included Gudigwa, Seronga, Ngarange and Kauxwi in the eastern part of the Panhandle. On the western part of the Panhandle, meetings were held at Shakawe, Samuchima and Sepopa.

Focus group discussions
FGDs were conducted at two levels, one being through the traditional stakeholder consultative workshops and the other approach comprising 6–15 discussants (Neuman, 2007). These followed the completion of a series of *Kgotla* meetings held within the Panhandle community. Four FGD meetings were held, one at Ngarange, comprising a heterogeneous group from within the community. Two other FGDs were held at Shakawe, the first being a homogeneous group of the San community (Bugakhwe and Khwee), the other being a heterogeneous group drawn from Shakawe and neighbouring localities. The motivation for hosting a homogeneous FGD with the San community was to cater for the language barrier and facilitate freedom of expression as they tend to shy away from public meetings and forums. The last FGD was held within Samuchima village, comprising fishers only. The attendance of the FGDs ranged from 8–15.

The FGDs helped in clarifying, extending, qualifying and probing issues raised during *Kgotla* meetings or identified in literature review and hence triangulation (Gill et al., 2008; Morgan & Spanish, 1984). They gave depth, insight and a rich understanding of phenomena or topics of interest.

Key informants
Initial consultative meetings were held with the COMPACT team comprising the Small Grants Programme (SGP) team, the Department of National Museums and Monuments and the Department of Environmental Affairs. Key informant interviews were held with Trust for Okavango Cultural and Development Initiatives (TOCaDI), non-governmental organisations (NGOs) working in the area, Ngamiland Council of Non-Governmental Organisations (NCONGO), local and traditional authorities, Village Development Committees (VDCs), fishers and Okavango Research Institute.

Stakeholder Analysis

A stakeholder analysis was conducted to identify individuals and groups with interest and influence in the COMPACT project within the Panhandle WHS of the Okavango Delta WHS. One of the principles of the COMPACT approach is participation; stakeholder participation at project inception, planning and implementation become imperative. Community-based initiatives, such as in the natural resources management programmes, require that stakeholders be actively involved in project decision-making processes to own both the project and the processes (Reed et al., 2009).

To ensure stakeholder participation from the onset, the stakeholder analysis employed was guided by the stakeholder-led stakeholder categorisation method (Reed et al., 2009). Through this approach, the stakeholders themselves classified the stakeholders to which they have created. This was done based on primary, secondary and tertiary stakeholder categorisation. A primary stakeholder was described as an intended direct beneficiary of a target project, i.e., those groups or individuals whose benefit will come through achieving the project purpose (Pomeroy & Douvere, 2008).

Accordingly, a secondary stakeholder has no direct benefit in the project or project area but is affected by what takes place in the form of activities at the landscape level. These form an integral part of the decision-making informing developments at the landscape level. These are often agencies which will be involved in the delivery of the project, which may be from the public sector, NGOs or the private sector. Tertiary or external stakeholders are those not directly involved but have an interest in the outcome of the project (Pomeroy & Douvere, 2008). This process resulted in the development of the stakeholder matrix (Table 21.3).

The exercise by the stakeholders revealed vast knowledge of diversity, interest and influence within the project area. The matrix of stakeholders provides a basis for project planning and implementation. All the different levels of stakeholders identified are important towards the successful implementation of the COMPACT project within the Panhandle area of the Okavango Delta WHS. The matrix further reveals a multiplicity of stakeholders, an important factor encapsulated within the landscape approach of the COMPACT model.

In addition to the above primary data collection methods, the following secondary data method was also applied:

Desktop Analysis

Literature relevant to the study area was reviewed to provide background information on the project site and documentation with a bearing on the intended project. This included both published and grey literature reposting mostly within government and academic institutions. The information included amongst others, policies and strategic documents that will guide the COMPACT project within the district and nationally, existing legislation, plans and other documents such as the ODMP of 2008, ODMP review report of 2013 and the Okavango Delta World Heritage Nomination Dossier 2013. Other key documents included the 2001 Okavango River Panhandle Management Plan and the 2009 Ngamiland Integrated Land Use Plan. International agreements or conventions were also reviewed, more so that the project site is listed under Ramsar and World Heritage Conventions. Policies and strategies relevant to the project and project area were reviewed to determine the policy environment within which the COMPACT project will work and be supported.

Table 21.3   Okavango Panhandle stakeholder matrix

Stakeholders – COMPACT Site Strategy: Panhandle

Primary	Secondary	Tertiary
Panhandle Communities (Morafe)*	**Ministries**	**Research Institutions/Projects**
Community Trusts	MENT	Okavango Research Institute
Okavango Jakotsha Community Trust (Jao, Ikoga, Etsha 3, Etsha 6, Etsha 13)	Local Government & Rural Development	BUAN
Okavango Community Trust (Seronga, Gunotsoga, Eretsha, Beetsha, Gudigwa)	**Non-government organisations**	KAZA
Bukhakhwe Cultural Conservation Trust	NCONGO	EcoExist
Tcheku Community Trust (Kaputura, Tovere, Kyeica)	TOCaDI	
Okavango Panhandle Community Trust (Shakawe, Mohembo West, Samochima, Nxamasere, Xhauga)	**Government Departments & Agencies**	Every River Has its Own People
Itekeng Community Trust (Ngarange, Mogotho, Sekondomboro, Xakao, Kauxwi, Mohembo East)	Department of Wildlife & National Parks	**International/Regional NGOs/Agencies/Organisations/ Conventions**
Teemashane Community Trust (Mohembo East, Mohembo West, Kauxwi, Xakao, Sekondomboro, Tobere, Ngarange, Mogotho, Shaikarawe)	Tawana Land Board	Conservation International
SETHAMOKA Community Trust (Sepopa, Tamachaa, Mowana, Kajaja)	Department of Mines	OKACOM
Okavango Polers Trust (Seronga)	Department of Environmental Affairs	UNDP
Makgobokgobo Youth Trust (Seronga)	Department of National Museums & Monuments	CITES
Okavango Panhandle Nature Conservation (Samochima, Xhauga, Nxamasere, Kajaja)	Botswana Tourism Organization	IUCN
Okavango Knowledge Trust	Department of Water Affairs/Water Utilities	UNESCO
Fishers Associations	Department of Waste Management	
Boiteko Syndicate	Department of Forestry & Range Resources	
	Department of Tourism	
Mohembo Fishers Association	**Technical Advisory Committee**	
Mmamasilo-a-noka	**Private Sector - Tourist Facilities**	
Ngororo	Shakawe River Lodge	
Okavango Horticulture Farmers Association	Drotskys Cabins	
OWMC	Xharo Lodge	

Stakeholders – COMPACT Site Strategy: Panhandle		
Primary	**Secondary**	**Tertiary**
	Jumbo Junction	
	Ditlhapi	
	Nxamesere Lodge	
	Swamp Stop	
	Guma Nguma Lodge	
	MmaPula (Beetsha)	
	HAWK	
	MoPhiri	
	The Boat	
	Sethabi	
	Shakawe Sands	

Note: * Morafe refers to community residents and members.
Source: Fieldwork.

312   *Handbook on tourism and conservation*

FINDINGS AND DISCUSSION

Stakeholders in COMPACT Implementation

Almost all the primary stakeholders are within the Panhandle area of the Okavango Delta WHS, while the majority of secondary and tertiary stakeholders are outside the project area. The positioning of the stakeholders within and outside the project area underscores the landscape approach espoused by the COMPACT model. Project implementation, monitoring and evaluation must therefore engage all the different stakeholders to ensure that their relative interests and influence are taken into account. This will ensure project support, legitimacy and consequently effectiveness.

The stakeholder matrix (Table 21.3) also shows that the COMPACT programme will be implemented within the context of existing CBNRM stakeholders. This is advantageous for the programme as it will benefit from the synergies, lessons and experiences already learned through the CBNRM programme implementation. As noted earlier, the COMPACT programme is a form of CBNRM, only that COMPACT take place within a WHS. The similarities of the stakeholders between the two programmes will facilitate smooth implementation as most of the stakeholders have learned to collaborate.

Institutional and Organisational Context for COMPACT Implementation

COMPACT implementation in the Panhandle area of the Okavango Delta WHS will take place within an already existing administrative and institutional framework operating at the national, district and local (site) levels. These include government institutions and structures (central and local government), civil society organisations and research institutions. Some of the organisations are operating at regional and international levels. This provides for potential synergies in terms of implementation, facilitation, monitoring, evaluation and joint funding for effective project coordination and management.

Government institutions

The Ministry of Environment and Tourism (MET) is the custodian of the World Heritage Convention of 1972 in Botswana (MENT, 2022). The key implementing Ministry for the COMPACT site strategy will therefore be the MET. Generally, the Ministry is mandated with the coordination of policies, strategies and programmes that promote conservation and tourism development in Botswana. The Ministry's strategic agenda is to protect and conserve the environment and promote investment opportunities to derive maximum socio-economic benefits from natural resources.

There are seven Departments under the MET (Table 21.4). These departments are directly or indirectly involved in CBNRM development as well as the promotion of community-based tourism and conservation initiatives in the country.

Since the COMPACT replication project is a type of CBNRM in a WHS, these Departments will continue to spearhead the implementation of COMPACT projects in the Panhandle area of the Okavango Delta WHS.

The MET will be operating at the national level, while the various Departments will operate at the district and local levels. The Department of Wildlife and National Parks (DWMP) has offices in Shakawe and Seronga, and the Department of Meteorological Services and

Co-management of world heritage sites for community benefit   313

Table 21.4   *Institutions and organisations under the MET*

Department/organisation	Sectorial responsibility
Ministry of Environment and Tourism	
Department of National Museum and Monuments	Protection, preservation and promotion of cultural and natural heritage for sustainable utilisation
Department of Environmental Affairs	Coordination of environmental policies and strategies, and promotion of environmentally sound based projects for the environmental conservation and protection
Department of Wildlife and National Parks	Conservation and management of wildlife and fisheries resources and their habitats
Department of Tourism	Coordination, management and promotion of sustainable tourism development through formulation, monitoring and implementation of policies and strategies that ensure sustainable tourism development
Department of Meteorological Services	Weather, climate information services and climate change
Department of Forestry and Range Resources	Conservation, protection and management of forestry and range resources
Department of Waste Management and Pollution Control	Waste, sanitation, pollution prevention and control
Parastatals	
Botswana Tourism Organisation	Tourism product development, marketing, grading and investment promotion in the tourism sector
Autonomous	
Forest Conservation Botswana	Management of the Tropical Forest Conservation Fund on behalf of the Ministry, promotion of sustainable use and conservation of forests by offering grants to relevant activities designed to conserve, maintain and restore the forests of Botswana
Other key role players	
Ministry of Land Management, Water and Sanitation Services	
Tawana Land Board	Land authority: Land use planning, administration and management
Department of Water & Sanitation	Water resource management and planning, and sanitation management. The International Waters Unit because of the transboundary nature of the Okavango River Basin
Ministry of Local Government and Rural Development	
North West District Council	Political leadership fostering local democracy and governance and service provision at the district level
Tribal Administration	Tribal leadership and administration

Source: MENT, 2022; MLMW, 2022; MLGRD, 2022.

Department of Forestry and Range Resources (DFRR) in Shakawe. Others such as the Department of National Museum and Monuments, Department of Tourism, Department of Environmental Affairs and Department of Waste Management and Pollution Control (DWMPC) have regional offices in Maun only. However, the administrative responsibilities of these regional offices cover the entire district, together with the Botswana Tourism Organization (BTO), the lone parastatal with relevance to the project and project area. The Forest Conservation of Botswana (FCB) is an NGO and has previously commissioned a consultancy for the development of a management plan and atlas for a proposed community forest reserve in Shaikarawe – situated 12 km west of Mohembo west. (FCB, 2015). The FCB also funded the development of a community forest nursery at Etsha 6 to promote the regeneration of important plant species used in basketry.

314   *Handbook on tourism and conservation*

Table 21.5   *Technical Advisory Committee membership*

Institution	Roles
Ministry of Environment and Tourism	**Key roles of TACs**
Department of Wildlife & National Parks	- Advise CBOs and district authorities regarding investments in
Department of Environmental Affairs	commercial utilisation of community-managed natural resources
Department of Forestry and Range Resources	at the district level
Department of Tourism	- Monitoring and implementation of CBNRM in the district
Department of National Museums & Monuments	- Facilitate and monitor joint venture arrangements between the
Botswana Tourism Organisation	private sector and CBOs
Land Authority	- Provide technical advice to CBOs on Trust operations,
Department of Lands	tendering procedures and government policies on CBNRM
Land Board	- Conflict resolution and mediation
District Administration	
North West District Council	
Economic Planner/Tourism Officer	
Tribal Administration	

Source: MEWT, 2008.

Technical Advisory Committee

The Technical Advisory Committee (TAC) is a district-based structure "charged with the responsibility of overseeing the implementation of CBNRM" (MEWT, 2008, p. 47). The Committee is chaired by the District Officer responsible for Development issues. CBNRM is cross-cutting and as such, the designation of the District Administration as the chair of the TAC is strategically meant to institutionalise the coordination of the programme at a higher level. The TACs have as one of their key activities the promotion of CBNRM and providing advisory services to the CBOs.

The TAC is composed of Departments mostly from the MET, and other Ministries indicated in Table 21.5. The Ministry of Land Management, Water and Sanitation Services, through the Tawana Land Board (TLB) and the Department of Water and Sanitation, has offices in Shakawe and Seronga. The Ministry of Local Government and Rural Development, with its portfolio responsibility focused on local governance and rural development, is widely spread with different offices in the Panhandle area. The Tribal Administration is found in most settlements, though at varied levels of authority, depending on the size of the village or settlement. A concentration of offices servicing the Panhandle area under the North West District Council is in Shakawe and Seronga, while others such as the Department of Social and Community Development are spread throughout the Panhandle communities. These are responsible for the provision of social welfare and community development services amongst the communities.

The TAC is a sub-committee of the District Land Use Planning Unit (DLUPU), a unit tasked with advisory roles in land-use matters within the District. The TAC is made up of representation from various departments and institutions within the district (Table 21.5).

Partner Organisations and Institutions

United Nations Development Programme

The UNDP, through the Global Environment Facility (GEF) Small Grants Programme (SGP), continues to be a partner in supporting national priorities and actions towards driving the sustainable development agenda in Botswana. The programme is guided by its Country

Programme Strategy for Operational Phase 7, which also targeted the Panhandle area (UNDP GEF-SGP, 2019). The aim of the Country Programme Strategy (CPS) is to promote and support innovative, inclusive and scalable initiatives, and foster multi-stakeholder partnerships at the local level to tackle global environmental issues in priority landscapes. The strategy document includes the Panhandle of the Okavango Delta as one of the priority landscapes, the others being the Makgadikgadi Framework Management Plan and the Bobirwa landscapes.

The assistance and support to be offered to the Panhandle landscape under the CPS will target community-based conservation of threatened ecosystems and species, sustainable agriculture and fisheries, and food security. The CPS identified four strategic initiatives that will be pursued under the Operational Phase 7 for the three landscapes; (i) community-based conservation of threatened ecosystems and species, (ii) sustainable agriculture and fisheries, and food security, (iii) low carbon energy access co-benefits, and (iv) local to global coalitions for chemicals and waste management (UNDP GEF-SGP, 2019).

The development of the COMPACT site strategy for the Panhandle of the Okavango Delta WHS is funded through the UNDP GEF-SGP facility. The CPS indicates that 70 per cent of the grant resources allocated for Operational Phase 7 will be reserved for the three selected landscapes (UNDP GEF-SGP, 2019). For the Panhandle area of the Okavango Delta WHS, the funding will provide small grants for community-based conservation and livelihood projects as well as targeted capacity-building, networking and exchange activities. The grants will be open to NGOs, research institutions as well as Community Trusts in the area to support community-based activities intended to strengthen biodiversity conservation in and around the target WHS.

The Permanent Okavango River Basin Water Commission
The Permanent Okavango River Basin Water Commission (OKACOM) is implementing a GEF-funded project, aimed at strengthening joint management and cooperative decision-making on the sustainable utilisation of natural resources to address livelihoods and socio-economic development challenges in the Cubango-Okavango River Basin (UNDP GEF, 2016). The four-year project, executed by the UNDP, is supporting the socio-economic development of the basin communities while sustaining the health of the basin ecosystem. The project was developed by the Strategic Action Programme (SAP) for the sustainable development and management of the Cubango-Okavango River Basin (OKACOM, 2011). The SAP builds on the knowledge collected through the Transboundary Diagnostic Analysis (TDA) which informed on priority areas for sustainable development and management of the Okavango River Basin. The project targets the three riparian states of the Okavango River Basin: Angola, Botswana and Namibia.

The project has three components, namely, (i) the basin development and management framework, (ii) environmentally conscious livelihoods and socio-economic development, and (iii) Integrated Water Resource Management (UNDP GEF, 2016). The first and last components generally target the three riparian states' water-related institutions and structures, with the first one having more focus on the OKACOM. The second component on the environmentally conscious livelihoods and socio-economic development has three sub-themes derived from the SAP: (i) community-based tourism, (ii) fisheries co-management, and (iii) conservation agriculture for improved food security at the local level (UNDP GEF, 2016; UNDP, 2019). The first two sub-themes are implemented in Namibia and Angola while the

316　*Handbook on tourism and conservation*

conservation agriculture for improved food security at the local level is undertaken in the three riparian states.

Specific to Botswana, community-based climate change adaptation demonstration projects are being conducted within the Maun area (100 km radius) and the Panhandle, specifically Shakawe. (OKACOM, 2019; NCONGO, 2020). The Ngamiland Council of Non-Governmental Organizations (NCONGO), a local NGO, was designated as the Botswana partner to implement the demonstration interventions. The demonstration projects are focused on horticultural production to link tourism and other higher market demands available in the area (OKACOM, 2019). To this effect, the NCONGO capacity has been enhanced through the provision of a horticulture specialist, with the key role facilitating and building farmers' horticulture capacity, mentorship and conducting market engagements.

The project has identified 20 champion demonstration farmers, comprising 55 per cent males and 45 per cent females. Thirty per cent of these farmers are from the COMPACT replication project area, comprising the youth, three males and three females (NCONGO, 2020). It is envisaged that these projects will maximise horticultural production while ensuring consistent supplies of quality farm products. By so doing, the projects will also contribute to improving the socio-economic status of local communities (UNDP, 2019).

The project is offering capacity development to farmers through mentoring and training in different agricultural practices targeting the tourism industry market. The project also assists the selected champion farmers by drilling and equipping well points and water reticulation for them, providing storage water tanks, shade nets, drip systems and supply inputs such as seeds (NCONGO, 2020). The project has contracted the Okavango Research Institute, University of Botswana, to undertake baseline assessments, monitoring and evaluation frameworks, and to perform the Social and Environmental Screening for the demonstration projects. The Institute also conducted soil analysis for the demonstration projects.

The Ngamiland Council of Non-Governmental Organisations
The NCONGO is an umbrella body of Ngamiland-based NGOs/CBOs and was established in 2008. The organisation, based in Maun, works with local stakeholders and communities on issues of CBNRM, NGO/CBO capacity-building, advocacy and socio-economic development. It represents an expanding membership of over 50 NGOs/CBOs within the Ngamiland district. NCONGO, as a volunteer-driven organisation, coordinates the efforts and advocates and lobbies for the needs of CBOs. It also facilitates networking and communication between the regions' non-state actors and strengthens the links with the government to enable real change. NCONGO members in the Panhandle include the Okavango Community Trust, Okavango Panhandle Community Trust, Okavango Poler's Trust, Sethamoka Trust, Tcheku Community Trust and the TOCaDI Trust.

NCONGO projects in the Panhandle include the following:

- *Maun/Okavango Delta Horticulture Demonstration Project*: The UNDP, GEF and the Governments of the Republic of Angola, Botswana and Namibia are providing technical and financial support to OKACOM to implement the "Support to the Cubango-Okavango River Basin (CORB) Strategic Action Programme (SAP) Implementation" project. Through this project, OKACOM intends to contribute to improving the socio-economic status of local communities across the basin with minimum adverse impacts on the protection of the basin ecosystem. The demonstration project focuses on enhancing horticultural produce in Maun/Shakawe through climate-smart practices to link horticultural production with the up-market tourism value chain and other local markets.

- In 2019, the Ministry of Agricultural Development and Food Security conducted joint assessments with the OKACOM and its demonstration projects implementing partner, the NCONGO. The joint assessment set agreed on criteria to identify and select demonstration farmers to form part of the OKACOM Horticulture Demonstration farmers to demonstrate how horticulture production can be enhanced and linked to higher value markets including the tourism markets. NCONGO is the local implementing agency supporting the implementation processes of the Maun/Okavango Delta horticulture demonstration project. The project started with 20 farmers. In the Panhandle area (i.e., Shakawe area) there are six farmers whose farms qualified as demonstration sites. NCONGO is providing guidance and service to a horticultural specialist. The goal is to maximise production while ensuring consistent supplies of quality farm products hence contributing to improving the socio-economic status of local communities.
- *Capacity-building*: NCONGO also provide workshop-based technical training to CBOs including those at the Panhandle. Some of the thematic focus of this capacity-building workshop includes CBNRM and governance, the relevance of CBNRM as a rights-based economic strategy, building an inclusive wildlife economy and sustainable wildlife management.
- *Advocacy and lobbying*: Advocacy remains a critical role played by NCONGO in the District and nationally. NCONGO thereby provide CBNRM CBOs with a group platform to network and share experiences and success through the District CBNRM Forum.

Trust for Okavango Cultural and Development Initiatives
TOCaDI was established in Shakawe in 1998 with a mission to build the capacity of the San and other minority groups in the Okavango sub-district to respond to their challenges. This is done to enable the marginalised groups to plan and implement their own culturally sensitive and sustainable programmes that will lead to an improvement in their quality of life. TOCaDI serves people in the Okavango sub-district, with a reach to about 21 villages and settlements situated on both the eastern and western sides of the Panhandle. The TOCaDI works to empower communities to become self-reliant and improve their standards of living. As part of the Kuru Family of Organisations, TOCaDI works to address the needs of the San people considered to be a marginalised community.

TOCaDI's current focus is on issues such as land and water rights, natural resource management, community health and cultural development. It also promotes income-generating opportunities by utilising and maintaining the natural resources in the area. It works in close collaboration with the Letloa Land, Livelihood and Heritage Resource Centre in Shakawe as well as with several Community Trusts to strengthen and train them and help them to mainstream activities.

TOCaDI's community development activities include the drilling and equipping of boreholes for cattle syndicates, the harvesting of thatching grass and interventions on health, such as HIV/AIDS and tuberculosis. Other projects proposed through the TOCaDI include the climate-smart water solution for rural communities of Okavango, harvesting thatching grass, tending vegetable gardens and fisheries and facilitating eco-tourism projects such as the Teemacane Cultural Hiking Trail.

318 *Handbook on tourism and conservation*

TOCaDI also continues to play an advocacy role for the San people in the District. Their impacts are more significant so that the Trust has networks nationally, regionally and internationally.

EcoExist

EcoExist seeks to reduce conflict and foster coexistence between elephants and people using an approach that connects science with practice (EcoExist, 2021). EcoExist focuses on applied research, land use planning, crop-raiding mitigation, agricultural experiment and innovation, and tourism development. This is achieved by gathering social, ecological and economic data to analyse the causes and consequences of human–elephant conflicts. The project intends to find and facilitate solutions that work for both species, that is, the people and the elephants.

The EcoExist project empowers farmers with practical, affordable and effective tools to minimise crop raiding and reduce conflicts with elephants. It is envisaged that in the long term, collaborations with local, national and international groups will be created as an enabling environment for a range of policies and programmes that tackle the root causes of human–elephant conflicts. In this way, the project addresses the human–elephant conflict in ways that may be modelled throughout Botswana and beyond.

EcoExist works with communities in the Eastern Panhandle of the Okavango Delta. The region has heightened competition for access to water, food and space. The projects undertaken by EcoExist in the eastern Panhandle are implemented within the scope of themes such as following the herds, planning a shared space, learning together to protect fields, harvesting early and boosting yields and building an elephant economy (EcoExist, 2021).

To attain 'the elephant economy', the EcoExist has forged links and collaboration with local artisans and artists with talents in music, dance, storytelling and crafts – and private companies to develop and market micro-enterprises for elephant-themed, elephant-friendly products and handicrafts, along with small-scale community-based tourism experiences for visitors. A Panhandle Cultural Fair showcasing local cultural arts and products primarily with elephant themes, identifying talents and helping market the Panhandle as an elephant-based destination was hosted in the area (EcoExist, 2015). The Fair aimed to celebrate the people, arts and cultures of the Eastern Okavango Panhandle and showcase the talents and experiences of people who live with elephants.

The Ngamiland District CBNRM Forum

The Ngamiland District CBNRM Forum was established in 1999 to create a platform for stakeholder dialogue, cooperation and coordination of CBNRM and related activities within the district. The Forum draws membership from different stakeholders from the district dealing with CBNRM. These include among others the CBNRM CBOs (Trusts), NGOs, government institutions and parastatals, the private sector engaged in tourism, Tribal Administration and Village Development Committees and academia. The motivation for the establishment of the Forum was to foster an enabling environment for CBNRM implementation within districts by promoting openness, trust, understanding and knowledge sharing among CBNRM stakeholders. The forum meets to lobby government CBNRM policy and implementation. All the Ngamiland CBNRM CBOs are members of the Ngamiland District CBNRM Forum.

Implications for Institutional and Organisational Context to COMPACT Programme

The COMPACT replication project at the Panhandle of the Okavango Delta WHS will be implemented within an enabling institutional and organisational support network that is already in place. Government institutions have been actively involved in the area and apart from the respective relevant Ministry's portfolio responsibilities, they have through the TAC assisted local communities towards forming Trusts to benefit through the CBNRM. This structure facilitates the much-needed concerted effort towards guiding the COMPACT replication project in the area.

The other partner organisations also in the area are an added advantage in undertaking complimentary and synergistic projects and programmes. Most policies and legislation within the MENT have advocated for local community involvement, NGOs and private sector participation and input in interventions geared towards natural resources management and conservation. These partner organisations in the form of NGOs, research institutes and forums are needed for joint implementation of the COMPACT replication project for effectiveness and impact. It is envisaged that with all these stakeholders taking part in the Local Consultative Body, and working in synergies, the aspirations of increasing the effectiveness of biodiversity conservation in the WHS while helping the livelihoods of local people to improve will be realised in the long term.

The collaboration within and across government sectors (e.g., wildlife, agriculture, land use planning, human–wildlife conflicts, cultural and heritage resources, and tourism) and with local communities and researchers is essential for building a long-term, effective strategy that will contribute to sustainable development. This will foster peace and security, care for nature, cultural diversity, inclusive social and economic development and the quality of life of communities, whilst fully respecting the Outstanding Universal Value of a WHS. Through this, the overall long-term project goals of (i) establishing the COMPACT community conservation programme constituting a permanent structure to support the involvement of local communities in the conservation of the WHS, (ii) providing small grants for community-based conservation and livelihood projects, and (iii) providing targeted capacity-building, networking and exchange activities can be achieved.

The biggest challenge within government institutions has been implementation and coordination. The ODMP midterm review found that there was no concerted effort to overcome the key problems hampering the implementation of the plan. It cited several issues regarding coordination among planning authorities as one of the areas of concern. The establishment of the Local Consultative Body, which will bring together all stakeholders, as part of the COMPACT project will go a long way towards ensuring effective project implementation. Regarding NGOs and academic institutions, funding is and has always been a challenge for them. It is commendable that there is funding available that can be accessed by civil society and academia to carry on interventions needed under this project for three years.

POLICY AND LEGISLATIVE FRAMEWORK

The policies, legislation, plans and other significant regional and international protocols and conventions with relevance to the development of the COMPACT site strategy for the

320   *Handbook on tourism and conservation*

Panhandle at the Okavango delta WHS were explored and reviewed. The review was done based on context, content, processes and actors.

Context

In terms of policy context, the review revealed that the policy and legislative framework formulation was initiated to address the challenges and threats facing natural resources conservation and management and the need for broad stakeholder involvement and participation in the same. In terms of COMPACT project implementation in the WHS, policies such as the CBNRM policy of 2007 become very relevant and contextual. The policy specifically refers to the country's culture as well as natural heritage and places great importance on the active involvement of host communities. The COMPACT project, which is CBNRM in a WHS will therefore be appropriately implemented through the guidance of this policy. However, as has been noted with CBNRM implementation in Botswana, there is a need for the approval of CBNRM guidelines, CBNRM Act and CBNRM strategy to give dedicated guidance and regulatory framework to CBNRM in general. This will help avoid erratic and haphazard decision-making which have characterised CBNRM implementation in the recent past.

Still, in context, the policy and legislative environment in Botswana are supported by the fact that most environmental-related agencies are housed under the same parent Ministry, being MET, making implementation smoother. The Ministry of Environment, Natural Resources, Conservation and Tourism is the custodian of the World Heritage Convention in Botswana, and hence directly responsible for the implementation of the COMPACT replication project using policies housed within its varied departments and agencies.

The policies reviewed within the context of the country, region and other international treaties are all relevant in terms of context to guide the COMPACT replication project. It is upon the implementers to see how best to harness the supportive policy and legislative framework for effective and efficient COMPACT replication project implementation. Concerns have been raised in the past that the problem with Botswana's policy and legislative framework is not the lack of appropriate policy and regulatory framework, but rather implementation and/ or enforcement bottlenecks. For effective and efficient implementation of COMPACT, practitioners and decision-makers must not adopt a 'business-as-usual' attitude but rather must take advantage of the enabling policy environment for the benefit of conservation and the people.

Content

Most policies reviewed had strategies and implementation plans, monitoring and evaluation statements or frameworks, which under normal circumstance must facilitate smooth and easy execution. Some policies such as the 2007 CBNRM Policy indicated that guidelines and legislative instruments will be developed to support CBNRM implementation. The revised Tourism Policy development was done simultaneously with its implementation framework. The implementation framework highlighted key actions and milestones to attaining the same. The policy further indicated that the monitoring and evaluation tools at sector level to facilitate effective delivery, more noting that tourism is a cross-cutting industry. However, with these good policy intentions, practice has shown that lack of dedicated resources to implement the policies based on the timelines is a challenge. For example, both the CBNRM guidelines and Bill are still at a draft stage, about 15 years since approval of the CBNRM policy. However, plans are advanced that both the guidelines and the Bill will be finalised, as drafts have long been prepared.

The time lag between policy approval and implementation of main actions, which are often overtaken by time and events, necessitates that policies and other strategic documents be reviewed or revised with a view to updating their implementation plans and providing requisite resources. Policy approvals must also be aligned to requisite resources so that approval is followed by immediate implementation.

Processes

A pertinent concern with policies within the MENT is that the policy review process takes a long time. Some policies, having been acknowledged that they are outdated, have long been reviewed but are not yet approved. A case in point is the Wildlife Conservation Policy of 1987, which was revised in 2013 and approved a year later in 2014. While the CBNRM strategy is recent in terms of drafting, approval must be expedited together with other policies pending approval. Some policies and strategies such as the Wildlife Conservation Policy, the Tourism Policy and the National Conservation Strategy have outlived their time, as they were formulated almost three decades ago.

There is a need to review these policy documents to align them to the current challenges and threats and infuse them with emerging biodiversity and global issues. Some policies, such as the Land Policy, were recently reviewed and approved within five years to align with the demands of the times. The same must apply to conservation and tourism-related policies. An advantage of legislation is that some has been amended frequently over the last five years such as the Fish Protection regulations of 2016.

Actors

Most of the policies, legislative instruments and strategic documents reviewed lie within one parent Ministry, which removes the challenge of fragmentation. The added advantage of the policies reviewed is that they embrace the multi-stakeholder approach in policy implementation as espoused by the WHC. Most of the policies specifically highlight the important role of multi-stakeholder participation and involvement in policy formulation and implementation. The policies and legislations even advance further by making provisions for the different actors or stakeholders to have committees that will give them a voice. The structures are provided such that they are at national, district, sub-district and local level settings. Most policies specifically highlight the importance of local and indigenous communities, NGO sector and private sector participation in their implementation. In terms of actors, the policies are robust and inclusive.

322 *Handbook on tourism and conservation*

THE GOVERNANCE STRUCTURE PROPOSED FOR THE COMPACT PROJECT IMPLEMENTATION

Modalities of Implementation

The modalities for implementation of the COMPACT Replication Project at the Panhandle of Okavango Delta WHS Strategy will follow the general guiding principles which advocates for "a decentralised, democratic and transparent process" (Brown & Hay-Edie, 2014, p. 27). The governance structure will have the following:

Local Coordinator

The Local Coordinator will serve the role of planning and implementing the programme and serve as a key link between local communities and different stakeholders and the COMPACT decision-making structures. According to Brown and Hay-Edie (2014), the Coordinator will also manage the small grants portfolio for COMPACT, as well as spearhead various capacity-building activities. While currently this role will continue to be served by the National Coordinator, Global Environment Facility Small Grants Programme based in Gaborone, future recruitment may be possible for a Coordinator outside the system and locally positioned within the Panhandle area of the Okavango Delta WHS.

Local Consultative Body

The Local Consultative Body (LCB) assists in ensuring that dialogue, coordination and consensus-building take place among key stakeholders at the level of the protected area (Brown & Hay-Edie, 2014), in this case, the Panhandle of the Okavango Delta WHS. The LCB recommends grant proposals to the SGP National Steering Committee as well as representing a set of key stakeholders in the landscape.

The guiding general prescription of the character of the LCB is as follows:

- Representative of the diverse actors concerned with the site and surrounding landscape: It may include representation from agencies responsible for the management of the site (in this case Department of National Museums and Monuments and Department of Environmental Affairs), the traditional leadership from within the target communities, local NGOs, Research Institute, local and central government, the private sector and donors.
- Voluntary: Service to the LCB will be voluntary, without expectation of compensation or allowance, though reimbursements of certain expenses such as transport may be considered on a case-by-case basis.
- Independent: Members should serve in their capacity as individuals, or as representatives of a community, organisation or business, but not as a representative of a political or administrative entity.
- Active: Members should be prepared to be actively involved beyond simply attending periodic meetings. They must be actively engaged in project review, site monitoring, capacity-building workshops and exchanges, and may serve as mentors to community groups.

Co-management of world heritage sites for community benefit   323

- Long-standing: The consistency ensured by a long-term structure is important. At the same time, the membership should change regularly, according to fixed terms of service, to bring in fresh perspectives.

During the Stakeholder Engagement Workshop, the stakeholders resolved that the LCB for the Panhandle at the Okavango Delta WHS shall consist of the following:

- Two members from each Community Trust in the area.
- Representatives from local fisher associations or groups.
- Indigenous groups in the area will be represented by TOCaDI, a local NGO.
- Traditional leadership will be represented through their respective oversight responsibilities, i.e., *dikgosi* (chiefs) heading a cluster of communities will represent them accordingly.
- Local research institute, currently the Okavango Research Institute, University of Botswana.
- Representation from local tourism facilities (HATAB will be consulted on this).
- Local and central government representative with portfolio responsibility in the area. In this case, the TAC working with CBNRM trusts in the area will serve the purpose.

The membership to the LCB will be reviewed from time to time to cater for the changing stakeholder landscape and new and emerging user groups.

National Steering Committee
Brown and Hay-Edie (2014) indicate that the National Steering Committee (NSC) is part of the SGP structure in each of the countries where it operates. It is a multi-stakeholder body operating at the national level and responsible for final decisions regarding small grants financed by the GEF. It offers an additional layer of neutrality and rigour to the review and approval of grants recommended by the Local Coordinator and the LCB operating at the national level. Membership in the NSC is also guided by the characteristics of representativeness, voluntarism, independence and active consistency. The existing NSC operating in Botswana at any time, being part of the SGP structure, shall continue to serve this purpose.

The structures indicated above are summarised in Figure 21.3.

CONCLUSION

This chapter examined the prospects of COMPACT programme implementation within the Panhandle area of the Okavango Delta WHS. Specifically, the chapter (i) explored the individuals and groups with interest and influence over the Panhandle area of the Okavango Delta WHS, (ii) examined available institutional and policy frameworks to facilitate COMPACT programme implementation and (iii) recommended governance structure to support COMPACT implementation.

Communities in the Panhandle have formed Trusts to benefit from the CBNRM programme within the region, and the implementation of the COMPACT will facilitate community participation and benefit. The stakeholder analysis matrix has revealed a multiplicity of stakeholders having both interests and influence in the COMPACT programme. The multiple stakeholders help advance the landscape approach espoused by the COMPACT model. The stakeholder matrix also mirrors CBNRM stakeholders, showing that the two similar programmes share groups and individuals involved in their planning and implementation. This similarity helps

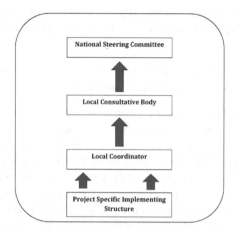

Source: Adapted from Brown & Hay-Edie (2014).

Figure 21.3   Proposed structures for implementation of the Panhandle of the Okavango Delta WHS COMPACT site strategy

the COMPACT programme as it will be implemented within a familiar context, in terms of groups and individuals that have been collaborating in one way or another. The stakeholders are almost at a par in terms of awareness and dynamics in stakeholder issues and management. To ensure the adaptive management approach principle of the COMPACT model, it is necessary to continue monitoring stakeholders' positions, in terms of interests and influence, as these are dynamic.

It is further recommended that stakeholder analysis must be viewed as an iterative process whereby additional stakeholders may continue to emerge and must be factored into the process. While the stakeholder-led stakeholder categorisation method used in this study allows near exhaustion of stakeholders, their position in terms of interests and influence is dynamic, i.e., not static but changes with time and circumstances (Reed et al., 2009). It is therefore important that stakeholder analysis be continually undertaken to identify emerging stakeholders and changing interests and power amongst them throughout project implementation.

The study has also shown that COMPACT replication project implementation will take place in a good institutional and organisational environment. There are existing government institutions with portfolio responsibilities direct or indirect in the area. There are also structures at District and sub-district levels such as the TAC whose key responsibilities and task is CBNRM implementation and monitoring. An advantage of the COMPACT replication project implementation in the Ngamiland district is that the District is home to numerous CBNRM Trusts. This has made the structures in place conversant with prevailing CBNRM issues and hence ready to extend their experience and expertise in the COMPACT replication project, seen as the CBNRM in a Word Heritage Site.

There are also other actors at District and local levels working on the site or with projects in the area. Some of these actors, such as the NGO community, have networks and project partners and donors working at regional and international levels. The NGO community also helps in advocacy and lobbying at the grassroots level. Others, such as the ORI and EcoExist, that are engaged in applied research offer COMPACT project implementation opportunities with strong synergies in areas such as training, capacity-building, research and monitoring. The presence of broad-based partners and actors in the District and project area can ensure effective and efficient COMPACT implementation within the WHS property. Formation of the Local Consultative Body that will bring all the stakeholders and partners together will be necessary at all stages of the project.

Policy and legislative review have also revealed multiple policies and relevant legislation about the natural resources sector. The policy environment is enabling and adequate, with most policies targeted to the same actors. The fact that most policies are housed within the same parent Ministry helps to smooth the implementation process. Most natural resources management-related institutions, implementing different sectoral policies are within one parent Ministry. Since the COMPACT replication project is of great interest to the World Heritage Convention and not only as a national initiative, coordination must continue to come from the highest office within the Ministry for maximal effect, impact and coordination.

Policy review revealed some shortcomings such as protracted policy formulation process, outdated policies and dependence on draft policy instruments. There are also key strategic documents that are still pending approval, such as the CBNRM guidelines, CBNRM Act and CBNRM strategy. Approval of these key legislative and strategic documents will strengthen CBNRM implementation.

ACKNOWLEDGEMENTS

This work was supported, and permission granted to publish through the UNDP Global Environment Facility – Small Grants Programme (GEF-SGP), working in collaboration with Government of Botswana. Part of the findings presented in this chapter came from a consultancy project entitled Engaging Local Communities in the Conservation and Management of the World Heritage Sites in Africa: Project No. 2019/029. The Consultancy developed planning frameworks for the implementation of the Community Management of Protected Areas for Conservation (COMPACT) Replication Project at the Panhandle of the Okavango Delta WHS, Botswana.

NOTES

1. NGx – Wildlife Management Areas (WMA) in Botswana are subdivided into Controlled Hunting Areas (CHA), and each CHA is designated a number within a specific district. The number allotted to a CHA is preceded by an acronym representing the first two letters of the district. In this case, all CHAs starting with NG denote that they are in the Ngamiland district, hence 'NG'. Therefore, NG7 denotes a seventh (7th) CHA in the Ngamiland district.
2. A *Kgotla* is a traditional public meeting place where customary judicial matters are handled, and public consultations take place within the community on a wide range of issues.

REFERENCES

Arts, B., Buizer, M., Horlings, L., Ingram, V., Van Oosten, C., & Opdam, P. (2017). Landscape approaches: A state-of-the-art review. *Annual Review of Environment and Resources, 42*, 439–463.

326 *Handbook on tourism and conservation*

Ary, D., Jacobs, L. C., Irvine, C. K. S., & Walker, D. (2018). *Introduction to Research in Education.* Boston: Cengage Learning.

Baird, J., Plummer, R., Schultz, L., Armitage, D., & Bodin, Ö. (2018). Integrating conservation and sustainable development through adaptive co-management in UNESCO Biosphere Reserves. *Conservation and Society, 16*(4), 409–419.

Bock, J. (1998). Economic development and cultural change among the Okavango Delta peoples of Botswana. *Botswana Notes & Records, 30*(1), 27–44.

Brown, J., & Hay-Edie, T. (2014). *Engaging Communities in Stewardship of World Heritage: A Methodology Based on COMPACT Experience.* World Heritage Paper Series No. 40. Paris: UNESCO.

Centre for Applied Research (2016). *2016 Review of Community-Based Natural Resource Management in Botswana.* Report prepared for Southern African Environmental Programme (SAREP).

Department of Environmental Affairs (2008). *The Okavango Delta Management Plan.* www.okacom .org/site-documents/project-reports/odmp-documents/okavango-delta-management-plan/view, accessed 28 June 2020.

DWNP (2019). *Okavango Sub-District Community Technical Advisory Committee Report.* DWNP, Maun.

EcoExist (2015). *A Cultural Fair Celebrating People.* www.ecoexistproject.org/reporting-back/blog/ cultural-fair-celebrating-people-elephants-eastern-okavango-panhandle/, accessed 5 May 2021.

EcoExist (2021). *The EcoExist Project: Reducing Conflict and Fostering Coexistence between Elephants and People.* www.ecoexistproject.org/.

Forest Conservation Botswana (2015). *Forest Conservation Botswana 2014/15 Annual Report.* FCB, Gaborone.

Gill, P., Stewart, K., Treasure, E., & Chadwick, B. (2008). Methods of data collection in qualitative research: Interviews and focus groups. *British Dental Journal, 204*(6), 291-295.

Government of Botswana (2007). *Community Based Natural Resources Management Policy.* Government Paper no. 2. Gaborone, Ministry of Environment, Wildlife and Tourism/Government Printers, Gaborone.

Kgathi, D. L., Kniveton, D., Ringrose, S., Turton, A. R., Vanderpost, C. H. M., Lundqvist, J., & Seely, M. (2006). The Okavango: A river supporting its people, environment and economic development. *Journal of Hydrology, 331*(1–2), 3–17.

Larsen, P. B. (2012). *IUCN, World Heritage and Evaluation Processes Related to Communities and Rights: An Independent Review.* IUCN, Gland.

Matswiri, G. M. (2017). Two in one: Explaining the management of the Okavango Delta World Heritage Site, Botswana. Master's thesis, University of Cape Town.

MENT (2022). *Okavango Delta Management Plan – 2021–2028.* Government of Botswana, Gaborone.

MEWT (2008). *Community Based Natural Resources Management: Practitioners Manual.* Department of Wildlife and National Parks, Gaborone.

MLGRD (2022). Ministry of Local Government and Rural Development. www.gov.bw/ministries/ ministry-local-government-and-rural-development, accessed 27 December 2022.

MLH (2009). *Ngamiland Integrated Land Use Plan.* Department of Lands. Gaborone.

MLHE (2001). *Okavango River Panhandle Management Plan.* Tawana Land Board. Gaborone.

MLMW (2022). Ministry of Land Management, Water and Sanitation Services. www.gov.bw/ministries/ ministry-land-management-water-and-sanitation-services, accessed 27 December 2022.

Morgan, D. L., & Spanish, M. T. (1984). Focus groups: A new tool for qualitative research. *Qualitative Sociology, 7*(3), 253–270.

Murphy, J. (2010). Does community involvement in conservation provide an alternative to fortress conservation? http://lionalert.org/page/article-community-involvement, accessed 20 June 2022.

Murray, C., & Marmorek, D. R. (2004). Adaptive management: A spoonful of rigour helps the uncertainty go down. Proceedings of the 16th Annual Society for Ecological Restoration Conference, Victoria, BC.

NCONGO (2020). *NCONGO Projects at the Panhandle.* NCONGO, Maun.

Neuman, L. W. (2007). *Social Research Methods*, 6th edition. Delhi: Pearson Education India.

NWDC (2007). *Ngamiland Tourism Development Plan.* North West District Council, Maun.

ODMP-Department of Environmental Affairs (2008). *The Okavango Delta Management Plan*. www .okacom.org/site-documents/project-reports/odmp-documents/okavango-delta-management-plan/ view, accessed 24 November 2022.

OKACOM (2019). *Project Implementation Review – Okavango Strategic Action Programme Implementation*. OKACOM, Gaborone.

ORI GIS Lab (2019). Okavango River Basin map. University of Botswana, Maun.

Pomeroy, R., & Douvere, F. (2008). The engagement of stakeholders in the marine spatial planning process. *Marine Policy*, *32*(5), 816–822.

Reed, M. S., Graves, A., Dandy, N., Posthumus, H., Hubacek, K., Morris, J., ... Stringer, L. C. (2009). Who's in and why? A typology of stakeholder analysis methods for natural resource management. *Journal of Environmental Management*, *90*(5), 1933–1949.

Songhurst, A. C. (2011). Competition between people and elephants in the Okavango Delta Panhandle, Botswana. PhD thesis, Imperial College London.

Statistics Botswana (2011). *Botswana Population and Housing Census*. Statistics Botswana, Gaborone.

Statistics Botswana (2015). *Ngami West Sub District Population and Housing Census 2011: Selected Indicators for Villages and Localities*. Statistics Botswana, Gaborone.

Statistics Botswana (2022). *Population & Housing Census 2022: Population of Cities, Towns, Villages and Associated Localities*. Statistics Botswana, Gaborone.

UNDP (2019). *The GEF/SGP Country Programme Strategy for Operational Phase 7 – Botswana*. Department of Environmental Affairs, Gaborone.

UNDP GEF (2016). *Support to the Cubango-Okavango River Basin Strategic Action Programme Implementation: Project Document*. UNDP, Gaborone.

UNDP GEF-SGP (2019). *The GEF SGP Country Programme Strategy for Operational Phase 7*. UNDP, Gaborone.

UNESCO (2019a). *Okavango Delta. World Heritage Convention*. https://whc.unesco.org/en/list/1432/ #:~:text=The%20inscribed%20World%20Heritage%20property,sands%20of%20the%20Kalahari %20Basin, accessed 10 January 2022.

UNESCO (2019b). *Operational Guidelines for the Implementation of the World Heritage Convention*. Paris: UNESCO.

USAID SAREP (2013). *Mid-term Review and Gap Analysis of the Okavango Delta Management Plan: Scoping and Gap Analysis Report*. USAID SAREP, Gaborone.

VanderPost, C., Baise, M., & Mahloane, P. (2003). *Community Based Natural Resources of Bugakhwe and Anikhwe in the Okavango Panhandle in Botswana*. IUCN/SNV CBNRM Support Programme in collaboration with the Teemacane Trust.

World Heritage Committee (2014). Decisions adopted by the World Heritage Committee at its 38th Session (Doha 2014). Decision 38 COM WHC-14/38. COM 16.

PART V

CONCLUSION

22. The interlinkage between tourism, environmental conservation, and natural resource management: a synthesis

Oluwatoyin D. Kolawole, Joseph E. Mbaiwa, Wame L. Hambira and Emmanuel Mogende

OVERVIEW

The tourism sector is one of the major income earners for national governments in continental Africa and the rest of the world (Rogerson, 2007). The sector continues to gain popularity as people's wanderlust progressively soars as it has over the last several decades. This *Handbook*, therefore, aimed to underscore the interconnectedness of tourism, environment, and sustainable natural resource use. Its ultimate goal was to identify strategic and innovative pathways for enhancing environmental conservation through sustainable tourism. Without any doubt, the compilation pointed readers' attention to the interdependent nature of the relationship existing between tourism and biodiversity conservation in every part of the world. The *Handbook* thus emphasised the role of communities in natural resource governance by applying Ostrom's social-ecological systems (SES) theory (Ostrom, 2009) as its analytical framework for reaching a consensus on divergent viewpoints on the subject within the context of global environmental change and emerging governance issues. The book, thus, highlights equity and environmental justice issues in relation to how destination communities have benefited from tourism activities in southern Africa and elsewhere.

EMERGING ISSUES AND CHALLENGES

Diverse issues on tourism and conservation emerged from the book. More importantly, a number of pertinent development challenges converge in relation to community exclusion and marginalisation from Protected Area (PA) arrangements; rural livelihoods and human–wildlife conflicts (HWC); climate change adaptation in the context of nature-based tourism; the subversive nature of market forces on conservation values; and institutional failures. Condensed, each of these identified issues will be highlighted as follows:

1. *Exclusion and marginalisation of local communities from PA arrangements*: Throughout the book, many authors agreed that communities that are contiguous to or situated within wildlife management areas (WMAs) or PAs continue to witness some forms of marginalisation and exclusion. In other words, they have little or no voice in determining how the natural resources in their local environments are managed. In most cases, local communities lack well mobilised civil society organisations that can fight for their rights when government undermines them. In another vein, Ramutsindela (2016, p. 27) argues that

330 *Handbook on tourism and conservation*

international agreements often 'privilege one form of knowledge system about nature conservation over the other which leads to the condemnation of the ways in which marginalised groups relate to and interact with their environment'. Not only are these communities marginalised. They also bear the consequences of conservation (see, for instance, in this volume Mzimela and Moyo; Blackie; Mbaiwa and Tafa; Gontse et al.). More often than not, the imbalanced power relations between the government or wealthy, rent-seeking individuals and local communities and the associated exclusion from decision-making rights over resource access and use may have engendered communities' negative perceptions of tourism objectives and the hostilities they exhibit towards such initiatives in the end. Suffice it to say that people's disenchantment with the conservation agenda is born out of the deprivation meted out to them in relation to land grabbing activities of tourism companies without any adequate compensation to meet the affected people's needs. To assert themselves, they use all sorts of strategies ranging from passive resistance to poaching in counteracting unfavourable government policies (see Scott, 1989). It is noteworthy that misdemeanour associated with certain infractions arises when prolonged deprivations are not adequately addressed and promptly minimised among any people (see Rotimi, 1997).

2. *The precarious situation of rural livelihoods in wildlife areas*: Closely related to community exclusion from the conservation agenda is the ever-present jeopardy witnessed in rural livelihoods in and around PAs due to livestock depredation and crop raiding by carnivores and elephants, respectively. Those who bear the brunt of conservation are local, resource-poor people that live in proximity to wildlife most especially where the latter's populations are not well managed. For instance, the burgeoning population of elephants in certain areas do cause untold hardships for hapless, arable farmers (see, for example, Gontse et al., this volume). The damage caused by wildlife is immediately problematic for any conservation agenda as this might engender increased intolerance for wildlife by affected farmers in the HWC hotspots (Graham, 2004; Gupta, 2013). The situation becomes aggravated when local farmers and other local stakeholders perceive wildlife as government owned and for which their livelihoods are in jeopardy without any adequate compensation in return for any loss and damage to lives and property.

3. *Climate change adaptation and nature-based tourism*: Nature-based tourism relies mainly on the natural environment and sceneries, which are, indeed, vulnerable to climate change. It is noteworthy, however, that the complex interrelationships between tourism activities and natural ecosystems do engender alterations in the natural environment, which could, in turn, induce climate change. It is thus apparent that tourism could impact people's natural, cultural, and social-economic assets. Authors' viewpoints on the need to devise innovative adaptation strategies to climate change for the purpose of enhancing sustainable tourism (Tervo-Kankare, this volume; El-Masry, this volume), and the need for a careful planning in land use and land cover change (LULCC) to lessen the negative impact of tourism on the environment (Gondo and Kolawole, this volume) point in the same direction. As changing climate continues to impact on biodiversity conservation in PAs leading to alterations in both cultural and natural landscapes (Aloia et al., 2019), there is an urgent need for some readjustment by all stakeholders to adequately respond to the changes being witnessed in the environment. In other words, the vulnerability of tourist destinations to climate change would warrant pragmatic steps or actions needed to equalise or lessen the phenomenon's negative impact on the aesthetic values of natural and cultural landscapes (Tervo-Kankare, this volume). Global warming, which is one of the effects of climate change, continues to

Tourism, environmental conservation, and natural resource management: a synthesis 331

threaten iconic tourist destinations on the African continent and elsewhere as witnessed in the decline in flora and fauna and other similar geographic features (see Dube, this volume).

4. *The subversive nature of market forces on conservation agendas*: The pervasiveness of capitalism and its associated power imbalances between the rich and poor may have undermined the intention of tourism or agrotourism in its regenerative inclination towards nature conservation. Prioritising profits over all other things including the natural ecosystems is counterproductive to real term development. Unlike growth that pushes for economic efficiency and expansion, real term development places premium on environmental and human wellbeing, ensuring the realisation of human potential. This is the position of ecological economics. When financially powerful individuals or multinational companies (MNCs) undercut local communities because of the latter's powerlessness, conservation will, perhaps, be a secondary concern in the end. Indeed, adopting tourism as an avenue to conservation may not necessarily be as straightforward as one would think. Finding a balance between the need to marketise and conserve natural resources as well as cultural heritage could be a daunting task. To be sure, Andéhn and L'Espoir Decosta (this volume) eloquently presented the scenario as a Catch-22 situation and a conundrum of some sort – being between Scylla and Charybdis, which indeed is about choosing the lesser of two evils. One way out of the quagmire could be through the implementation of conservation programmes that empower local communities to assume leadership roles in preserving local resources only if the local populace is somehow immune to the influence of market forces (see also Andéhn and L'Espoir Decosta, this volume), and where they themselves do not metamorphose into being bourgeoisie in the long run.

5. *Institutional failures*: Institutional frameworks that prioritise international agreements over and above local people's interests and needs continue to run counterproductive to conservation agendas (Mogende and Ramutsindela, 2022). Although international agreements are not legally binding, developing countries are often coerced to implement initiatives that do not suit local desires partly because they may lose material benefits such as donor funding and foreign aid that are associated with being a member of such agreements. Perceiving local communities as constituting a barrier to wildlife conservation and creating an apprehensive atmosphere among hapless, local communities in an attempt to entrench the commodification and territorialisation of wildlife resources are, indeed, anti-development. For example, institutionalised commodification of wildlife resources continues to attract foreign tourism businesses, which seemingly dictate and control wildlife conservation and tourism policies in Botswana (see, for instance, Suping, this volume). While the CBNRM was designed to enhance community empowerment and participation in natural resources governance (Mogende and Kolawole, 2016), it has not realised its objectives because local-level capacity development is still largely non-existent in communities, who therefore lack the ability to compete favourably with foreign tourism investors. Besides, international and regional agreements may have grossly constrained the government's ability to implement impactful interventionist development initiatives in relation to natural resources conservation (Suping, this volume; Solway 2009).

332 *Handbook on tourism and conservation*

POLICY ISSUES

Power and politics play a crucial role in policy development. Transforming structures and processes in a bid to enhance people's access to natural resources might be hamstrung by political issues (Ashley and Carney, 1999). More often than not, rent seekers and powerful individuals pursue their own agenda to thwart the realisation of the common good. In this context, policy development in natural resources conservation must, therefore, transcend *personal interests* and *business as usual* or even crisis situations (see Grindle and Thomas, 1991) to achieve the goal of creating an impactful development at the grassroots level. Overcoming the Leviathan in administrative reforms will mean that state reformers need to make transparent, strategic choices in policy design that are easy to relate to by all stakeholders (see, for instance, Heredia and Schneider, 2003).

The following policy issues emanated from the book:

- *Putting communities first*: Placing people at the centre of wildlife conservation is the only solution for sustainable environmental conservation. Throughout, most authors agreed that local communities have been disenfranchised from the management of and access to natural resources. National governments have a duty to act in good faith to promptly reverse the trend and with all sincerity of purpose ensure that local people are placed at the centre of any conservation agenda in tourist destination areas. While it is not as simplistic as it seems, the avoidance of too much emphasis on commercialisation and commodification at the expense of local beneficiaries will go a long way in enhancing the realisation of sustainable tourism and environmental conservation. In response to the necessary criteria for evolving SES, there is a need to establish adaptive governance frameworks whereby actors draw on existing institutional, cultural, and social arrangements to (re)shape institutions in response to the changing complex situations. Reshaping the institutional landscape that places communities at the centre of resource management attempts to offer mechanisms that broker better economic growth, social equity, and sustainable development (Hall, 2007). Thus, economic benefits accruing from tourism MNCs must trickle down to the affected communities with a view to improving the latter's quality of life (Hambira, 2020).
- *Climate change adaptation in tourist destination areas*: It is no longer news that climate change and variability have become an albatross, which humanity continues to contend with in the twenty-first century. Paradoxically, the hapless, poor majority are those who bear the brunt of extreme weather conditions in developing countries. Clearly nature-based tourism will continue to be impacted if proactive and climate smart initiatives are not properly activated in tourist destination areas. Approaches that would enable de-risking of climate change scenarios, which impact on both flora and fauna (including aquatic resources) need practical applications. In littoral communities, policymakers and government programmes need to focus attention on beach nourishment programmes if only to avoid jeopardising tourism and livelihoods activities in the affected areas.
- *Product development and marketing strategies of agrotourism*: It is becoming increasingly necessary that tourist destination communities devise appropriate technologies to develop innovative products that can enhance marketability of local products and by that means improve people's socio-economic status as they receive foreign visitors in their local environments. One of the promising avenues to achieve this may be through agrotourism where business owners could create cultural identity in the development of local products and

Tourism, environmental conservation, and natural resource management: a synthesis   333

cuisines. Value additions to artefacts and other similar products may help to market and further entrench the cultural identity of local people. People are encouraged to contribute to environmental conservation when they genuinely reap the gains of tourism activities in their areas.

- *Capacity development*: Situating communities at the centre of tourism and conservation would mean that they need to be empowered to gain managerial and entrepreneurial skills in nature-based or cultural tourism. Therefore, policy development geared towards functional education and training will be necessary to propel effective community participation in CBNRM activities tailored towards tourism and environmental conservation.
- *Safety and security*: Policy issues in safety and biosecurity cannot be wished away most especially within the context of agrotourism in which local farm facilities are exposed to international tourists who are likely to come in contact with farm animals and other agro-allied products. The recent COVID-19 pandemic serves as a wake-up call for stakeholders in the industry to promptly devise appropriate mechanisms that could impede the spread of zoonotic diseases and the like across space and time.

All that said, the *Handbook on Tourism and Conservation* intended to identify strategic and innovative pathways for enhancing environmental conservation through sustainable tourism. We believed that this book has achieved its objective. The contributions in this volume will be handy for policymakers and relevant institutions in the implementation of tourism policies that equally place pertinent issues in tourism and environmental conservation on the same pedestal, and which in turn could enhance wider participation in tourism development by all stakeholders in the sector. The need to craft policies that are uniquely relevant for the current dispensation will go a long way in enhancing the attainment of both sustainable tourism and environmental conservation meant for the good of all. While this book may not have provided any uniquely new information on the subject, it has contributed additional literatures to the body of knowledge on tourism and conservation issues in southern Africa and beyond.

REFERENCES

D'Aloia, C. C., Naujokaitis-Lewis, I., Blackford, C., Chu, C., Curtis, J. M. R., Darling, E., … Fortin, M.-J. (2019). Coupled networks of permanent protected areas and dynamic conservation areas for biodiversity conservation under climate change. *Frontiers in Ecology and Evolution*, 7(27). https://doi.org/10.3389/fevo.2019.00027.

Ashley, C., & Carney, D. (1999). *Sustainable Livelihoods: Lessons from Early Experience*. London: Department of International Development.

Graham, H. (2004). The ecology of conservation of lions: Human wildlife conflict in semi-arid Botswana. PhD dissertation, University of Oxford.

Grindle, M. S., & Thomas, J. W. (1991). *Public Choices and Policy Change: The Political Economy of Reform in Developing Countries*. Baltimore: Johns Hopkins University Press.

Gupta, A. C. (2013). Elephants, safety nets and agrarian culture: Understanding human-wildlife conflict and rural livelihoods around Chobe National Park, Botswana. *Journal of Political Ecology*, 20, 238–254.

Hall, A. (2007). Global governance in a globalizing world. In A. R. Turton, J. Hattingh, D. Roux, M. Classen, G. A. Maree, & W. F. Strydom (eds), *Governance as a Trialogue: Government-Society-Science in Transition* (pp. 29–38). Berlin: Springer Verlag.

Hambira, W. L. (2020). A review of community social upliftment practices by tourism multinational companies in Botswana. In M. T. Stone, M. Lenao, & N. Moswete (eds), *Natural Resources, Tourism*

334 *Handbook on tourism and conservation*

and *Community Livelihoods in Southern Africa: Challenges of Sustainable Development* (pp. 52–63). London: Routledge.

Heredia, B., & Schneider, B. R. (2003). The political economy of administrative reform in developing countries. In B. R. Schneider & B. Heredia (eds), *Reinventing Leviathan: The Politics of Administrative Reform in Developing Countries* (pp. 1–22). Coral Gables, FL: North-South Center Press at the University of Miami.

Mogende, E., & Kolawole, O. D. (2016). Dynamics of local governance in natural resources conservation in the Okavango Delta, Botswana. *Natural Resources Forum*, 40, 93–102.

Mogende, E., & Ramutsindela, M. (2022). The violence of greening the state in Africa. In M. Ramutsindela, F. Matose, & T. Mushonga (eds), *The Violence of Conservation in Africa: State Militarisation and Alternatives* (pp. 37–52). Cheltenham, UK and Northampton, MA, USA: Edward Elgar Publishing.

Ostrom, E. (2009). A general framework for analysing sustainability of socio-ecological systems. *Science*, 325(5939), 419–422.

Ramutsindela, M. (2016). Political dynamics of human-environment relations. In M. Ramutsindela, G. Miescher, & M. Boehi (eds), *The Politics of Nature and Science in Southern Africa* (pp. 20–36). Basel: Basler Afrika Bibliographien.

Rogerson, C. M. (2007). Reviewing Africa in the global tourism economy. *Development Southern Africa*, 24(3), 361–379.

Rotimi, O. (1997). Before Sharia again. *The Guardian*, Lagos, Nigeria, 1 September.

Scott, J. C. (1989). Everyday forms of resistance. *The Copenhagen Journal of Asian Studies*, 4, 33–62.

Solway, J. (2009). Human rights and NGO 'wrongs': Conflict diamonds, culture wars and the 'Bushman Question'. *Africa: Journal of the International African Institute*, 79(3), 321–346.

Index

Abarca Alpízar, F. 150
Adamov, T. 214, 215
adaptive management 307, 324
Addo, K. A. 94
adventure tourism 4
Africa
    climate change effects in 112
    cultural tourists in 279
    poverty and policies of African governments 268
    tourism and climate change vulnerabilities in 86–100
    Tourism Climate Index (TCI) for 104–6
    tourism industry and risk of climate change in 101
Afriski, Lesotho 89, 91, 97
agency 82
agriculture
    arable 306–307
    dryland farming 46, 48–9, 306–7
    flood-recession (*molapo*) farming 46, 48, 306–7
    monoculture 170
    in Zimbabwe 205
agritourism 5–6, 147–9, 169
    benefits of 210–13, 217
    challenges facing development of 213–17
    definition of 204
    development of, in Zimbabwe 5, 204–21
    European Union policies 157
    in Fortín, Veracruz, Mexico 147–63
    growth in 205, 208, 216
    and market-driven conservation 164–75
    in the multifunctionality of the landscape 150
    product development and marketing strategies of 332–3
    and regeneration 165–6
    technical assistance, education and training 156
    typologies of 158
agroforestry 31
Aguilar-Rivera, Noé 5
Akinyemi, F. O. 198
'all inclusive' resorts 167
Allison, E. H. 180
Alpine region 76
Alternative Agrifood Networks 5
AmaNgwane community 13, 17
AmaSwazi community 13, 17

AmaZizi community 13, 17, 20
Ammirato, S. 212
Andéhn, Mikael 5, 331
Andereck, K. L. 38
Angola 119, 302
Antarctica 76
Ariffin, A. R. M. 210
Aronson, J. 284
artificial water points (AWPs) 131
Ashley, C. 33, 56
Asmelash, A. G. 211
Australia
    climate change adaptation plans in 82
    Great Barrier Reef 115
    weather extremes in 103
    wetlands 78

Babbie, E. R. 60
Badimo, Dandy 7
Baipai, Rudorwashe 5
Bajgier-Kowalska, M. 212, 215
Bakgalagadi community 269
Balearic Islands, Mallorca wetland 78, 79
Bandura, A. 11, 12, 17
Banerjee, G. 266
Barbieri, C. 151, 215
Basarwa tribe, Botswana 30–31, 32, 55, 225, 229
Basso, K. H. 280
beach tourism 4, 87
Bebbington, A. 184
Belsky, J. M. 267
Bender, B. 295
Berkes, F. 185
Bhammar, H. 18, 20
biodiversity loss 119
birds
    bird sanctuary, at Nata 63
    vulnerability to climate change 97
Black, R. 191–2
Black South Africans, views of conservation 17
Blackie, I. R. 6, 244
Blanco, M. 156
Blau, P. M. 57
Boateng, I. 94
Bolaane, M. 225
Bolsen, T. 125
Bonsai 160
Boro River 198
Botswana 86, 117

336 *Handbook on tourism and conservation*

artificial water points (AWPs) 131
attitudes of local people towards conservation in 56
basket making in 184
CBNRM projects in 40–41, 227, 232, 239
CBTOs in 251–2
Chobe Enclave 11
CITES 232
contribution of tourism to economy in 25–6, 37
Country Programme Strategy (CPS) 315
Covid-19 pandemic and 26, 196–7, 227
crop damage by elephants in 43
cultural landscapes 279
Department of Environmental Affairs (DEA) 226, 232
Department of Forestry and Range Resources (DFRR) 313
Department of Meteorological Services 312
Department of Wildlife and National Parks (DWNP) 45, 134, 225, 240, 312
District Land Use Planning Unit (DLUPU) 314
drought in 115–16
ecotourism in 54–6, 264
electric fences along Boteti River 44, 47–8
elephant population in 37, 43, 44
exogenous tourism enterprises in 227
Forest Conservation of Botswana (FCB) 313
forest reserves in 267
GDP and tourism in 196
Hospitality and Tourism Association 196
hunting ban 228
ICT adoption in 249, 258
Integrated Support Programme for Arable Agriculture Development (ISPAAD) 42
international cooperation in 6, 224, 230
Ipelegeng (drought relief programme) 40
land use and land cover in 5, 190, 191, 193–9
legal system 256
local media articles on climate change 119
map of 283
media coverage of tourism 119–23
Ministry of Environment and Tourism (MET) 312, 313
Mohembo Bridge 198
National Parks and Game Reserves Regulations 2000 49
natural resources of 223, 226
Ngamiland District 237
North-East District (NED) 3, 54–71
participation in international and regional environmental agreements 232
poverty in 268–269

private tourism enterprises 227
rainfall in 117
sovereignty of the state in Botswana's WMAs and TFCAs 231–2
state-making in 228–9
Tawana Land Board 306
Technical Advisory Committee 314
Technical Assistance Committees 241
threat of Covid-19 to tourism in 196–7
tourism industry in 115
Tourism Satellite Account 240
Tribal Land Act (TLA) 240
Wildlife Conservation and National Parks Act 225
wildlife conservation in 223, 224, 231
Wildlife Conservation Policy (WCP) 225
Wildlife Management Areas (WMAs) 228–9
wildlife tourism in 194, 226
see also Goo-Moremi Cultural Landscape; northern Botswana
Botswana Communications Regulatory Authority (BOCRA) 249
Botswana Press Agency 116
Botswana Tourism Organisation 197, 244, 245, 313
Brines, J. 38
Brockington, D. 224, 225, 231
Brown, J. 322, 323
Brysiewicz, P. 87
Buckles, D. 54
Budowski, G. 264
buffalo (*Syncerus caffer*) 48, 136
Bugakhwe group 28, 225, 305
Burgess, R. G. 60
Butler, R. 267
Bwana, M. A. 209
Bwindi Impenetrable National Park, Uganda 18

Cameroon 102
Campbell, J. M. 215
Campbell-Smith, G. 47
Canada
    glaciers in 81
    tourism industry and risk of climate change in 101
    Waterton Lakes National Park 132
capacity development 331, 333
capital 178–9
    built 179, 180, 183
    cultural 179, 180, 184
    financial 179, 180, 183
    human 178–179, 180, 184
    natural 179, 180, 183
    political 179, 180, 182
    social 178, 180, 182

Index 337

capitalism 171, 172
carbon dioxide 124
carbon sinks 109
Cassidy, L. 49, 243–4, 252
Cavric, B. I. 199
CBNRM *see* Community-Based Natural Resource Management (CBNRM)
CBT *see* community-based tourism
CBTOs *see* community-based tourism organisations (CBTOs)
CECT *see* Chobe Enclave Community Trust (CECT)
Cele, H. M. S. 18
Central Asia, tourism industry and risk of climate change in 101
Central Kalahari Game Reserve 224
Chambers, R. 180
Chang, Y. C. 87
Chase, L. 204, 211, 216
Chatterjee, B. 265, 266
Chen, A. A. 89
Chen, F. J. 64
Chen, Y. 210
Chenje, M. 55
Chikuta, Oliver 5
Chimanimani National Park 95
China, Chinghai Biosphere Mountain reserve 64
Chiutsi, S. 199
Chobe Enclave, Botswana 33, 237, 253, 255, 257
Chobe Enclave Community Trust (CECT) 252, 253, 257–8
Chobe Forest Reserve 255
Chobe Linyanti River systems 115
Chobe National Park, Botswana 130, 131, 133–4, 255
    wildlife migratory corridors, between Namibia and 140
Chobe River 255
Choi, H. C. 38
Chula Vista camping 153
Churchill, Canada 78–9
Ciolac, R. 216
CITES (Convention on International Trade in Endangered Species of Wild Fauna and Flora) 223, 230, 232
civil wars 268
climate change 4, 101–14, 140
    adaptation for 73–85, 97, 107, 109–12, 332
    consequences for the tourism sector 104–8
    contribution (footprint) of tourism to 108–9
    definition of 102
    effects in Africa 112
    impacts in protected areas 77–79
    Intergovernmental Panel on Climate Change (IPPC) 102, 103, 112

literature reviews on protected areas 75–6
    media articles on 118, 119
    nature-based tourism and adaptations for 330–31
    vulnerabilities of African tourist destinations to 86–100
Climate Change 2021: The Physical Science Basis 102
climate change communication 115–29
    issues in 123–5
    recommendations 125–6
    research on 116
climate change sceptics 123
climate variability 130, 132
coastal tourism, sea level rise and 92–5
coffee 148–9, 170
Colding, J. 288
collaboration
    in decision-making among users of protected areas 82
    in management of protected areas 18
collective efficacy 3, 11–13
Collett, D. 55
Colorado, ecotourism in 55
communities, role in natural resource governance 329
community-based initiatives 300, 301, 308–9
Community-Based Natural Resource Management (CBNRM) 6, 39, 65, 132, 133, 140, 251, 314, 331
    Act and regulations 244–5
    and COMPACT 301–2
    implementation of 242–4, 245
    poverty reduction and 237–48
    projects 40, 49
    Support Association of Botswana (CSABO) 244
community-based organisations (CBOs) 241, 242–3
community-based tourism 11, 14, 17, 21, 250
community-based tourism organisations (CBTOs) 249
    in Africa 249
    in Botswana 251–2
    ICT and 249–63
community capitals framework 177, 178–80, 185
community development 133, 141
community empowerment 14, 21, 31, 33
Community Management of Protected Areas Conservation *see* COMPACT programme
Community Natural Resources Management Lease (head lease) 239, 244, 251
community power 18–20
community sustainability 133
COMPACT programme 300–301, 307, 319, 324

338 *Handbook on tourism and conservation*

CBNRM and 301–2
governance structure for implementation of
322–3
policy and legislative framework 319–21,
325
see also Okavango Panhandle
conflicts
land-use conflict resolution 59
between resource-dependent communities
and PA/tourism development
managers 16
conservation
conservation areas in Africa 55
as counteracting excesses of tourism 171
ecotourism as a tool for 54
implications of human-elephant conflicts for
rural livelihoods and 37–53
local people's perceptions of 56, 65–6
relationship between PA-based tourism and
16
subversive nature of market forces on
conservation agendas 331
see also wildlife conservation
Constanza, R. 179
Controlled Hunting Areas (CHAs) 239, 306
Convention on International Trade in Endangered
Species of Wild Fauna and Flora (CITES)
223, 230, 232
Conway, G. R. 180
coral reefs 92
bleaching of 81, 92, 111
in North Africa 92
Corbin, J. 118
corruption 268
Covid-19 pandemic 115, 130, 155, 164, 226, 333
impact on tourism 25, 86
and tourism in Botswana 26, 196–7, 227
Cox, M. 253
Creswell, J. W. 13
Cristina, M. 215
crop damage
electric fences to minimise 44, 45, 47, 48
and elephants 43
implications for livelihoods 46
implications for wildlife conservation 47
by wildlife 38, 40, 43
Crossley, M. N. 197
Cuadrado, E. 18
cultural heritage 7
cultural landscapes
globalisation and commodification of 279
sense of place and tourism in 279–99
cultural tourism 21, 32, 164, 166, 167–8, 172, 279
empowerment through 20–21
in Okavango Delta, Botswana 184

small-scale culture-based tourism initiatives
176–88
culture, marketization of 167–9
Cunningham, S. J. 88
cyclones 95–96, 97

D'Amour, D. 48
Darkoh, M. B. 55, 198
Das, M. 265, 266
Dawson, J. 79
dependency syndrome 275
desertification 111
destination communities, and natural resources
conservation 6–7
digital image processing 190
Dikgosi (village chiefs) 271
Ditladi village 59
Djibouti 102
Downing, T. A. 90
droughts 111, 115
Dube, K. 4, 80, 82, 94, 95, 96
Dumitras, D. E. 212
Dutra, L. 77, 80, 82

Earth Resources Technology Satellite (Landsat)
193
East Asia, weather extremes 103
EcoExist 318
ecological economics 331
ecotourism 32, 227
in Africa 264
alleviation of poverty through 264–78
in Botswana 54–6
costs and benefits of 267
definition of 55
improving livelihoods through 265–7
in Kweneng West 269–70, 274
local people's perceptions of 64
Quebec Declaration on Ecotourism 2002 265
residents' perception of Tachila Nature
Reserve 54–71
education 268
Edwards, J. 178–9
Egypt, reefs and tourism in 92
El Niño events 92
elephants (*Loxondanta africana*) 136
die-off in Okavango Delta 140
human–elephant conflicts 37–53, 231, 318
implications of elephant crop damage for
livelihoods 46
elite capture 241
Elkington, John 206
Ellis, F. 179, 180
Emery, M. 178, 180, 184
empowerment

through cultural tourism resources 20–21
through tourism development 14
enclave tourism 225
Engels, Friedrich 171
Engleton, Abigail Lillian 7
environment
climate change communication for
sustainable environmental
conservation 115–29
cultural tourism and 20
impact of agritourism on 5
environmental change 131–2, 140, 141
effects on wildlife-based tourism 137–8
implications for community development
138
perceptions of tour guides on 136–7
role of climate in 136–7
environmental justice 1, 6, 329
Erlingsson, C. 87
Ethiopia
Abijata-Shalla Lakes National Park 130
relocation of people from Nechasar National
Park 55
Etongo, D. 94
Europe, weather extremes 103
European Union, agritourism policies in 157
Ex Hacienda Monte Blanco, Santa Lucía 159
exclusion
of hunter-gatherers 239–42
of local communities 237
extreme weather
adaptation actions/techniques 111
effects on tourism sector 107

Faccioli, M. 78
farming *see* agriculture
Feld, S. 280
Ferguson, J. 231
Finland 82, 130
fires
global warming and 88
on mountain tourist areas in Africa 90
Fisichelli, N. A. 79
Fitchett, J. M. 92, 115, 116
Flanigan, S. 151
Flora, C. 178, 180, 184
Flora, J. 178
Folke, C. 288
Folorunso, C. O. 195
food production 109
food waste 109
foot and mouth disease (FMD) 217
forest-based tourism 3, 25–36
forests
community empowerment through 31, 33

contribution of forest resources to rural
livelihoods 30–31
local people's attitudes towards forest
reserves 64
mangrove forests 109
Fortín, Veracruz, Mexico
agriculture in 148–9, 151
agritourism in 5, 147–63
criminal incidence in 157
flower cultivation in 149, 151, 152
la Feria de la Flor (the Flower Fair) 159
management and entrepreneurship in 156–7
map of 150
Metlac canyon 154
non-working farms 159
product development and marketing
strategies in 156
public policies in agritourism 157
safety and security in 157
surveys of producers of 152
tourist attractions of 152, 153
tourist services 154–5
working farms 158–9
fortress conservation 56
framing theory 117
Francistown, Botswana 57

Gandiwa, Edson 5
Garekae, Hesekia 3
Garmin handheld global positioning system
(GPS) 193
geographic information systems (GIS) 88, 190–91
geospatial technologies 193
Getz, D. 55
Ghana
rising sea levels 94
tourism climate change index (TCI) 104, 106
Gil, C. 147
glaciers, melting of 111
Glaser, M. 252
Glasson, J. 295
Global Distribution Systems (GDSs) 250
Global Environment Facility Small Grants
Programme 314, 322
Global South 4
global warming 4, 86, 330–31
impact on
coastal and ocean tourism 91–2
mountain tourism 89–91
tenting accommodation 96
and water-based tourism resorts 96
Goffman, Erving 117
Gohori, O. 11, 17
Golnik, B. 217
Gonarezhou National Park 96

340   *Handbook on tourism and conservation*

Gondo, Reniko 5
Gontse, Kenalekgosi 3
Goo-Moremi Cultural Landscape 279–97
Goo-Moremi Gorge 282
    ancestor worship at 286–7, 296
    community views on the use of 285
    drilling for water at 293
    fencing of Motlhodi (water spring) 292–3
    Kgosi Tshekedi Khama road and car saga
        292
    Sir Seretse Khama Alarm Stone 294–5
    tourism projects and culture 293–5
Goo-Moremi Village 281, 282, 293
    development projects 291
    residents' sense of place, place attachment
        and identity 280–81
    role of ancestors for 287
    taboos and social norms 288–91
Good, K. 239
Google Earth 88, 90, 91, 193
Gorongosa National Park 95
Gössling, S. 74, 101, 132
governance systems 2
Graham, H. 38
greenhouse gas emissions 108, 110
Groulx, M. 74, 83
Grove, R. 55
Guliyev, S. 212
Gumede, T. K. 19
Gupta, A. C. 48
Gupta, S. 268
Gutiérrez-Montes, I. 178, 184
Guvamombe, I. 206

Hall, D. 190
Hall, H. 38
Hambira, W. L. 7, 96, 124, 125, 132, 226
Hamilton, Clive 171
Hashem, H. 280
Hay-Edie, T. 322, 323
Helama, S. 90
Hemp, A. 90
Hernández-Sampieri, R. 150
Hilmi, N. 92
Hinch, T. 267
Homans, G. C. 38, 56
Honey, M. 55
Hoogendoorn, G. 91, 92, 115, 116, 125
hospitality tourism 166, 167
Huberman, A. M. 256
Hubona, Zibanani 294
human–elephant (*Loxodonta africana*) conflicts
    231, 318
    implications for rural livelihoods and wildlife
        conservation 37–53

human–lion (*Panthera leo*) conflicts 38
human–orangutan (*Pongo abelii*) conflicts 47
human–wildlife conflict (HWC) 37, 38, 55, 131,
    140
hydrophytes 196, 197

Iceland, glacier tourism in 82
identity 279, 280, 281, 282, 283, 284, 285, 291,
    295, 296
Igoe, J. 224, 225, 231
Ilesi, Kakamega County, Kenya 181
    clay harvesting site in 183
    Ilesi Clay Potters Women Group 177, 182–4,
        185
    micro and small-scale tourism initiatives
        (MSTIs) 176–7
Imbaya, Beatrice 5
impala (*Aepyceros melampus*) 136
India
    agritourism in 209
    fatal injuries of elephants in 39
Indian Ocean, coral bleaching in 92, 93
indigenous knowledge 10, 15, 16, 20
information and communication technology (ICT)
    6, 249, 250–51
infrastructure
    impact of weather extremes on 78
    tourism 97
institutions
    and international agreements 331
    natural resources conservation policies 332
Integrated Conservation and Development
    Projects (ICDP) 300
Intergovernmental Panel on Climate Change
    (IPPC) 102, 103, 112
    Special Report on Ocean and Cryosphere 91
International Union for Conservation of Nature
    (IUCN) 3, 73, 300
Irandu, E. M. 56
iSimangaliso Wetland Park 19
Isukha community, Kenya 177, 181
Itekeng Community Trust 306
ivory
    ban on sale of 232
    poaching of 140

Jacobs, B. 79
Johnson, P. 55
Joint Venture Partnerships 6, 251, 252, 258
Juukan Gorge cave shelter, Western Australia
    166, 168, 169, 170–71

Kahn, M. 280
Kakamega County, Kenya 181
Kakamega forest 181, 184

Kakamega National Park 181
Kangas, K. 76, 78
Karampela, S. 210
Kariba, Lake 96
Kashe, K. 48
Kavango-Zambezi Trans-frontier Conservation
    Area (KAZA) 231
Kegamba, J. J. 17
Keiner, M. 199
Kenya
    GDP of 181
    Kisimu Region 209
    Ministry of Culture, Sports and Social
        Services 183
    Ministry of Tourism and Wildlife 176
    National Bureau of Statistics' Gross County
        Product Report 2019 181
    opposition to the lifting of the ban on the sale
        of ivory, 232
    poisoning and spearing of lions in 39
    pottery production by women's groups in 5,
        177, 182–4, 185
    small-scale culture-based tourism initiatives
        in 176–88
    tourism climate change index 104, 106
Kenya, Mount 89–90, 97
Kern, T. 38
Kgalagadi Transfrontier Park 17
Kgathi, D. L. 197
Khairabadi, O. 209
Khama, Kgosi Tshekedi 292, 294
Khama, Sir Seretse 292
Khumaga village, Botswana 3, 39–40, 44
    arable farming at 41–43
    DWNP officers 45, 49–50
    elephant crop damage in 43–7, 50
Kibaara, O. N. 176
Kilimanjaro, Mount 89, 90–91, 97
Kilinc, C. C. 212, 216
Kilungu, H. 90
Kirilenko, A. P. 116, 119
Kiyaga-Mulindwa, D. 286
Klitgaard, R. 268
Kolawole, Oluwatoyin 5
Kotler, P. 170
Koutra, C. 178–179
Kruger National Park 19, 95, 96
Kubickova, M. 215
Kumah, M. 209
Kumar, S. 211
Kweneng West, Botswana
    access to natural resources 275
    community efforts to rise out of poverty in
        273
    ecotourism in 269–70, 274

poverty alleviation programmes 275
poverty in 271–2
role of the government in alleviating poverty
    in 272–3

Lagos, Nigeria 195
Lamie, R. D. 204
land use and land cover (LULC) 191, 193–9
land use and land cover change (LULCC) 5, 190,
    191, 193, 194–8, 330
Landsat programme 193
landscape, influence of environmental and
    climate change on 137
Lane, B. 189
last chance tourism (LCT) 78–79
Latin America, agricultural development in 156
Le Roux, W. 239–40
Lee, J. 124
legislations, governing natural resources 243,
    245, 246
Leh, O. L. H. 215, 216
Leininger, M. M. 284
Lemieux, C. J. 78, 79
Lenzen, M. 108
Lepetu, Joyce 3
L'Espoir Decosta, Patrick 5, 331
Letloa Trust 29
Levy, S. 170
Libya 102
Lindsay, W. K. 55
Littrel, M. A. 178
Liu, S. 214
Liu, T. 77, 80
livelihoods *see* rural livelihoods; sustainable
    livelihoods
living culture, encounter with 169–71
local communities, exclusion from protected
    areas 329–30
Luo, X. 39
Lupi, C. 210

Ma, S. D. 116, 119
Mabibibi, M. A. 19
McGinnis, M. D. 2
Mackenzie, A. F. D. 64
Maetzold, J. A. 210
Magole, L. 48
Magole, L. I. 244
maize (*Zea mays L.*) 46, 48
Makgadikgadi Pan National Park 38, 39, 40
Makuleke community 19, 20
maladaptation, climate change 81, 82
Malaysia, agritourism and crimes in 216
Malinowski, B. 38
Mallorca, wetland in 78, 79

342 *Handbook on tourism and conservation*

Man and Biosphere Programs (MABPs) 300
Mana Pools National Park, Zimbabwe 199
Manatsha, B. T. 63, 65
Marekia, E. N. 55
marginalised communities, forest-based tourism
and 25–36
market forces 331
marketization 165, 172
 as challenge to experiential value potential of
 tourism 166–7
 of place 168
Maruping, M. 291
Marx, Karl 171, 172
Mashame, G. 198
Masry, Esraa El- 4
mass tourism destinations 166
Masunga, Gaseitsiwe Smollie 7
Mathayomchan, B. 87
Mather, A. A. 94
Matheson, W. 199
Matshelagabedi village 59
Matteucci, X. 15, 16
Mauritius 92, 93, 94
Mauro, P. 268
Mauss, M. 38
Mavondo, F. 140
Mbaiwa, J. E. 3, 6, 56, 177, 184, 185, 195, 196,
 197, 198, 226, 244
 on ancestor worship at Goo-Moremi Village
 294
 on CBNRM programmes in Botswana 40
 on ecotourism 54
 on enclave tourism in Botswana 225
 on use of fences to separate wildlife and
 livestock 48
Mbaiwa, O. I. 48
Mediterranean countries
 tourism industry and risk of climate change
 in 101
 weather extremes 103
Merwe, P. Van Der 210, 213
Mexico 130
 agritourism in 147–63
Middle East, tourism industry and risk of climate
 change in 101
Mieczkowski, Z. 87, 102
Miles, M. B. 256
Miller, K. 38
Miller, R. G. 88
Mitchell, J. 33
Mmegi OnlineI (news outlet) 120
Mmualefe, L. C. 197
Moganane, B. O. 55
Mogende, E. 26, 131, 331
Molek-Kozakowska, K. 120

Molokwane, Thekiso 6
Mombasa, Kenya 94
Monametsi, N. F. 38, 45
Moraru, R.-A. 214, 215
Moremi Game Reserve 56, 224, 225
Morocco, tourism climate change index (TCI)
 104, 106
Moses, O. 96
Mosie, Ineelo 7
Moswete, N. 140
Moyo, I. 3, 18
Mozambique channel 92
Mpolokang, Maduo 4
Mudzengi, B. K. 199
Murphree, M. W. 55, 56
Murphy, P. E. 250
Murray, I. 38
Musakwa, W. 77
music and dance 274, 275
Mutanga, C. N. 5, 56
Mwenya, A. 49, 56
Myanmar (Burma) 55
Mzimela, Jabulile Happyness 3

Namibia 119
Namibian Association of CBNRM 245
Nan, W. 120
Napa Valley, California, USA 170
NASA 90, 102
Nata, Botswana, bird sanctuary 63, 64
National Aeronautics and Space Administration
 (NASA) 193
national parks, temperature anomalies at 88
natural resources
 centralising of, by government 243
 commodification of 226–8
 conservation of 6–7
 poverty reduction and community-based
 237–48
natural vegetation 199
nature-based tourism 1
 and adaptations for climate change 330–31
 affected by storm damage 4
 participation of locals in 16
nature bonding 281, 282
Nechasar National Park, Ethiopia 55
Nepal, conservation in 39
Nett, R. 60
new public management 165
New Scientist 121
New Zealand
 climate change adaptation plans 81
 Indigenous Māori cultural values and
 wisdom 172
Ngami, Lake 115

Ngamiland Council of Non-Governmental
    Organisations (NCONGO) 316–17
Ngamiland District, Botswana 28, 196, 304, 305
    CBNRM Forum 242, 244, 318
    Tourism Development Plan 305
Ngamiland Integrated Land Use Plan 305, 306
Ngazy, N. 92
Ngoma Safari Lodge 257
Ngwenya, Barbara N. 7
Nhamo, G. 80, 82, 96
Nicholson-Cole, S. 121
Nisbet, M. 117
Nithi Falls, Mount Kenya 90
Nkone, Agnes 6
non-timber forest products (NTFPs) 31
northern Botswana 4
    affected by climate change 120
    CBNRM programme in 301
    effects of environmental change in 130–45
    human–wildlife conflict in 37
    road networks in 195
    see also Botswana
Northern Fennoscandia 76
Nthiga, Rita 5
Nuriyeva, K. 212
Nyaupane, G. P. 178, 179, 180, 185
Nyongesa, K. W. 79, 90
Nzama, A. T. 19

Ochna pulchra (Monyelenyele) 30
Ogra, M. 37
oil, production from trees 31
OKACOM 315–316, 317
Okavango Delta, Botswana 26, 28, 34, 115, 130,
    131, 192, 198, 266, 301, 304, 305
    CBNRM in 229
    drying of the 123
    ecotourism in 264
    Game Park 192
    land use and land cover (LULC) in 193–9
    Management Plan 301
    reliance of marginalised communites on
        forest products 32
    vulnerability to climate change 96
    as a World Heritage Site 230, 303
Okavango Panhandle 28, 302–3, 304
    agriculture in 306–7
    CBOs 306
    communities 305
    COMPACT implementation in 312–13
    Cultural Fair 318
    ethnic groups living in 304
    livelihood activities of communities in 305–7
    Maun/Okavango Delta Horticulture
        Demonstration Project 316–17

stakeholders 310–11, 312
tourism in the 305
see also COMPACT programme
Okavango Research Institute, University of
    Botswana 316
Okavango River Basin 119, 302, 303, 315
Okello, M. 48
O'Neill, S. 121
Ostrom, E. 2, 253, 329
Otago Peninsula 130
Ouma, Y. O. 89

Palapye village 198
Paris Agreement 109–10
Paris syndrome 168
parks, economic change and planning future of
    80–81
PAs *see* protected areas
Pasquinelli, C. 170
Patayamatebele village 58
Peluso, N. 231
Perdue, R. R. 55
Pérez-Olmos, Karina Nicole 5
Perkins, Jeremy 4
Permanent Okavango River Basin Water
    Commission (OKACOM) 315–16, 317
Peru, Ministry of Foreign Trade and Tourism of
    150
Pfouts, J. H. 38
Philippines, agritourism in 210, 215
Phillip, S. 151, 158
piloncillo (product of sugarcane juice) 149
place attachment 279, 280, 281, 282, 283, 295
plant succession 196
Poberezhskaya, M. 118, 119
policy issues 332–3
    capacity development 333
    climate change adaptation in tourist
        destination areas 332
    product development and marketing
        strategies of agrotourism 332–3
    safety and security 333
Poteete, A. R. 245
pottery craft 5, 182, 185
poverty 239, 241, 245
    alleviation of 6, 269, 270, 272–3, 275
    causes of 267–9
    indicators of 271–2
Power, A. 241
Pradhan, M. S. 240
precipitation, effects on tourism sector 107
Preston-Whyte, R. A. 116
Pretty, J. N. 179
private tourism 226
Priyanka, S. 209

344 *Handbook on tourism and conservation*

pro-poor tourism 3, 33, 34
Prosser, R. 56
protected areas 4, 37, 55
  absence of community involvement in
      management of 17–18
  in Africa 10, 55
  benefits of PA-based tourism 10
  in Botswana 240
  community empowerment through tourism
      in 15–16
  exclusion of local communities from 329–30
  exertion of pressure on communities by
      10–11
  impacts of climate change in 77–9
  management of 19, 21
  recreation in 75–6
  role in natural resources conservation 3
  rural livelihoods in 330
  tourism
  and climate change adaptation in 73–85
  and conservation in 10–24

Raey, M. El- 92
Ramberg, L. 198
Rammotokara, Gorata 3
Ramutsindela, M. 329–30
re-regulation 226
recentralisation, of natural resources management
      246
Red Sea, coral bleaching 92
regenerative tourism 165
relationships, evaluation of 57
remote sensing (RS) 190, 191
resource systems 2
resource units 2
Réunion Islands 93
Rhodesia *see* Zimbabwe
Ribot, J. C. 245
Riera, A. 78
Rihoy, E. 56
Ringrose, S. 199
Riveros, H. 156
Roberts, L. 190
Robinson, R. N. 195
Rodney, W. 55
Rogerson, C. M. 204
Rogerson, J. M. 204
Roman, M. 217
Rotherham, I. D. 189
rural livelihoods 329
  dependence on natural resources 54
  implications of human-elephant conflicts for
      37–53
  in wildlife areas 330
rural tourism

attributes of urban and 190
definition of 189
environmental impact of 189–203
rurality 189
Rusnak, G. 54
Russia, media coverage of climate change in 118

Saarinen, J. 133
safari-based holiday packages 214
Safarov, B. 198
Sagoe-Addy, K. 94
St Lucia Wetland Park 95–6
Sakuze, L. K. 177
Sakwa, J. 48
San community 239, 269
  challenges facing 29–30
  removal from the Moremi Game Reserve,
      Botswana 55
  social exclusion of, in Botswana 240
San Juan Hill (Antennas Hill) 152, 153
Sankuyo Tshwaragano Management Trust
      (STMT) 243, 251
Santawani Lodge, Cultural Village 266
Savé Valley Conservancy game reserve 96
Scheyvens, R. 14, 17
Schilling, B. J. 217
Schinziophyton rautenii (Mongongo) 30, 31, 32
Scoones, I. 180
Scotland, ecotourism development in 55
Scott, D. 101, 132
sea level rise
  adaptation actions/techniques 111
  and coastal tourism in Africa 92–5
  effects on tourism sector 107
sea surface temperatures 92
Sehuhula, Siamisang 6
self-efficacy 11
Sen, A. 182, 184
sense of place, tourism and 279–99
Senses of Place (Feld and Basso) 280
Serengeti National Park, variability of rainfall
      seasons in 78
Seronga, Botswana 304
SESF *see* Social-Ecological Systems Framework
      (SESF)
Seychelles 92, 93, 94
Sgroi, F. 27
Sh, D. O. 198
Shackleton, C. M. 27
Shackley, M. 279
Shaikarawe community 27, 28–9, 30
Shaikarawe Community Trust 34
Shaikarawe Forest Conservation Area (SFCA) 29
Shakawe, Botswana 192, 193, 196, 304
  growth of tourism in 195

land use in 191, 194, 198, 199
Shapiro, M. A. 125
shopping, contribution (footprint) to climate
    change 109
Sibitane, Z. 92
Sife, A. S. 119
Simin region, Iran 209
Simpson, John 230
Siphambe, Gladys 6
Siyao, P. O. 119
Sjoberg, G. 60
Skelton, Paul 123
Skondomboro settlement 198
Slaper, T. F. 207
small island developing states (SIDS) 92
    sea level rises in 93
    tourism industry and risk of climate change
        in 101
snowfall
    adaptation actions for variability in 111
    effects of rising temperatures on 106
social-ecological systems 1, 329
Social-Ecological Systems Framework (SESF) 2,
    117, 252–4, 256
social exchange theory 3, 38–9, 49, 56–7, 66, 67
social exclusion 240, 241
social justice 6, 14
social life, exclusion of people from 268
Socio-Cognitive Theory 11
Son, J.-Y. 38
sorghum 46
South Africa 86, 88, 102, 130
    acts to dispossess Black Africans of their
        native land 10
    co-management approach post-apartheid 15
    environmental change in 132
    Natal Colonial Government 10
    National Environmental Management:
        Protected Areas Act 57 2003 17
    policies on communities in and around PAs
        in 12, 21
    tourism climate change index (TCI) 104, 106
South Asia, tourism industry and risk of climate
    change in 101
South Korea, reports on climate change in 124
Southern Africa, tourism and climate change in
    116
Southern African Development Community
    (SADC), Protocol on Wildlife
    Conservation and Law Enforcement 223,
    230, 231
Sowa, J. 244
space, and place 166, 171
stakeholders

and climate change adaptation in protected
    areas 73
and measurement of the 3Ps 207
perceptions of agritourism 208, 210, 217
primary 309
secondary 309
state-making 228–9
Stedman, R. C. 280
Steele, F. 281
Stone, M. T. 11, 134, 140, 178, 179, 180, 185
Stone, P. R. 64
Strategic Action Programme (SAP) 315
Strauss, A. 118
Streifeneder, T. 204, 215, 216
Stretch, D. D. 94
Stronza, A. L. 177, 197, 198
Sub-Saharan Africa
    effects of rising temperatures on species in
        national parks 106
    food waste 109
sugarcane 148
sun-and-sand mass tourism 166, 167
sun-lust and wanderlust conceptual framework
    191–2
Suping, Kekgaoditse 6
Survival International 229
sustainability, Triple Bottom Line approach to
    206–7, 217
sustainable development 157, 185
    Agenda for Sustainable Development 2030
        110
Sustainable Development Goals (SDGs) 92, 110,
    132
sustainable livelihoods 26, 141
    among clay potters women group, in Kenya
        177
    role of conservation and forestry in 27
Sustainable Livelihoods Framework (SLF) 131,
    133, 177
sustainable tourism 14, 169

Tachila Nature Reserve, Botswana 3, 54–71
Taecharungroj, V. 87
Tafa, Unabo 3
Taiwan 78, 80
Tanbi Wetland National Park, Gambia 81
Tanzania 86
    environmental change in 132
    establishment of PAs in 16
Tateishi, R. 89
Tati Company 57, 65
Tatsugoro Bonsai Museum 152, 153, 159
Taylor, M. 242
Tcheku Community Trust (TCT) 306
temperature rise

346 *Handbook on tourism and conservation*

adaptation actions/techniques 111
effects on tourism sector 106
territorialization 224, 225
Tervo-Kankare, Kaarina 4
Tew, C. 151
Thailand
dive industry 115
Mu Ko Surin National Park 81
Thakadu, O. T. 3, 4, 7
Thamalakane River 122
Thapelo, K. 48
thatching grass 31
Thouless, C. 48
Tolan, S. 120
Tolvanen, A. 76, 78
tour guides 134
tour operators 210
tourism
contribution to the global economy 25
environmental conservation, natural resource
management and 329–34
micro and small-scale tourism initiatives
176–88
power imbalances in 171–2
Tourism Climate Index (TCI) 4, 87, 102
for Africa 104–6
model 112
tourists, decision-making process of 192
trans-frontier conservation areas (TFCAs) 230–32
Transboundary Diagnostic Analysis (TDA) 315
Trentino, Italy 55
TripAdvisor 87, 89, 90
Trust for Okavango Cultural & Development
Initiatives (TOCADI) 29, 317–18
Tshekedi, Kgosi 292
Tshipa, Sharon 4
Tsodilo Hills, Botswana 192
Tugade, L. O. 210, 211, 214
Tule lagoon 152
turtles 92
Tuscany, Italy 170
Twumasi-Ampofo, K. 94

uKhahlamba-Drakensberg Park (UDP), South
Africa 13, 17–21
United Nations, Sustainable Development Goals
(UN SDGs) 92, 110, 132
United Nations Development Programme 314–15
United Nations Educational, Scientific and
Cultural Organisation (UNESCO) 300
World Heritage Convention 303
World Heritage Sites 192
United Nations Environment Programme (UNEP)
74, 110, 230

World Conservation Monitoring Centre
(WCMC) 73
United Nations Framework Convention on
Climate Change (UNFCCC) 110
Climate Neutral Now initiative 101
United Nations World Tourism Organization
(UNWTO) 73, 74, 101, 110, 178
Urioste-Stone, S. M. De 78, 79, 83
USA 78
climate change adaptation plans 82
National Oceanic and Atmospheric
Administration (NOAA) 87
Uzbekistan 198

Vacik, H. 79, 90
Van der Merwe, P. 11, 17
Van der Watt, H. 14, 17
Van Wilgen, N. J. 88
Van Zyl, C. 204, 210, 213
Vandergeest, P. 231
Varisco, C. A. 151
Venice, Italy 166, 168
Victoria Falls 89, 96, 115
Victoria, Lake 181
Vietnam 130
Villanueva-González, Carlos Enrique 5

Walker, K. P. 55
Wall, Geoffrey 73
Wang, L. 80
water availability, and wildlife-based tourism
137–8
water shortages, climate change adaptation
actions for 111
waterbuck (*Kobus ellipsiprymnus*) 136
Watson, H. K. 116
weather extremes 78, 103
Weiler, B. 191–192
Welling, J. 79
Werbner, R. 286
Western Europe, tourism industry and risk of
climate change in 101
wetlands
in Australia 78
in Mallorca 78, 79
Ramsar Convention on Wetlands 303
White, R. 286, 289
wildlife, negative attitudes of local communities
towards 49
wildlife conservation
in Botswana 223, 224, 231
human–elephant conflicts and implications
for rural livelihoods and 37–53
implications for crop damage 47
putting communities first in 332

see also conservation
Wildlife Conservation Policy, 1987 321
Wildlife Management Areas (WMAs) 223, 226, 231–2, 240, 302, 329
wildlife tourism 132, 194, 226
Willcocks, L. 38
Wilson, S. 196
Wilson, W. J. 241
wines 170
Wolf, I. D. 16
Woodroffe, R. 48
World Economic Forum 115, 244
World Ecotourism Summit, 2002 264
World Heritage Committee 300
World Heritage Convention 300
world heritage sites, co-management of, for community benefit 300–327
World Meteorological Organization 86, 92
World Travel & Tourism Council (WTTC) 101, 115, 176
World Wildlife Fund (WWF) 300

Yamagishi, K. 214, 215
Yamane, Taro, sampling size formula 61
Yang, J. T. 195
Yildirim, G. 212, 216

Zacal, R. G. 210, 217
Zafirovski, M. 38, 57
Zanzibar 92
Zimbabwe 56, 77, 78, 86
   agritourism in 5, 204–221
   climate change adaptation plans 82
   land audit programme 206
   Lands Department 216
   Manicaland province 11, 17, 208, 209
   Mashonaland West province 208, 209, 212
   regions of 207–8
   Special Agriculture (Maize) Production Programme 206